Bugs Rule!

Bugs Rule!

An Introduction to the World of Insects

WHITNEY CRANSHAW AND RICHARD REDAK

Princeton University Press
Princeton and Oxford

Published by Princeton University Press, 41 William Street, Princeton,
New Jersey 08540

In the United Kingdom: Princeton University Press, 6 Oxford Street, Woodstock,
Oxfordshire OX20 1TW

press.princeton.edu

Cover photograph: Milkweed longhorn by David Leatherman.

ISBN 978-0-691-12495-7

Library of Congress Control Number: 2013934688

British Library Cataloging-in-Publication Data is available

This book has been composed in Sabon LT Std by SPi Publisher Services, Pondicherry, India.

Printed on acid-free paper.

Printed in the United States of America

10 9 8 7 6 5 4 3

Dedication

This book is dedicated to Bill Cranshaw, whose boundless curiosity about the world—and his passion for books—will always inspire.

Contents

Preface

Bugs Rule! An Introduction to the World of Insects was originally conceived to meet the needs of an entomology course taught to nonscience majors—the type of course where you have only one shot to get across everything-you-need-to-know-about-insects to a group of students with little previous exposure to the subject. And although such students rarely have the opportunity to take more advanced courses in entomology, it is always hoped that their initial exposure will open them up to a lifetime of learning about the subject.

There have been many influences affecting the development of this book. Probably the greatest have been entomology books that approached the subject by often talking about individual insects or insect groups,—their life history, and how they intersect with humans. *Life on a Little Known Plant*, by Howard Ensign Evans, will always remain a classic in this regard, able to open the world of insects by his use of fascinating details combined with an enjoyable, highly readable style. More recently, there have been a substantial number of contributions by May R. Berenbaum, through both her columns and books (e.g., *Ninety-Nine Gnats, Nits, and Nibblers*; *Bugs in the System: Insects and Their Impact on Human Affairs*) that have helped familiarize and popularize all manner of insect subjects. Also of great influence was the emergence of some highly successful entomology courses around the country that engaged large numbers of students, such as what Tom Turpin achieved at Purdue University, accompanied by his *Insect Appreciation Digest*.

The authors have also reviewed and considered most of the books that have been developed to teach a more standard entomology course. Many do an excellent job in teaching the basics of entomology needed for students with a biology orientation. In our opinion, *The Insects: An Outline of Entomology* by Gullan and Cranston (now in its fourth edition) is a particularly thorough and well-organized book for teaching entomology. *Borror and DeLong's Introduction to the Study of Insects* (now in its seventh edition) by Johnson and Triplehorn is also considered to be a standard and has been used by the authors as the guide for present taxonomy—a constantly shifting and debated subject. Insight into phylogenetic relationships and insect evolution are particularly well handled in *Evolution of Insects*, by David Grimaldi and Michael S. Engel.

Drawing from all these resources and many more, *Bugs Rule!* has its own approach to the subject. Overall, the book is more natural history oriented than are standard entomology textbooks. Many of the "basics" are present—such as anatomy, physiology, systematics—but these are given a fairly compressed treatment within introductory chapters, and later referenced as appropriate when specific insect groups are discussed. The great majority of the book is an introduction to the cast of characters—the various insect orders with a bit of emphasis on some of their more important or poignant (to our eyes) members.

Some features of this book differ considerably from what one finds in most entomology textbooks. Perhaps most notable is the fairly extensive treatment given to the noninsect arthropods—crustaceans, myriapods, and particularly the arachnids (spiders, mites, scorpions, etc.). Most people lacking a formal course in entomology appear to have a somewhat vague idea about the arthropods, considering them all to be "bugs" of some sort. And many—a great many—of the questions that they typically have about "bugs" involve arachnids. So this book tries to include a bit more information about these fascinating animals that typically get sidestepped.

Also likely to be a bit controversial are a few topics that respond to common questions students have expressed. "Do insects sleep?" and "Do insects feel pain?" come immediately to mind. We recognize that there remains considerable debate about such matters. As a result they are typically ignored as textbook subjects; we chose to attempt a coverage of the topics.

Both the authors have long taught the type of introductory courses for nonscience majors that *Bugs Rule!* is designed to support. We hope it will be useful to others who are in the business of increasing entomological (in a very broad sense) literacy or who just want an introduction of their own to the fascinating life of arthropods.

Acknowledgments

So many have assisted in so many ways to help bring this book to completion.

The extensive use of images—photographs, figure drawings, and other illustrations—was critical to the production of this project, and scores of people have contributed in this area. Credits have been provided with each illustration, but there are some individuals who have made exceptional contributions and, we feel, deserve special recognition and thanks. This starts with the three photographers that are most heavily represented within these pages—Brian Valentine (Lord V), Tom Murray, and Jim Kalisch, each an amazingly talented and productive photographer of all types of arthropods. One will also find Gary Alpert, Scott Bauer, Joseph Berger, Mark Chappell, William Ciesla, Tom Coleman, Jillian Cowles, Susan Ellis, Ken Gray, David Leatherman, Gerald Lenhard, Sturgis McKeever, Herbert A. Pase, Jr., S. Dean Rider, and James Solomon well represented within this book. Then there were a few individuals we were able to go to repeatedly for help with the images we found most difficult to locate—Alex Wild (Hymenoptera and rock crawlers), Scott Turner (termites), Lynn McCutcheon (arachnids), David Shetlar (all manner of oddities), and Jack Kelly Clark (the California connection).

Matt Leatherman was our main illustrator, to whom we turned repeatedly for help in providing critical illustrations, particularly in the early chapters. Illustrations related to medical/veterinary entomology rely heavily on those provided by Scott Charlesworth with Purdue University and the Centers for Disease Control. The historical illustrations produced by the Department of Entomology within the Smithsonian Institution, primarily penned by Art Cushman, were also used extensively.

Boris Kondratieff provided great guidance to the original development of the text, others in review, including Mike Rust, Wayne Brewer, Paula Cushing, Mike Rust, Tom Weissling, Richard Zack, and anonymous reviewers. All comments are appreciated and helped improve the book.

As this project finally moved to end stages (following too many years of undue tardiness by the authors!), the folks in Princeton University Press worked magic. Terri O'Prey took over to keep production on schedule, with Larissa Klein assisting. The primary editing and all layouts were coordinated by Gunabala Saladi and her staff at SPi Global. Throughout, from the original conception of the project and through the rocky periods, Robert Kirk shepherded the process.

THANK YOU ALL!

Whitney and Rick

Bugs Rule!

Introduction to the Arthropods

What Is an Arthropod?

The subjects of this book are the arthropods that live among us, primarily the insects but also some of their relatives, such as arachnids, millipedes, centipedes, and a few crustaceans. When formally classified, these animals are placed in the phylum Arthropoda, which comprises a huge number of species with a tremendous diversity of forms and habits. Nonetheless, all arthropods share certain features that together define them as a distinct form of life:

(a)

(b)

(c)

FIGURE 1-1
Three representative arthropods. (a) Dragonfly (insect), (b) julid millipede, and (c) windscorpion (arachnid). All show the basic external features of arthropods including an exoskeleton, segmentation of the body, jointed appendages, and a body design that is bilaterally symmetrical. Photograph of the dragonfly courtesy of Brian Valentine; photograph of the millipede courtesy of Jim Kalisch/ University of Nebraska; photograph of the windscorpion by Jack Kelly Clark and provided courtesy of the University of California IPM Program.

- All arthropods have a body supported by a hardened external skeleton (**exoskeleton**), a reverse type of engineering compared to our internal skeleton. To allow growth, this exoskeleton must be periodically shed, and a new one rebuilt.
- The body of an arthropod is divided into segments, a feature shared by some other animal groups, such as earthworms (phylum Annelida) and velvet worms (phylum Onychophora).
- The appendages of arthropods—their legs, antennae, and mouthparts—are jointed. This is the feature that defines the phylum. (In Greek, arthropod means "jointed foot.")
- Internally, the nerve cord runs along the lower (ventral) part of the body and is not enclosed in a protective spinal column. These features contrast with those found in phylum Chordata to which we belong.
- Blood is moved by the aid of a tube-like heart, located along the back (dorsal) part of the body.
- The overall body arrangement is **bilaterally symmetrical**, so that, if the body were cut through the center from head to tail, the two halves would be a mirror image of one another.

The Diversity and Abundance of Arthropods

The arthropods are, by far, the most diverse life form on the planet. Insects alone, with approximately 970,000 known species, comprise over one-half of all kinds of life known to occur on the planet. Yet despite the impressive numbers, these reflect only "known species," ones that have been suitably described in the scientific literature and accepted as distinct species. This number represents only a small fraction of the number of species estimated to be present on the planet today. This number is also a tiny fraction of all the insects that ever were on the planet. It has been suggested that perhaps 95% of all insects that have ever existed, since their first appearance some 400 million years ago (mya), are now extinct.

Today, the number of insect species thought to occur is often estimated at about four to five million species. The great majority of these, at least 80%, remain unknown to science so far. Progress is being made to close this gap, with over 7,000 new insect

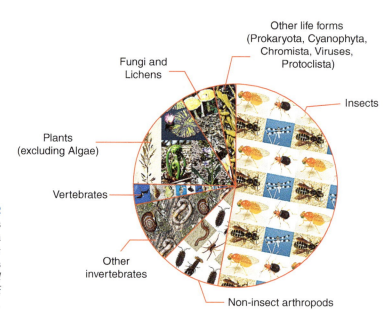

FIGURE 1-2
The relative number of different kinds of life forms known on Earth, based on the number of known species. Of the approximately 1.9 million presently recognized species, just over half are insects. Figures based on *Numbers of Living Species in Australia and the World*, 2nd ed. (2009). Photographs courtesy of Tom Murray.

species being described annually, over 20 per day on average. At this rate of new discovery, impressive as it is, perhaps we can expect a full catalog of the five million insects to be ready in about 550 years or so.

A much more difficult question to answer is "How abundant are insects and other arthropods in terms of total population numbers?" One of the problems is that the overwhelming number of arthropods are minute and live in soil. For example, one of the first attempts at counting all of the arthropods in a sample of soil was done in an English pasture during November 1943. About 2.5 billion arthropods were estimated per hectare, with mites comprising some 62% and springtails 23% of the total number. On the basis of surveys such as this it has been estimated that the insects, springtails, mites, and other land-dwelling arthropods outnumber humans by as much as 250 million to 1. Furthermore, these arthropods collectively comprise over 80% of the total biomass of the terrestrial animals, far outweighing all the other land dwellers such as earthworms, reptiles, birds, and mammals.

(a)

(b)

FIGURE 1-3
(a and b) Springtails and soil-dwelling mites are the most abundant kinds of animal life on the planet. A billion or more may typically be found in a hectare of fields, pasture, or lawn. Photographs courtesy of Brian Valentine.

The Many Roles of Arthropods

> If all mankind were to disappear, the world would regenerate back to the rich state of equilibrium that existed ten thousand years ago. If insects were to vanish, the environment would collapse into chaos. (E. O. Wilson, *The Diversity of Life*)

Although small in size, arthropods, in their tremendous numbers, collectively account for the most biomass of all land animals. In the Amazon rain forest, the weight of just one family of insects, the ants, is estimated to be four times more than all the mammals, birds, fish, reptiles, and amphibians combined. Furthermore, the roles of arthropods in ecosystems are myriad, but central to the functioning of planet Earth:

Pollination of flowering plants. Insects are essential to the pollination of most flowering plants, and many of the flowering plants are the result of **coevolution** with their insect pollinators. The tremendous variety of flower types reflect different ways that plants have evolved to more efficiently attract pollinators. In response, new species of insects have arisen to better exploit these sources of nectar and pollen. In addition to native plants, essentially all fruits, vegetables, and many of the forage crops (e.g., clover, alfalfa) are dependent on insects to produce seed.

FIGURE 1-4
A leafcutter bee pollinating sweet pea. Many plants are dependent on insects for their pollination. Photograph by Whitney Cranshaw/Colorado State University.

FIGURE 1-5
Blow flies colonizing fresh carrion. Insects help in the decomposition of dead plant and animal matter and have central roles in the cycling of nutrients in natural systems. Photograph by Whitney Cranshaw/Colorado State University.

Recycling plant and animal matter. Many insects develop by feeding on dead plant matter, dead animal matter, or animal dung. In this role, they function as **macrodecomposers** that are in the first-line "clean-up crew" essential to the recovery and recycling of nutrients. Through insect feeding, these substances are broken down into much smaller particles and partially digested, which greatly accelerates the process of decay that frees the nutrients to nourish later generations of plants. In the absence of insects, nutrient-recycling systems break down and organic matter accumulates.

Soil formation and mixing. The great majority of terrestrial arthropods live within the soil. These animals help to turn the soil and incorporate organic matter and nutrients. The impacts of these activities can be very dramatic, with some of the social insects (e.g., the ants and termites) moving and mixing tremendous amounts of soil as they tunnel. These processes are critical to soil formation and the maintenance of soil fertility. Without these insects, plant growth would be reduced and restricted.

Centrality to animal food chains. Through their feeding activities, plant-feeding insects (about 25% of the species on the planet) convert plant biomass to animal biomass. In turn, these creatures serve as the primary source of food for other insects (another 25% of the planet's species) and for many birds, fish, and mammals

FIGURE 1-6
Mound-building termites are central to soil formation and mixing in large areas of Africa. Photograph courtesy of USDA APHIS PPQ Archives/Bugwood.org.

that are, in turn, food for yet still more animals. Thus, plant-eating insects are the critical link between plants and much of the rest of animal life on Earth (including humans).

FIGURE 1-7
Many types of wildlife utilize insects as an important part of their diet. Photograph courtesy of David Leatherman.

Maintenance of plant communities. Although the effects of large plant-feeding mammals are conspicuous, it is the activities of insects that most often determine what plant life is present. Insects do this in many ways, including feeding on plants (**phytophagy**), feeding on seeds, pollination, and dispersing seeds.

FIGURE 1-8
Through their foraging activities, leafcutting ants can have dramatic effects on the kinds of plant life that occur. Photograph courtesy of Ronald F. Billings/Texas Forest Service/Bugwood.org.

Unfortunately, most people recognize only those arthropods that are directly and immediately affecting human activities. These species are often considered negatively, as competitors, because of their ability to cause several types of harm—destruction of crops, damage to stored products or structures, transmission of plant and animal pathogens, and stings or bites— or merely some degree of annoyance. Those that do affect us in these ways are judged to be "pests," a subjective and very flexible term that is defined by how much impact they are personally perceived to have. It is important to keep in mind that only a tiny fraction of all arthropods are ever elevated to this infamous status. A listing of all insects worldwide that are considered pests for one reason or another would include fewer than 10,000 species, approximately 1% of the total number of *known* insect species. A list of species that are directly beneficial to humans may be larger by an order of magnitude.

All too often people try to separate the insect world into "good bugs" and "bad bugs." Alternatively, one often hears the question "What good is this insect/scorpion/spider?" These types of categorizations fail to recognize the tremendous importance of the arthropods to the functioning of this planet, usually in ways we little understand. It is perhaps important to keep in mind the words of pioneer conservationist/naturalist John Muir: "When we try to pick out anything by itself, we find it hitched to everything else in the Universe."

Insects are neither good nor bad. They are, along with all other extant life forms, a representation of the latest expression of what has evolved on Earth.

Classification of the Arthropods

In the classification of biological organisms, all life forms are grouped according to how related they are, usually based on physical features. Within this organization, all life forms are arranged in a series of subgroupings that become increasingly specialized. This science of classification is known as **taxonomy** and is conducted by specialists known as taxonomists. Closely associated with taxonomy, and often guiding the classification arrangements, is the science of **systematics** that seeks to determine the relatedness of different life forms. Systematists make extensive use of the fossil records of extinct species along with all

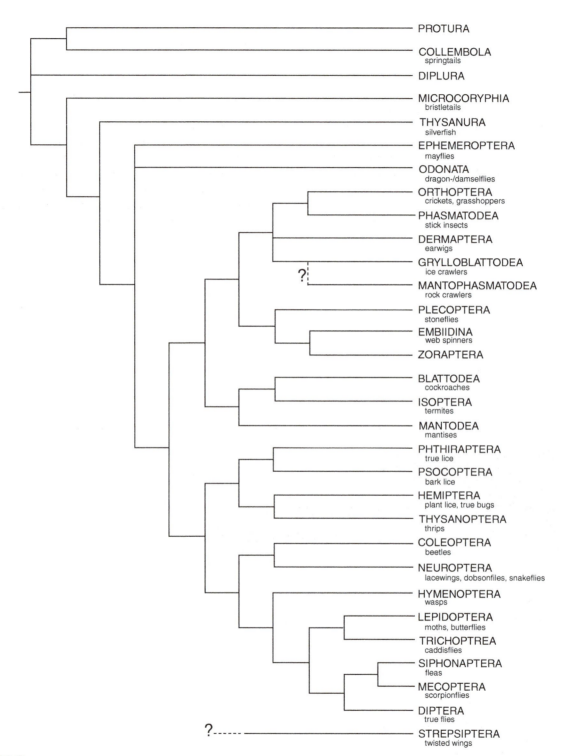

FIGURE 1-9

A diagram of a proposed phylogenetic relationship of the insect orders. Phylogenetics is the study of the evolutionary relations between organisms, and in a diagram such as this the orders that are most closely clumped are most closely related. Phylogenetics is a very active field that has been greatly aided by genetics. With new information, evolutionary relationships and taxonomic groupings are constantly being reevaluated, and changes in taxonomic arrangement are frequent. (Modified from Grimaldi and Engel, 2005.)

manner of biological features of present (**extant**) species. Increasingly, genetics also guides systematics. The powerful new tools that allow sequencing of genes are producing a revolution in the systematics of insects (and many other organisms) during which we are seeing many taxonomic arrangements being modified and many new species being recognized.

Using this system, all animals are classed together within the broadest type of grouping, a **kingdom**—specifically the kingdom Animalia. The kingdom containing all animals is next subdivided into various **phyla** (singular, **phylum**), one of which is Arthropoda—the arthropods that are the focus of this book. (Examples of some other animal phyla include Annelida, the segmented worms; Nematoda, the round worms; Mollusca, the mollusks; and Chordata, the animals with a hollow, ventral nerve cord, which includes humans.) In turn, a phylum is subdivided into sections, each known as a **class**. Four classes of arthropods (millipedes, centipedes, arachnids, hexapods/insects) are the primary focus of this

book. Also discussed, in part, are a group of arthropod classes collectively known as crustaceans (subphylum Crustacea).

The classes are subdivided into **orders**. For example, once you have identified something as an insect (from the class Hexapoda), the next grouping is the order of insects where it has been placed. Butterflies and moths, insects that have characteristic scale-covered wings, are placed by taxonomists in the order Lepidoptera. Beetles that have a hardened front pair of wings are in the order Coleoptera, while the flies, with their unique single pair of wings, are classified in the order Diptera. Because of differences in how scientists classify the insects, you may see some differences in the number of and names for the orders among the many books describing insect life and in their names. The classification system used for this book follows that of the 7th edition of *Borror and DeLong's Introduction to the Study of Insects* (2005), which lists in the class Hexapoda 28 orders of insects and 3 orders of entognathous hexapods.

(a)

(b)

(c)

FIGURE 1-10

(a–c) Representatives of three insect orders: sulfur butterfly (Lepidoptera), blatellid cockroach (Blattodea), and ground beetle (Coleoptera). Photographs courtesy of David Cappaert/Michigan State University/Bugwood.org, Ken Gray/Oregon State University, and Jim Kalisch/University of Nebraska, respectively.

Orders are subdivided into **families,** scientific names usually ending in "idae." For example, the beetles (order Coleoptera) are divided into scores of families, including lady beetles (Coccinellidae), weevils and bark beetles (Curculionidae), and leaf beetles (Chrysomelidae). Each family is divided into **genera** (singular **genus**), and each genus into various **species.**

(a)   (b)   (c)

FIGURE 1-11

(a–c) Representatives of three families within the order Coleoptera (beetles): lady beetle (Coccinellidae), weevil (Curculionidae), and darkling beetle (Tenebrionidae). Photographs courtesy of Whitney Cranshaw/Colorado State University, Brian Valentine, and Jim Kalisch/University of Nebraska, respectively.

Each species of insect, as well as all other life forms, has its own **scientific name.** This name is constructed by combining the genus name and what is known as the **specific epithet.** The genus name is capitalized, the specific epithet is not, and both are written in italics. For example, the scientific name of the house fly is *Musca domestica* and that of the tomato hornworm is *Manduca quinquemaculata.* The idea of giving each species a scientific name that is universally recognized was formalized by Carolus Linnaeus (sometimes Anglicized as Carl Linnaeus), a Swedish physician and biologist (1707–1778). The outline he developed, published in the book *Systema Naturae* (1st edition 1735), was revolutionary and remains the fundamental framework whereby all living organisms are classified, based on shared features.

(a)   (b)   (c)

FIGURE 1-12

(a–c) Representatives of three different species within the beetle family Coccinellidae (lady beetles): *Hippodamia parenthesis, Harmonia axyridis, Coleomegilla maculata.* Photographs by Whitney Cranshaw/Colorado State University.

Since each scientific name has two parts, it is described as **binomial nomenclature**. This has become the universally recognized standard for discussing the identity of different organisms in a world that shares few other common languages. In the formal naming of an organism, the person who originally described it is also placed at the end of the name. Therefore in the scientific literature the house fly would be *Musca domestica* Linnaeus and the tomato hornworm *Manduca quinquemaculata* (Haworth), recognizing that these two insects were originally described by Linnaeus and Haworth, respectively. In this book, the descriptor names are left out for simplification, not to diminish in any way the contributions of those who took it upon themselves to first identify the insect as being a unique species.

Several mnemonic phrases have been developed to help reinforce learning of the basic taxonomic groups—kingdom, phylum, class, order, family, genus, species—including the following:

King Philip cuts open five green snakes.
Kings play cards on fat green stools.
Kings play chess on Fridays, generally speaking.
King Philip cried out—"for goodness sake!"

FIGURE 1-13
The principles that guide the classification of living organisms was first formalized in the book *Systema Naturae*, written by Carolus Linnaeus (1707–1778). Painting by Alexander Roslin.

TABLE 1-1 Taxonomic position of the honey bee, *Apis mellifera* Linnaeus, and the southern black widow, *Latrodectus mactans* (Fabricius).[‡]

HONEY BEE (COMMON NAME)[*]	SOUTHERN BLACK WIDOW (COMMON NAME)[†]
 **FIGURE 1-14** The honey bee, *Apis mellifera* Linnaeus. Photograph courtesy of Joseph Berger/Bugwood.org. Phylum—Arthropoda Class—Hexapoda	**FIGURE 1-15** The southern black widow, *Latrodectus mactans* (Fabricius). Photograph courtesy of Clemson University/Bugwood.org. Phylum—Arthropoda Class—Arachnida *(continued)*

TABLE 1-1

HONEY BEE (COMMON NAME)*	SOUTHERN BLACK WIDOW (COMMON NAME)†
Order—Hymenoptera	Order—Araneae
Family—Apidae	Family—Theridiidae
Genus—*Apis*	Genus—*Latrodectus*
Species—*mellifera*	Species—*mactans*
Original descriptor—Linnaeus	Original descriptor—Fabricius ‡

*Common name accepted by the Entomological Society of America.
†Common name accepted by the American Arachnological Society.
‡The original 1775 description by Fabricius used the genus name *Aranea*. In later revisions, the southern black widow was placed in a different genus (*Latrodectus*). This change from the original is indicated by the parentheses surrounding the name of the original descriptor.

It must be recognized that whatever type of classification is used, it is a human construct and thus subject to change. Orders, families, and even classes of organisms may be rearranged following revisions made by taxonomists as new information becomes available through discoveries of new species, better fossil records, and the use of modern molecular genetic techniques.

As our understanding of how different organisms are related has increased, additional groupings have been required. These are most often created by the prefix "sub" or "super." For example, a subclass is a division of a class but will still contain within it one or more orders of the class. A superfamily will contain one or more families within the same order. The taxonomic arrangement used for this book is presented in table 1-2.

TABLE 1-2 Primary taxonomic divisions of the phylum Arthropoda. Orders have been included for the terrestrial or freshwater arthropods that are the focus of this book (classes Arachnida and Hexapoda; subphylum Crustacea in brief).

Subphylum Trilobita—trilobites (fossils only)
Subphylum Chelicerata
    Class Merostomata—eurypterids (fossils only) and horseshoe crabs
    Class Arachnida—arachnids
        Order Scorpiones—scorpions
        Order Palpigradi—micro whipscorpions
        Order Thelyphonida (Uropygi)—whipscorpions
        Order Schizomida—shorttailed whipscorpions
        Order Amblypygi—tailless whipscorpions, whipspiders
        Order Araneae—spiders
        Order Ricinulei—hooded tickspiders
        Order Opiliones—harvestmen, daddy longlegs
        Order Acari—mites and ticks
        Order Pseudoscorpiones—pseudoscorpions
        Order Solifugae—windscorpions, sunspiders
    Class Pycnogonida—sea spiders
Subphylum Crustacea—crustaceans
    Class Cephalocarida

(*continued*)

TABLE 1-2

Class Branchiopoda
  Order Anostraca—fairy shrimp
  Order Notostraca—tadpole shrimp
  Order Conchostraca—clam shrimp
  Order Cladocera—water fleas
Class Ostracoda
Class Copepoda
Class Mystacocarida
Class Remipedia
Class Tantulocarida
Class Branchiura
Class Cirripedia
Class Malacostraca
  Order Amphipoda—amphipods
  Order Isopoda—isopods
  Order Stomatopoda—mantis shrimp
  Order Decapoda—lobsters, crayfish, crabs, shrimp
Subphylum Atelocerata
  Class Diplopoda*—millipedes
  Class Chilopoda*—centipedes
  Class Pauropoda*—pauropods
  Class Symphyla*—symphylans
  Class Hexapoda—hexapods (includes insects)
    Subclass Entognatha†
      Order Protura—proturans
      Order Diplura—diplurans
      Order Collembola—springtails
    Subclass Insecta—insects
      Order Microcoryphia‡—jumping bristletails
      Order Thysanura‡—silverfish
      Order Ephemeroptera—mayflies
      Order Odonata—dragonflies and damselflies
      Order Orthoptera—grasshoppers, crickets, katydids
      Order Phasmatodea—walkingsticks and leaf insects
      Order Grylloblattodea—rock crawlers
      Order Mantophasmatodea—heelwalkers or gladiators
      Order Dermaptera—earwigs
      Order Plecoptera—stoneflies
      Order Embidiina—webspinners
      Order Zoraptera—zorapterans, angel insects
      Order Isoptera—termites
      Order Mantodea—mantids
      Order Blattodea—cockroaches
      Order Hemiptera—true bugs, cicadas, hoppers, psyllids, whiteflies, aphids, and scale insects
      Order Thysanoptera—thrips
      Order Psocoptera—psocids
      Order Phthiraptera—lice

(continued)

TABLE 1-2

Order Coleoptera—beetles
Order Neuroptera—alderflies, dobsonflies, fishflies, snakeflies, lacewings, antlions, and owlflies
Order Hymenoptera—sawflies, parasitic wasps, ants, wasps, and bees
Order Trichoptera—caddisflies
Order Lepidoptera—butterflies and moths
Order Siphonaptera—fleas
Order Mecoptera—scorpionflies
Order Strepsiptera—twisted-wing parasites
Order Diptera—flies

*Arthropods that are often referred to as Myriapoda, the myriapods.
†The classification of the various entognathous hexapods is subject to debate, although each of the three groups is considered distinct. Some classification schemes consider them as separate subclasses or even classes.
‡The orders Microcoryphia and Thysanura (alternately named as Archaeognatha and Zygentoma) consist of insects with primitive features that originated before the development of wings. As such they are sometimes considered together as the Apterygota, in contrast with the other insect orders (Pterygota) that have physical features associated with wings. This arrangement is subject to debate, as many other features of the Thysanura indicate that they are more closely related to the insects that developed wings than to the Microcoryphia.

Common Name or Scientific Name?

Some insect orders and families, and many individual species of insects, have a **common name**. This is the familiar insect name in English, in contrast to the more formal **scientific name**. For example "beetles" is the common name for the order Coleoptera, "swallowtails" is the common name for the butterflies within the family Papilionidae, and "house fly" is the common name for the insect *Musca domestica*.

Scientific names are universal; they are the same in every country. That is their utility and appeal—although names of genera and even families are sometimes rearranged when new information (now usually genetics) leads to taxonomic revisions. However, most people find it easier to learn and use the common names when discussing local insects. Unfortunately, such common names may be used for very different insects in different locations, thus leading to some confusion. For example, an insect formally known as the armyworm (*Mythimna unipuncta*) is a common pest caterpillar of grain crops in much of the central United States. However, when outbreaks of the forest tent caterpillar (*Malacosoma disstria*) occur in forests of northern Minnesota and Wisconsin and the caterpillars are seen marching across roads, this very different insect is called an "armyworm" and elsewhere other caterpillars seen in bands are often referred as "armyworms." Similarly, an odd group of insects known as Jerusalem crickets are known locally by a wide variety of names such as "children of the earth," "old baldheaded man," and "potato bugs." (In turn, a great number of other generally round-bodied arthropods are known as "potato bugs," including pillbugs and the Colorado potato beetle.) Therefore the use of formally accepted common names provides a means to discuss and write about insects in a manner that allows the identification of the species to be consistently recognized.

The Entomological Society of America attempts to standardize the common names of insects used in the United States in the publication *Common Names of Insects and Related Organisms*. Common names of arachnids are similarly codified by the American Arachnological Society. Around the world, similar publications have been developed by various professional organizations committed to the study of arthropods.

(a)

(b)

FIGURE 1-16

(a) The armyworm, *Mythima unipuncta*, and (b) the forest tent caterpillar, *Malacosoma disstria*, each have formalized common names through the Entomological Society of America. The armyworm is a pest of grain crops, and the forest tent caterpillar feeds on various deciduous trees. During outbreaks, forest tent caterpillars are sometimes referred to as "armyworms," which can cause confusion as to the species in question. Photographs by Frank Peairs and Whitney Cranshaw/Colorado State University.

(a)

(b)

(c)

FIGURE 1-17

Jerusalem crickets (a), *Stenopelmatus* spp., may locally be called by many different names including "children of the earth," "old baldheaded man," and "potato bug." Among the other arthropods that are sometimes called "potato bugs" are (b) pillbugs, *Armadillidium vulgare*, and (c) the Colorado potato beetle, *Leptinotarsa decemlineata*. Photographs by Ken Gray/Oregon State University, Whitney Cranshaw/Colorado State University, and David Cappaert/Michigan State University/Bugwood.org, respectively.

2

What One Sees on the Outside—External Features of Insects

The Exoskeleton

Surrounding any arthropod is a hard, tough watertight covering known as the **integument**. Unlike the integument (skin) of humans, the integument of an arthropod is also the primary organ of structural support, providing the **exoskeleton**. The exoskeleton is an immense organ that makes up much of the mass of most arthropods. It not only covers the exterior body but also lines much of the digestive tract, all of the respiratory system, the reproductive system, and the dermal glands and ducts. One may consider the entire insect to be coated inside and out with an exoskeleton.

FIGURE 2-1

An adult dog-day cicada, *Tibicen* sp., emerging from the exoskeleton of the last nymphal stage. The discarded exoskeleton of the previous stage consists of hardened (sclerotized) areas, i.e., the exocuticle, which are in the form of plates. Photograph courtesy of Jim Kalisch/University of Nebraska.

The exoskeleton is composed of many plates (**sclerites**) that may be fused or joined together with flexible membranes. It is not smooth and uniformly thick and may fold inward at points, producing a plate (**apodeme**) or a point (**apophysis**) to which muscles and other tissues attach. Elsewhere, the exoskeleton may project outward and produce various structures, such as spines, protective shields, or prominent horns. The exoskeleton may be particularly thick and hardened at points where added strength is required, such as for jaws or for legs that grasp or dig. Thinner and more flexible areas of the exoskeleton allow for bending.

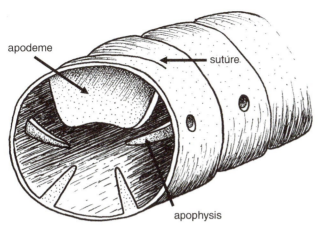

FIGURE 2-2

Cross-section of an insect showing areas of the exoskeleton that fold internally. An infolded plate (apodeme) or point (apophysis) serves as a point of attachment for muscles, much as does our internal skeleton. These infoldings may also provide internal reinforcement. Drawing by Matt Leatherman.

CONSTRUCTION OF THE EXOSKELETON

The exoskeleton is a complex structure that may vary in chemical composition, flexibility, and thickness at different areas of the body, supporting diverse functions. A primary constituent is **chitin,** a tough, durable polysaccharide similar to cellulose and starch found in plants. It is a long-chained molecule, often occurring as parallel bundles laid down as overlapping layers, somewhat like plywood.

Large amounts of cuticular proteins are interspersed, forming a protein–chitin matrix. Some of these proteins may undergo chemical reactions that produce cross-linkage, a process known as **sclerotization,** which results in stiffening of the exoskeleton, reduction of its water content, and resistance to degradation. Darkening of the exoskeleton occurs when many of the more common cuticular proteins undergo sclerotization.

The strength and flexibility of the exoskeleton are largely determined by the types of proteins present and their locations. There are other types of proteins that provide nonstructural functions. For example, **resilin** is a compressible protein present at points in the body to assist in jumping, flying, or other movements. Other proteins may help with wound healing or defense from invading organisms. Additionally, small amounts of lipids are present that, mixed with waxes, provide much of the waterproofing.

The exoskeleton is constructed in many layers, known as the **epicuticle, procuticle, epidermis,** and **basement membrane.** The basement membrane covers the interior of the exoskeleton. It is not composed of cells and simply moderates the movement of nutrients and waste materials into and away from the living cells of the exoskeleton, i.e., the **epidermis.** The epidermis, in turn, secretes the procuticle and epicuticle.

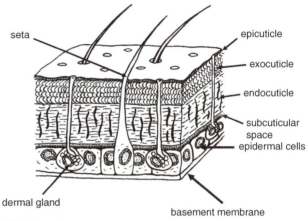

FIGURE 2-3

Exoskeleton layers of an insect. Drawing by Matt Leatherman.

Most of the exoskeleton is made up of the **procuticle,** which itself is composed of two layers: **endocuticle** and **exocuticle.** The endocuticle is the innermost layer of the procuticle and is relatively soft and light colored as it does not undergo sclerotization. Beyond the endocuticle is the exocuticle, hardened to varying degrees and usually darkened by sclerotization. The exocuticle provides most of the support and strength to the exoskeleton.

Covering the entire insect is the **epicuticle.** The epicuticle may be extremely thin, but it has a tremendously important function—protection from water loss. The epicuticle is constructed in layers of wax and cementing materials.

The exoskeleton is produced by the living, cellular epidermis. When a new exoskeleton is needed, the cells of the epidermis are stimulated to produce a sequence of events that results in shedding of the old exoskeleton and construction of a new larger one. This process is known as **molting.** Scattered throughout the insect are special cells of the epidermis (**trichogen cells**) that are capable of producing hairs (**setae**). Insect setae can have several functions. For example, dense hairs in insects may offer protection from predators or provide thermal insulation. Setae that are flattened are known as scales, such as those that cover the wings of moths and butterflies. Some setae are associated with underlying nerve cells and have sensory functions, such as detecting vibration or air movement.

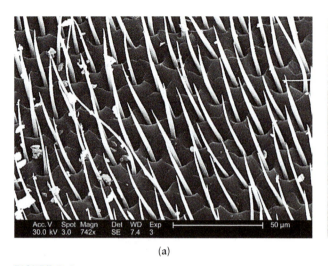 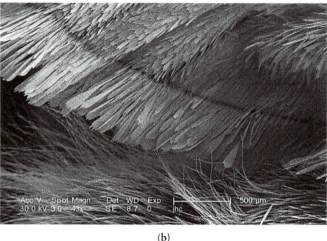

(a)   (b)

FIGURE 2-4

(a) Setae from the body of a honey bee and (b) scales from the wing of a moth. Scales are a special sort of flattened setae. Photographs courtesy of the Centers for Disease Control Images Library.

● Arthropods and Small Size

Currently among all animals there is a 13 order-of-magnitude size range, from the blue whale (100 t) to a tiny rotifer (0.01 mg). Among these creatures, insects and other terrestrial arthropods are relatively small; however, there remains a five orders of magnitude size range among the land-living arthropods. They are all on the small end.

As they are small, insects are affected by physical forces differently than are more massive animals such as vertebrates. One key relationship is that their surface areas are much larger in proportion to their masses. This is because the surface area of an object increases exponentially as a squared function (x^2) while mass increases exponentially as a cubic function (x^3). The differences in the relative amount of surface area to mass become tremendous when a typical insect is compared to even a small vertebrate, such as a rat or house cat. But even among the insects, there is a considerable range. For example, when a small leafhopper of about 3 mm is compared to a 12 mm modest-sized insect of similar proportion, such as a boxelder bug, we see a 16× (4^2) difference in surface area and a 64× difference (4^3) in mass—a 1:4 change in the surface

area to mass ratio. If we compare the same leafhopper to a somewhat larger insect, such as a 30 mm long grasshopper, the differences in surface area and mass would be 100× and 1000×, respectively. This produces a 10-fold change in the surface area to mass ratio.

Some problems come with being small. The comparatively large surface area increases problems associated with air resistance; smaller animals are relatively underpowered, which is why most small insects have a difficult time bucking even a modest headwind. Even more important, a large surface area increases water loss. The evolution of the waterproof exoskeleton was critical to allowing small animals, such as insects, to live in the dry areas of land.

Conversely, possessing a small mass provides several positive features. Small organisms are much less susceptible to impact forces that occur with a fall or collision. For example, an insect that is 1/10 the weight of a mouse and traveling at the same speed would impact with only 1/1000 the force. Also acting on the smaller animal is drag (resistance force) as it moves through either air or water. Because of an insect's relatively large surface area

(*continued*)

to mass ratio, these forces can result in slowing terminal velocity or even limit entirely the animal's ability to move through the medium. Together these forces (gravity and resistance) produce the very real effect that "the bigger they are the harder they fall." (Or, in the case of small insects, "the smaller they are the softer they fall.")

The relatively large surface area to mass ratio is also fundamental to many of the more notable activities that insects are able to engage in. For example, water striders that skate across ponds without sinking possess hairs at the tips of their legs, which help to maximize surface tension on water. This force is sufficient to offset the gravitational effects acting upon them; a heavier animal would sink. Modifications on the legs of flies provide adhesion forces that allow them to hang suspended from a ceiling or crawl on a vertical wall. Gravitation acting on a greater mass would exceed these forces, preventing such abilities in a much larger animal (Spiderman notwithstanding).

field." Because of the physical laws affecting larger animals, these analogies quickly break down, and a human-sized ant would likely be only as strong as a human or ape. Similarly, a kangaroo-sized grasshopper would likely not jump much farther than a kangaroo.

Perhaps the biggest advantage of being the size of insects is that the world is divided into so many ecological niches where they might establish themselves. A cow and a calf may need several hectares to support themselves in a rangeland. A large predator, such as a mountain lion, may range across several thousand hectares. A tiny vinegar fly can develop in moist spilled food in a floor crack, and there may be 1,000 or more different arthropod species that reside in a modest-sized yard with some landscaping. The small size allows arthropods to exploit different foods and live in different sites that cannot sustain a large animal.

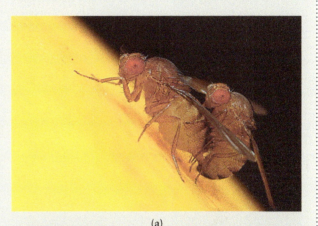

(a)

(b)

FIGURE 2-5
Small physical mass and adaptations that avoid breaking water tension allow water striders to overcome the effects of gravity so that they can move over the surface of water. Photograph by Whitney Cranshaw/Colorado State University.

The small mass also makes arthropods appear to be extraordinarily strong. This is because muscle strength is proportional to the cross-sectional surface area. Although a small insect is moving a small mass, the strength of the muscle is relatively large compared to the mass being moved. Therefore, insects can perform what appear to be extraordinary feats—carrying weights much heavier than their bodies or jumping distances that are a great many times their body lengths. As a result people often attempt analogies such as "if a man were as strong as an ant he could pull two boxcars" or "if a grasshopper were as large as a man it could jump the length of a football

FIGURE 2-6
(a and b) Vinegar flies of the genus *Drosophila* are very small and require little food for their larval development. Large numbers can develop in a bit of overripe fruit or the yeasts that settle in the bottom of a discarded beer bottle. Photograph of the adults courtesy of Brian Valentine.

CONSEQUENCES OF THE EXOSKELETON

The exoskeleton is a marvelous adaptation for small land-dwelling animals such as insects. Most importantly, it can provide a means to prevent water loss, a particularly critical concern for small animals that have a relatively large surface area to mass ratio. Further aiding water conservation are the infoldings of the exoskeleton that form the tracheal system. This allows for a means of moving oxygen directly to cells without much concurrent loss of moisture, in marked contrast to mammals that expel massive amounts of moisture from the lungs.

The rigid exoskeleton functions well to provide areas of support for muscles and other tissues. Precise internal modifications at different points can allow muscle systems for wings, legs, jaws, genitalia, and other structures that allow insects to move, feed, and reproduce. Externally, the exoskeleton can also be modified to produce different structures that allow an arthropod to enable specialized functions like the following: jaws to slice, suck fluids, or grind; legs to grasp, dig, or jump; wings to enable flight or provide protection; and myriad antennal forms that help allow the collection of chemicals, vibrations, or other critical signals of its environment.

FIGURE 2-7
Darkling beetles common in arid regions have very hard exoskeletons with thick waxy layers covering the surface. These features make the insects highly resistant to water loss and provide protection from many potential predators. Photograph by Whitney Cranshaw/Colorado State University.

The exoskeleton also provides an excellent defense against predators, parasites, and pathogens. Since it also extends to include the foregut and hindgut of the digestive system, it protects these sites from abrasion and pathogen entry. It also provides excellent shock-absorbing capabilities. The ability to withstand a blow or a fall is much greater for an animal with an exoskeleton than it is for an equal-sized animal with an endoskeleton.

The most significant downside of the exoskeleton is that it requires periodic replacement in order to allow growth. This is achieved during the process known as molting, discussed in detail in chapter 4. During this time when the old, hardened exoskeleton is discarded and the new one is constructed, the animal is soft, largely immobilized, and highly vulnerable.

FIGURE 2-8
A damselfly adult emerging from the exoskeleton of the last nymphal stage. During molting, the body is soft, and the insects are quite vulnerable. Photograph courtesy of Brain Valentine.

COLOR PRODUCTION IN INSECTS

From dull earth tones to spectacular iridescent reds, blues, greens, and violets, insects can produce a remarkable range of colors. These colors may serve many important functions:

Mate recognition. Many day-active insects such as butterflies and dragonflies use colors to help identify potential mates or rivals.

Warning colors. Bright colors and bold patterns are often used to signal a warning that the animal is capable of producing toxic chemicals or stings, or has other defense mechanisms. This is known as **aposematic coloration,** and certain warning colors (e.g., red and black, yellow and black) are displayed by a wide range of insects, thereby reinforcing these signal color patterns.

(a)

(b)

FIGURE 2-9

Certain bright colors and combinations, such as the yellow or orange and black of many stinging wasps, serve as a warning signal. Illustrated here is (a) a western yellowjacket, *Vespula pensylvanica*, and (b) a tarantula hawk, *Pepsis angustimarginata*; both produce a painful sting. Photographs by Whitney Cranshaw/Colorado State University.

Camouflage. Many insects have coloration that allows them to blend well with their background, allowing them to avoid being eaten by birds or other predators. Camouflage may also aid insects that are ambush hunters.

Startle markings. Brightly colored hindwings, or eyespot markings, may be suddenly displayed by insects seeking to confuse or deter predators.

Thermal regulation. Dark or light colors may be used as a mechanism to capture or reflect,

FIGURE 2-10

Eyespot markings on the wings of insects can be suddenly exposed to startle potential predators. Photograph courtesy of Howard Ensign Evans.

respectively, solar radiation to maintain body temperature (thermal homeostasis).

Color is produced in insects by one of two fundamental mechanisms. Some colors result from pigments incorporated into the exoskeleton. Other colors are produced by the structural features of the exoskeleton, which cause differences in how the various wavelengths of light bounce off the insect. Colors produced by pigments tend to be in the warmer wavelengths (reds, oranges, and yellows), while colors produced by structural features tend to be cooler (blues and greens), although there are many exceptions. Individual insects are not limited to using just one mechanism for creating color. Sometimes, both pigments and structural features are used in combination to create different colors and appearances. They are often used in tandem to enhance given colors (e.g., blues or iridescent violets).

Pigments are chemical compounds embedded in the exoskeleton or tissues immediately beneath the exoskeleton, which differentially absorb different wavelengths of light. Those wavelengths of light that are not absorbed by the pigments are allowed to pass through the tissue or are reflected back. For example, a red pigment on a grasshopper wing acts to differentially absorb some colors but reradiates a disproportionate amount in the red wavelengths. Where all visible light is absorbed and none reflected, the object

TABLE 2-1 Categories of pigments found in insects and the colors they produce.

PIGMENT CATEGORY	COLORS PRODUCED
Melanins	Black, brown, reddish brown
Pterins	White, red, yellow
Ommochromes	Red, brown, yellow
Tetrapyrroles	Green, blue
Papiliochromes	Reddish brown, yellow
Anthraquionones	Red, yellow
Carotenoids	Red, orange, yellow, green, blue
Flavonoids	Cream, yellow

appears black; when all visible light is reflected, white color is produced.

Insects use various types of pigments to produce colors and patterns (table 2-1). They are able to produce some during their normal metabolism. Others (e.g., flavonoids, carotenoids) are acquired from the food eaten during the immature stages.

Colors due to structural features are produced by the physical properties of the exoskeleton. Some molecular arrangements within the exoskeleton may allow longer wavelengths of light to pass through, while shorter wavelengths are reflected, producing blue color. Fine microsculpturing of the body surface or scales can cause different wavelengths of light to scatter in different directions by a phenomenon known as **diffraction**. When this occurs, color reflecting off the body of the insect will change depending on the angle of view.

Transmission of light may also be affected by the arrangement of chemical layers within the exoskeleton. Light waves of different lengths become separated as they enter the exoskeleton and are later reflected in a new arrangement. When such new arrangements of light waves are produced, the phenomenon is known as **interference**. Insect colors produced by light interference are iridescent or metallic.

FIGURE 2-11

Color in insects can be produced by two different mechanisms, one of which involves pigments. These grasshoppers, *Dactylotum bicolor*, show bright patterning produced by pigments. Photograph by Whitney Cranshaw/Colorado State University.

(a)

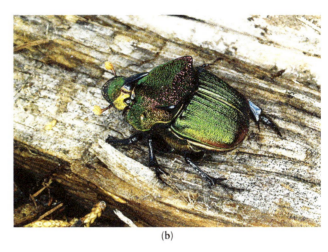
(b)

FIGURE 2-12
Scattering of light due to microsculpturing of the exoskeleton can produce many of the metallic and iridescent colors insects display, such as those produced on the wings of (a) a Morpho butterfly (*Morpho* sp.) and (b) a rainbow scarab (*Phanaeus vindex*). Photograph of rainbow scarab courtesy of Jim Kalisch/University of Nebraska. Photograph of the Morpho butterfly by Whitney Cranshaw/Colorado State University.

● Peppered Moth—An Example of Cryptic Coloration on the Move

The peppered moth, *Biston betularia* (Lepidoptera: Geometridae), native to Great Britain (and now found elsewhere, including Japan and the southeastern United States), is illustrative of how insects can evolve to changing environments. Our understanding of this insect benefits greatly from the long history of the English fascination with nature. This has been particularly true of the more "charismatic" species such as moths and butterflies, several of which have been studied for over 200 years.

The peppered moth is found in forested areas, and during the day rests on branches and trunks of trees. For protection from birds and other predators, it relies on camouflage. The **cryptic coloration** of the moths allow them to blend well against a background of bark and, particularly, crustose lichens, which are common on branches.

Wing colors and patterning of the peppered moth do vary. Light forms, which blend with lichens, predominated in early collections of the species; however, some dark forms (**melanistic**) began to be observed after 1848. The melanistic character of wing color is controlled by a few genes.

Incidence of the dark forms of this moth increased dramatically following the Industrial Revolution. Factories and the associated soot from the tremendous increase in coal burning rapidly transformed the English landscape. With extensive contamination by soot, the crustose lichens died out or were obscured. Lighter-patterned wings no longer provided effective camouflage for the peppered moth in industrialized areas. The darker morphs thrived on the now darker and lichen-denuded bark while the lighter forms were more susceptible to predation by birds. (Light morphs remained dominant in rural areas.) A series of experiments and observations by H.B.D. Kettlewell of Oxford in the late 1800s demonstrated the new advantage of the darker forms from bird predation.

Since the late 1800s, there have been dramatic changes in air quality and with it the return of lichens. Along with this the light-winged moths also returned as the predominant form. The dark morphs returned to being the more susceptible form to bird predation, while the lighter form thrived. The relatively brief rise and fall of the dark-winged form of this insect has been termed "industrial melanism" and is a good example of evolutionary processes.

(*continued*)

(a)

(b)

FIGURE 2-13

(a and b) The peppered moth, *Biston betularia*, has camouflage wing markings. Light forms of the peppered moth blend in well with the lichen on trees and rocks, and these forms predominated prior to industrialization in England. Where the background is darkened, as it happened following the pollution produced by heavy industrialization in England, dark forms of the moth become better camouflaged, and these forms predominated during the Industrial Revolution of the 1800s. When air pollution was reduced, the lighter forms again predominated. Photograph of the peppered moth courtesy of Lesley Simpson. Drawing by Matt Leatherman.

BODY SEGMENTATION AND TAGMOSIS

With even a casual glance it is apparent that arthropods are segmented animals, a feature they share with some other animals, such as the segmented worms of phylum Annelida and the velvet worms of phylum Onychophora. In certain places on the body, some segments coalesce to produce body regions that have distinct functions, a process known as tagmosis. The distinct body regions produced by the grouping and fusing of segments are called **tagmata** (singular **tagma**). In the insects, the body segments are consolidated into three tagmata: head (6 segments), the thorax (3 segments), and the abdomen (11 segments).

Different groupings of body segments occur among different classes of arthropods. For example, arachnids have a body that is arranged in two tagmata: **cephalothorax** and **abdomen**.

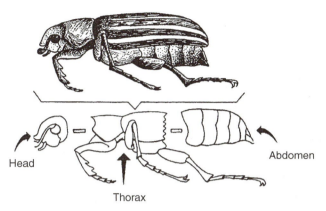

FIGURE 2-14
The primary body regions (tagmata) of a beetle—head, thorax, and abdomen. Drawing by Matt Leatherman.

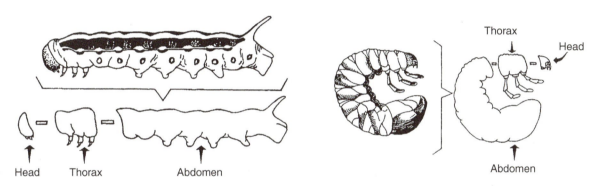

FIGURE 2-15
The primary body regions (tagmata) of the immature stage of a sphinx moth (hornworm) and a scarab beetle (white grub). Drawing by Matt Leatherman.

The Insect Head

The insect head is usually in the form of a hardened capsule produced by a near complete fusion of the first six segments. In addition, there are five associated pairs of appendages that are modified legs, constituting the mouthparts and antennae.

Most of the functions of the insect head are similar to that of the human head, allowing it to "go forward" to meet the environment. Much, but by no means all, of the sensory input comes through the external structures of the head, including eyes, antennae, and sensory palps associated with the mouthparts. Internally, the head contains masses of neurons (collectively known as the brain), which help integrate sensory input and send messages to the rest of the body. The head houses the mouthparts used to gather and ingest food.

FIGURE 2-16
The head of a katydid, an insect with prominent chewing mouthparts. Photograph by Whitney Cranshaw/Colorado State University.

INSECT MOUTHPARTS

Four pairs of appendages are associated with the mouthparts of an insect: the **labium**, **maxillae**, **mandibles**, and the **labrum**. In the majority of insects, these structures have evolved to enable insects to chew (See figure 2-17.) With insects that have chewing mouthparts the labrum is a fused, frontal pair of appendages, which forms a sort of an upper lip; a swollen area on the undersurface of the labrum is known as the **epipharynx**. The epipharynx may be inconspicuous in insects with chewing mouthparts but becomes more developed in some insects, such as mosquitoes.

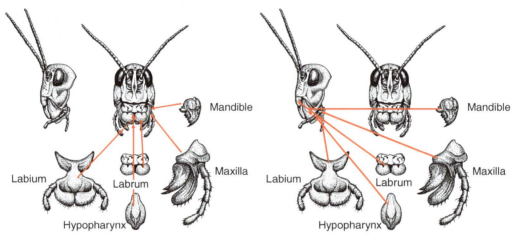

FIGURE 2-17

Chewing mouthparts of a grasshopper. In the front is the labrum, which appears as a flap. Underneath this are the mandibles, which are heavily sclerotized and used to cut food. The maxillae (singular **maxilla**) are underneath the mandibles, which help manipulate food and have associated sensory palps. The hindmost segment of the mouthparts is the labium, which covers the underside of the mouth and also has many sensory structures. A tongue-like hypopharynx is internal and helps move food into the pharynx of the gut. Drawings by Matt Leatherman.

FIGURE 2-18

Tiger beetles have chewing mouthparts that are modified for prey capture. Photograph courtesy of David Cappaert/Michigan State University/Bugwood.org.

Behind the labrum is a pair of opposable mandibles, which work horizontally and act as jaws. Often these are the most prominent structures of the mouthparts with cutting and grinding surfaces, which allow insects to cut up and chew food.

The appendages behind the mandibles are the maxillae, which are used primarily for food manipulation. In addition, these often have sensory appendages (**palps**) for taste, touch, smell, and temperature perception. A tongue-like **hypopharynx** projects between the maxillae and helps direct food into the oral cavity. The opening of the salivary ducts is often located near this structure, allowing saliva to be mixed with the food.

The hind structure of the mouthparts is the **labium**, formed by the fusion of the last pair of appendages. The labium forms a kind of lower lip, which helps in chewing. Sensory structures (palps) are often associated with the labium as well.

A great many insects have modified mouthpart structures, which allow them to feed in ways different from chewing. Individual parts may be greatly enlarged, greatly reduced, or have very different forms. Such different arrangements can allow insects to suck nectar from plants, withdraw blood from insects or other animals, or sponge up fluids. A summary of some of the different mouthpart arrangements and their functions is provided in table 2-2.

TABLE 2-2 Some derivations of insect mouthpart design and function.

INSECT EXAMPLE	FUNCTION OF MOUTHPARTS	MOUTHPART DESIGN
True bugs, aphids, hoppers, scale insects, whiteflies, psyllids, cicadas, etc.	Piercing and sucking fluids	Mandibles and maxillae are extremely elongated and form a stylet bundle. Fluids are removed and saliva introduced through two parallel channels of the interlocked maxillae. The labium is enlarged and supports the stylets but does not penetrate and fold when feeding.

FIGURE 2-19
The mouthparts of a cicada are highly elongated and modified to suck fluids. The food is withdrawn through a food canal made of the interlocking maxillae. Photograph by Whitney Cranshaw/Colorado State University.

(continued)

TABLE 2-2

INSECT EXAMPLE	FUNCTION OF MOUTHPARTS	MOUTHPART DESIGN
Mosquitoes	Piercing and sucking blood	The labrum, hypopharynx, mandibles, and maxillae form a stylet bundle, which pierces skin. Saliva is introduced through the hypopharynx, and the food canal runs between the labrum and hypopharynx. The labium is enlarged and supports the stylets but does not penetrate and fold during feeding. Palps of the maxillae are enlarged and have sensory functions.

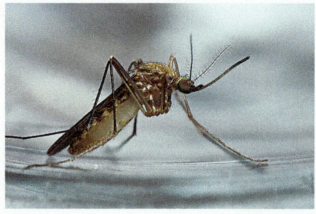

FIGURE 2-20

In the highly elongated mouthparts of a female mosquito, the blood is ingested through a food canal, which is composed of the labrum and hypopharynx. Photograph courtesy of Brian Valentine.

| Stable fly and tsetse fly | Piercing and sucking blood | The labium is greatly enlarged, tipped with hard teeth, and serves as the primary structure for piercing. The labrum and hypopharynx are elongated and thin and rest within the labium. Saliva is introduced through the hypopharynx, and the food canal runs between the labrum and hypopharynx. Mandibles and maxillae are extremely reduced. |

FIGURE 2-21

Stable flies suck on the blood of animals by use of a specially modified labium that can pierce skin. Photograph by Whitney Cranshaw/Colorado State University.

(continued)

TABLE 2-2

INSECT EXAMPLE	FUNCTION OF MOUTHPARTS	MOUTHPART DESIGN
House fly and blow fly	Sponging fluids	The labium is greatly enlarged and tipped with large fleshy lobes, which channel fluids. The labrum and hypopharynx are elongated and thin and rest in the labium. Saliva is introduced through the hypopharynx, and the food canal runs between the labrum and hypopharynx. Mandibles and maxillae are extremely reduced although the maxillary palps are prominent and have sensory functions.

FIGURE 2-22
Most of the mouthparts of a blow fly are highly reduced or absent, but the labium is greatly enlarged and functions to wick fluids. Photograph courtesy of Brian Valentine.

Lepidoptera	Sucking fluids	A portion of the maxillae (galea) is greatly elongated to form a straw-like mouthpart with a central food canal. This coils when not in use. Palps of the labia are well developed and have sensory functions. Other structures are reduced or absent.

FIGURE 2-23
Butterflies and moths are fluid feeders, using modified structures of the maxillae that produce a siphoning tube. Photograph by Whitney Cranshaw/Colorado State University.

(continued)

TABLE 2-2

INSECT EXAMPLE	FUNCTION OF MOUTHPARTS	MOUTHPART DESIGN
Thrips	Piercing and sucking fluids	Only one elongated mandible is present (the left), and it functions to penetrate the plant surface. The maxillae are thinner and longer, forming stylets that puncture underlying plant cells. The labium and labrum are greatly enlarged and form a cone, which is used to suck released fluids into the mouth.

FIGURE 2-24
Thrips have unique mouthparts that include only a single mandible (used to penetrate surface cells) and a pair of elongated maxillae. The labrum and labium are enlarged to form a cone. Photograph courtesy of John W. Dooley/ USDA APHIS PPQ/Bugwood.org.

INSECT EXAMPLE	FUNCTION OF MOUTHPARTS	MOUTHPART DESIGN
Honey bee and bumble bee	Chewing and lapping fluids	Mandibles are retained for chewing. The labium is greatly elongated and adapted to pick up fluids, which are then moved to the mouth.

FIGURE 2-25
Many bees, such as this bumble bee, have mouthparts that allow both chewing and sucking fluids. Photograph courtesy of Brian Valentine.

ANTENNAE

The antennae are sensory structures usually located near the anterior (front) portion of the head close to the compound eyes. Antennae are covered with sensory cells, which allow the insect to perceive odors, humidity, and vibration. They can also be used as touch organs and are sometimes called "feelers," although given their primary role, they are better described as "smellers."

FIGURE 2-26
Head of an aphid showing structures of the antennae, compound eye, and ocelli. Photograph courtesy of Brian Valentine.

FIGURE 2-27
Head of a wasp showing structures of the antennae, compound eye, and ocelli. Photograph courtesy of Brian Valentine.

All antennae are composed of three main parts: **scape**, **pedicel**, and multisegmented **flagellum**. The overall shape and relative size of the antennae vary tremendously among the insects. Long, thin, filamentous antennae are characteristic of many common insects such as crickets, cockroaches, earwigs, and some moths. Antennae of other insects may be club-like, tipped with knobs, or have an arrangement like a saw blade. Antennal forms may also be variable between sexes; the males of some moths have large plume-like antennae while females have smaller, thinner antennae.

ORGANS FOR LIGHT PERCEPTION

Several different types of structures for perceiving light can be found on the insect head. Most prominent are **compound eyes**, a pair of which is found on the head of almost all adult insects and many immature insects. (Species that live in soil and life stages of insects spent buried in soil or other dark substrates usually lack compound eyes.) Each compound eye is made up of many different light-sensing units called **ommatidia** (singular **ommatidium**). The largest number of ommatidia is found among dragonflies, which may possess 10,000 or more; low numbers of ommatidia, often only a few dozen or even fewer, occur among insects that spend much time in dark environments.

FIGURE 2-28
Many flies have very large compound eyes, which make them more visually adept among most of the insects. Photograph courtesy of Brian Valentine.

Most insects, in both adult and immature stages, also possess up to three **ocelli**, or "simple eyes," which are found on the top of the head, typically in a triangular arrangement. The ocelli cannot determine forms but are very sensitive to light intensity. The functions of the ocelli are often unknown, but can include recognition of changes in day length, which can be important signals for insect development and activity.

A third type of light-perceiving organ is the **stemmata**, which can be found on the sides of the head. These occur only in the immature stages (larva) of some insects that undergo complete metamorphosis, such as caterpillars, sawfly larvae, and beetle larvae. Their numbers are few, fewer than a dozen, and they provide individual points for light detection. By moving the head, the larval insect can use the stemmata to help direct its movements.

FIGURE 2-29
Stemmata are small light-sensing organs found on the heads of caterpillars. Photograph courtesy of Brian Valentine.

In some insects, light can be detected by other parts of the body. However, the most important light-perceiving organs, and the only ones capable of detecting images, occur on the head.

The Thorax

The **thorax** is the body region located behind the head and is composed of three segments: **prothorax**, **mesothorax**, and **metathorax**. Thick sclerites cover the thorax as plates. A sclerite found along the side of the body, a lateral sclerite, is referred to as a **pleuron**. A dorsal sclerite is known as a **notum**, while a ventral sclerite on the underside is a **sternum**. Altogether sclerites form a very strong box to which many muscles are attached internally.

The thorax is responsible for locomotion, allowing insects to walk, run, jump, swim, and fly. The legs occur on the thorax, with one pair on each of the three segments. In winged insects, the wings are found on the second and third segments.

INSECT LEGS

The basic structure of an insect leg includes five segments. At the base is the **coxa**, a segment that connects the leg to the thorax. Leading outward (distally) from the coxa are the **trochanter, femur, tibia**, and **tarsus**.

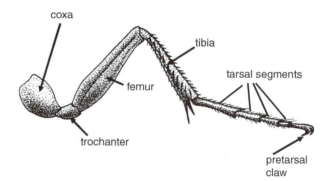

FIGURE 2-30
External structures of an insect leg adapted to jumping (saltatorial). The coxa is the segment attached to the thorax. The small trochanter is followed by the large, heavily muscled femur. The tibia is also usually quite prominent and may have spines as shown in this figure. Several segments make up the tarsus, and the leg is tipped with the pretarsus, which is usually clawed. Drawing by Matt Leatherman.

The **trochanter** is typically a small segment that allows certain kinds of leg flexing. The **femur** (plural **femora**) is often the largest and most prominent part of the leg, particularly in a jumping insect such as a grasshopper. The **tibia** is also prominent, particularly among insects that use it to help grasp prey or dig through soil.

FIGURE 2-31
Tarsal claws tip the legs of most insects. Photograph courtesy of the Centers for Disease Control Images Library.

FIGURE 2-32
Ground beetles have legs designed for walking or running (cursorial). Photograph courtesy of Jim Kalisch/University of Nebraska.

At the end of the insect leg is the **tarsus,** which itself is composed of three to five very small segments. At the end of the last tiny segment of the tarsus is the pretarsus, which functions as a landing pad or a set of claws. The **pretarsus** is where "the rubber meets the road" as far as insects are concerned and is modified to allow the insect to walk or land upon various substrates. The pretarsus of a fly allows it to walk up a wall or upside down on the ceiling.

Insects differ in their habits, and the form of their legs often reflects this. **Cursorial legs** are modified for running or walking and are the most common leg types found among insects. **Natatorial legs** are modified for swimming, producing a feathered oar-like form, used by beetles and bugs that spend their lives in water. **Saltatorial legs** allow jumping, provided by a greatly enlarged femur on the hindlegs. Such a leg type is common with insects such as grasshoppers, fleas, and some jumping beetles (flea beetles). **Raptorial legs** function to seize and hold onto prey. These typically occur on the front pair of legs on insects such as mantids and assassin bugs. Front legs may also be enlarged in the case of insects that spend much time digging. These **fossorial legs** are found in insects such as mole crickets and immature cicadas that develop below the ground.

FIGURE 2-33
Water boatmen have legs designed for swimming (natatorial). Photograph courtesy of Tom Murray.

FIGURE 2-34
Robust camel crickets have hindlegs designed for jumping (saltatorial). Photograph by Whitney Cranshaw/Colorado State University.

FIGURE 2-35
Ambush bugs have forelegs designed to help grasp prey (raptorial). Photograph courtesy of Tom Murray.

FIGURE 2-36
Mole crickets have front legs that are adapted for digging (fossorial). Photograph courtesy of David Cappaert/Michigan State University/Bugwood.org.

WINGS

Insect wings are novel structures produced by outgrowths of the exoskeleton. Insects have two pairs of wings, and they always originate on the sclerites of the second and third segments of the thorax (meso-thorax and metathorax). Each wing is composed of two layers of integument with thickened areas, called veins, which provide reinforcement as well as a means to pump fluids to the wings. The veins further help to provide strength and rigidity to the wing by their arrangement as a series of miniature hills and valleys, somewhat in the manner of a folded paper fan.

Just as leg modifications are seen in some species, various insect groups have modified wings to allow different functions and uses. A common adaptation is that the front wing is thickened so that it provides a protective function for the hindwing, often losing importance for flight. Wings that are used in flight may also have increasing or decreasing numbers of veins for support. These features are very distinctive of the insect orders; the suffix "ptera," meaning "wing," often figures prominently in the names used to identify the insect orders (e.g., Diptera, Coleoptera, Lepidoptera). A summary of some of the different wing types and their functions is provided in table 2-3.

TABLE 2-3 Some derivations of insect wing types.

INSECT EXAMPLE	FUNCTION OF WINGS
Mantids, roaches, and grasshoppers	The forewings (mesothoracic wing) are thickened and provide cover for the more membranous hindwings (metathoracic wing), which are used in flight. These are known as **tegmina**.

FIGURE 2-37
The forewings of a grasshopper are thickened and have protective functions (tegmina). Photograph by Whitney Cranshaw/Colorado State University.

(continued)

TABLE 2-3

INSECT EXAMPLE	FUNCTION OF WINGS
True bugs (Hemiptera, suborder Heteroptera)	Forewings are thickened and protective in function, but the distal (outer) area is more membranous, producing a **hemelytra**. The hindwings are membranous and used in flight.

FIGURE 2-38
Stink bugs and other "true bugs" in the order Hemiptera have unique forewings, which are thickened at the base and more membranous at the tip (hemelytra). Photograph by Whitney Cranshaw/Colorado State University.

Beetles	The forewings are thickened and hardened (**elytra**) and provide a shield over the body, which protects the hindwings.

FIGURE 2-39
The front wings of beetles are hardened (elytra) and function to shield the hindwings and body. Photograph courtesy of Jim Kalisch/University of Nebraska.

(continued)

TABLE 2-3

INSECT EXAMPLE	FUNCTION OF WINGS
Flies (house flies, mosquitoes, and other members of the order Diptera)	The forewings (mesothoracic wings) are used for flight, and the associated mesothoracic segment is greatly enlarged to support flight muscles. The hindwings (mesothoracic wings) are reduced to small balancing organs (**halteres**), which help to stabilize flight.

FIGURE 2-40
Flies have only a single pair of wings, which are attached to the middle section of the thorax (mesothorax). The hindwings are modified into small structures called halteres, which aid in flight stability. Photograph courtesy of Brian Valentine.

Butterflies and moths

The hindwings are smaller than the forewings. Scales cover the wings.

FIGURE 2-41
The wings of butterflies and moths are clothed in flattened setae (scales). Photograph by Whitney Cranshaw/Colorado State University.

(continued)

TABLE 2-3

INSECT EXAMPLE	FUNCTION OF WINGS
Bees and wasps	The hindwings are smaller than the forewings. A series of hooks on the front edge of the hindwings (hamuli) lock to the hind edge of the forewings so that the two pairs of wings work together in flight.

FIGURE 2-42
Wasp wings hook together in flight by means of small hooks along the leading edge of the hindwings. This is a cicada killer wasp subduing its prey, a dog-day cicada. Photograph courtesy of Nancy Hinkle/University of Georgia/ Bugwood.org.

Dragonflies and damselflies	Wings cannot fold over the body but may be very large and can project far from the body. The wings are membranous and have a very large number of bracing veins.

FIGURE 2-43
Dragonflies have large wings that are supported with numerous cross veins. Photograph courtesy of Brian Valentine.

(continued)

TABLE 2-3

INSECT EXAMPLE	FUNCTION OF WINGS
Thrips	Both pairs of wings appear to have a central solid core with radiating projections, which produce a feathered or fringed effect.

FIGURE 2-44
Thrips have wings that appear fringed. Photograph courtesy of Plant Protection Archives/Bugwood.org.

The Abdomen

The abdomen is composed of 11 segments, but some may be so reduced that they are not readily visible. The abdominal region is primarily important because various internal functions take place as it houses much of the digestive tract, organs for excretion, and the reproductive tract.

Among insects that live on land, there are often openings along the side of the abdomen known as

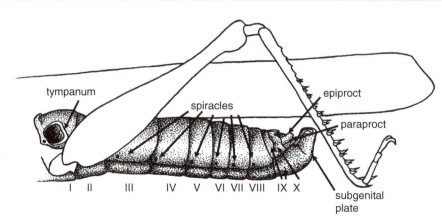

FIGURE 2-45
Primary features of the exterior of the abdomen of an insect, based on a generalized grasshopper. Drawing by Matt Leatherman.

spiracles. They are the entry ports that allow gas exchange, moving oxygen into the animal and waste gases out. The spiracles are usually found along the sides of the insect but may be located as a single tube at the tip for insects that develop in water. Many insects that develop in water (mayflies and caddisflies) have thin areas of the abdomen (gills) that allow oxygen to move into the body.

The terminal segment of the abdomen often bears a pair of tail-like appendages called **cerci,** which often have a sensory function. In adult insects, there are usually external structures of the abdomen used for copulation and egg laying. For example, males may have appendages that function to clasp females during mating and an **aedeagus** (penis) for copulation. Females may have a visible **ovipositor** that allows them to lay eggs into soil, plant stems, or other materials.

FIGURE 2-46
Sensory cerci are found at the tip of a stonefly nymph abdomen, and feathery gills occur concentrated along the sides of the thorax. Photograph courtesy of Tom Murray.

FIGURE 2-48
Male genitalia of a winter stonefly, *Allocapnia* sp., showing the claspers used during mating. Photograph courtesy of Tom Murray.

(a)   (b)

FIGURE 2-47
Stout cerci are a prominent feature of earwigs and can be used to help manipulate food. (a) The cerci of a male earwig are more bowed than those of (b) a female and can also be used to grasp the female during mating. Photographs courtesy of Tom Murray.

FIGURE 2-49
The bizarrely enlarged male genitalia of scorpionflies lend them their common name. Photograph courtesy of Tom Murray.

FIGURE 2-50

A long sword-like ovipositor is found among some katydids and crickets to insert eggs into soil. Photograph by Whitney Cranshaw/Colorado State University.

FIGURE 2-51

A highly elongated ovipositor adapted for drilling is found among parasitic wasps that lay eggs on hosts located within plants. Photograph courtesy of Brian Valentine.

3

The Internal Organization

Arthropods share the same basic needs for sustaining life as all animals: an ability to acquire food for energy and growth and oxygen for respiration processes required by living cells, a system to excrete waste, and some method of reproduction. However, the specific details in how these challenges are met by arthropods may differ considerably from how humans or more familiar animals function, often contributing to perceptions that arthropods are rather alien and bizarre life forms.

Gas Exchange

Insects, like humans, require oxygen to fuel their metabolism. However, because of their small size, insects are highly susceptible to desiccation and the process of gas exchange usually must be done with minimal water loss. This is achieved by most insects using an **open tracheal system**, which delivers oxygen directly to their tissues.

FIGURE 3-1
Side view of a caterpillar showing elliptical-shaped spiracles on segments of the prothorax and abdomen. Spiracles are openings where air moves into the body of the insect and is carried into the tracheal system to the cells. The spiracles may open and close, as needed. Photograph by Whitney Cranshaw/Colorado State University.

The open tracheal system directly connects all of the living tissues of the insect to its own supply of oxygen-containing air. The air is drawn first into the body through openings called **spiracles**. The spiracles are usually found along the side of the thorax and abdomen and are often modified with **valves** that can close when not in use, sealing the tracheal system.

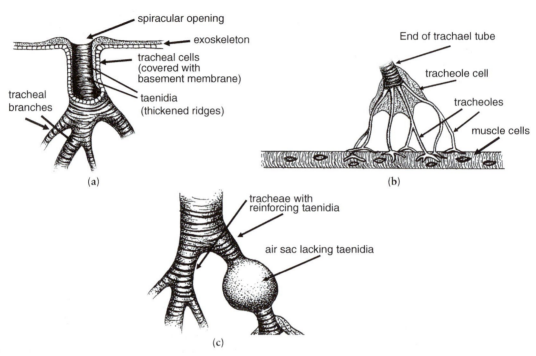

(a)

(b)

(c)

FIGURE 3-2
Spiracles are exterior openings of the tracheal system (a), which is primarily made of branching tracheae. The tracheae are reinforced with thickened ridges (taenidia) and are part of the exoskeleton that is replaced at each molt. At the end of the tracheal system (b) are tracheoles, which are extremely narrow (ca. 1 μm in diameter). The tips of the tracheoles lie immediately next to a cell, and oxygen moves from the tracheoles into the cells through passive diffusion. Within the tracheal system there may also be enlarged air sacs (c) that lack taenidia. The air sacs can have many functions, including increasing buoyancy, ventilation of the tracheal system, and sound production. Drawings by Matt Leatherman.

Once air passes through the spiracle, it enters the **trachea** (plural **tracheae**). The trachea is made of exoskeleton and is in the form of a tube that is reinforced along most areas by spiral thickenings known as **taenidia**. The tracheae branch repeatedly, narrowing with each branching and infiltrating throughout the body. At the end the tube becomes extremely narrow (approximately 1 μm in diameter) and the exoskeleton is thin, producing a blind sac called a **tracheole**. The tracheoles are located immediately adjacent to the living cells of the organism and oxygen can then diffuse through the thin-walled tracheoles and enter tissues.

At certain points along its length, the trachea may also expand to form **air sacs**, which lack the

FIGURE 3-3
A canna leafroller (*Calpodes ethlius*) has an unusually transparent body. The larger branches of its tracheal system can be seen radiating from the spiracle. Photograph courtesy of David Shetlar/The Ohio State University.

reinforcing taenidia, allowing the air sacs to expand and contract. These can provide many important functions. Alternating expansion and contraction of the air sacs can help increase the volume of air flowing through the tracheae. When air sacs expand, they also make the insect less dense, which can increase the insect's buoyancy in water or enable the insect to use less energy for flight. Air sacs are also used in sound production by some insects and the ability of air sacs to be compressed can provide some room for growth of internal organs.

Most of the movement of gases occurs through the process of **passive diffusion**, flowing along a concentration gradient from an area of high partial pressure (i.e., high concentration) to one of low partial pressure. Thus, oxygen will typically move from outside the insect, through the spiracles and down the tracheal system, to areas of the tracheoles where oxygen occurs at lower concentration, having been withdrawn for use in respiration by the adjacent cells. In larger, active insects (like grasshoppers), passive diffusion alone is not sufficient to deliver enough oxygen to meet the high respiratory demands of the tissues. These insects will also use a process known as **active ventilation** where rhythmic pumping of the abdominal muscles actively pushes air throughout the tracheal system and the spiracles alternately open and close. The air sacs are important in this pumping, allowing expulsion of oxygen-depleted gas and drawing in new air in a manner similar to a bellows.

Insects that live in water and those that live as internal parasites of other animals do not have access to atmospheric oxygen and must acquire it from the water or surrounding fluids. These insects have a **closed tracheal system**, which lacks external spiracles. They use a tracheal system that is entirely contained within their bodies. Most aquatic insects have areas known as **gills** covered by a very thin layer of exoskeleton. These are often quite large and feathery, providing a large surface area through which oxygen from the water diffuses to enter the tracheal system. Very small insects in moist environments and those that occur as internal parasites may have a very thin exoskeleton and can acquire oxygen directly through their cuticles.

Not all aquatic insects possess gills. Many have not only an open tracheal system with spiracles but also special adaptations that allow them to live underwater. Some, such as mosquito larvae, have a single pair of large spiracles at the tip of a long

FIGURE 3-4

Many aquatic insects bring oxygen into their bodies by means of gills, which are covered by a thin layer of exoskeleton through which oxygen can diffuse. In the illustrated stonefly nymph, the gills are present at the base of the thoracic legs. Once oxygen moves through the gills, it is transported via the tracheal system to the cells throughout the body. Photograph courtesy of Tom Murray.

snorkel-like projection at the end of the abdomen, allowing them to take in fresh air at the water surface. Other aquatic insects possess hairs at certain points of their bodies that help trap a bubble of air that covers the spiracles while the insect is underwater. As the insect consumes oxygen, the oxygen within the bubble becomes depleted (partial pressure of oxygen drops), causing fresh oxygen to diffuse into it from the surrounding water where oxygen concentration is higher. This "air bubble gill" is known as a **plastron** and is the insect equivalent of a scuba tank, allowing some insects to

FIGURE 3-5

Rattailed maggots, *Eristalis* spp., are fly larvae adapted to living in highly polluted, oxygen-depleted water. They acquire oxygen through a highly modified spiracle, which arises from the back of the larva and terminates at the end of a long tube. By extending the spiracle to the surface, they can acquire oxygen. Photograph courtesy of Jim Kalisch/University of Nebraska.

FIGURE 3-6
Several aquatic beetles and bugs can stay underwater for an extended period by the use of a plastron. This is a bubble of air that is held next to the body by means of air-trapping hairs, with the bubble covering the spiracles. As oxygen is depleted in the bubble, some replacement oxygen diffuses into the bubble. Photograph of a giant water scavenger beetle courtesy of Joseph Berger/Bugwood.org.

remain underwater for hours before they need to gather a new bubble with a quick surface visit.

Excreting carbon dioxide presents a somewhat different challenge than acquiring oxygen. Carbon dioxide diffuses through insect tissues approximately 19× faster than oxygen, largely due to its much greater solubility in water. This can often allow much of the carbon dioxide to diffuse through the exoskeleton. Simple diffusion through the exoskeleton is a mechanism through which Collembola (springtails) and some insects that inhabit very moist environments (i.e., develop as internal parasites) excrete waste carbon dioxide. In most insects, carbon dioxide, in addition to diffusion through the exoskeleton, is eliminated via the tracheal system, sometimes in the form of actively expelled bursts.

● Why Insects Can't Be Big

When considered among the animals, insects and other terrestrial arthropods are relatively small creatures. Even those found in the oceans rarely get very large, although some of the behemoths are commonly found on supermarket shelves, such as king crabs and lobsters. Most arthropods are less than 1 cm in length, and great size is not possible with these animals. Two arthropod features that contribute to their success as animals limit their potential size: the exoskeleton and their manner of air exchange.

The basic support structure of an exoskeleton resembles a tube. For small animals this provides a combination of strength and relatively light weight, avoiding the use of dense heavier support such as the bones of a mammal endoskeleton. Such an arrangement fails with larger organisms due to the effect of increased size on the relationship between surface area and body mass.

With each doubling in body size, the surface area increases by the square while body mass increases by the cube. For example, an insect that is 6 cm long would have a surface area four times that of the insect having the same shape that was half as long (3 cm). On the other hand, a larger insect would have a body mass that was eight times as heavy. The exoskeleton can be, and is, thickened in larger insects; however, at a certain point the exoskeleton can no longer support the body mass without becoming overly thick, heavy, and burdensome.

Another problem concerns molting or shedding the exoskeleton, which must be done repeatedly during the development of all arthropods. During the critical time when an arthropod is shedding the old exoskeleton and building a new one, the body is quite soft and vulnerable; the exoskeleton only becomes supportive once it has gone through the process of sclerotization and has hardened. An arthropod with a massive body would simply collapse during molting. (Arthropods found in water can support a somewhat larger body size due to buoyancy effects, but these are also ultimately limited by the same constraints of an exoskeleton.)

The size of arthropods is also limited by the way they acquire and distribute oxygen. Insects, and most terrestrial arthropods, achieve this through an elaborate tracheal system that involves infoldings of the exoskeleton that extensively branch until they ultimately reach every cell. This system has a great advantage in water conservation, but it depends largely on the process of **passive diffusion**. With passive diffusion, gases redistribute to be in a state of equilibrium, causing oxygen to move to areas of lower concentration, such as the respiring cells.

Relying on passive diffusion of gases works fine when the distance is short. It becomes much less efficient as the tubes of the tracheal system become long, requiring more time for oxygen to reach the tracheoles through which it is passed to cells. These cells thus become starved of

(continued)

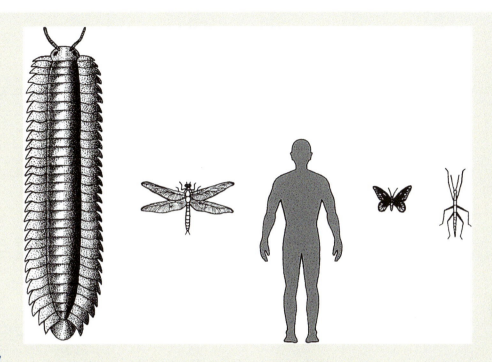

FIGURE 3-7

Relative size of a human in comparison to some extinct and extant arthropods. On the left is an ancestral diplopod, *Arthropleura*, which reached 2.6 m in length and thrived during the upper Carboniferous period, about 300 mya. Next to it is a representative of griffinflies, a distant relative of today's dragonflies, which had a wingspan of up to 75 cm wide and were present a bit later, during the late Permian period. Comparatively, today the widest wingspan found in a butterfly is about 31 cm, and the insect with the longest body is a walkingstick, which is about 55 cm long. Drawing by Matt Leatherman.

oxygen and cannot function. It is thought that one of the main reasons that some larger insects were found in the Carboniferous period, such as the giant griffinflies of the extinct order Protodonata, was that the oxygen concentrations were much higher at that time—estimated at perhaps 35% compared to the present level of 21%. Under these conditions it was a bit easier for sufficient oxygen to move to all cells of the insect's body.

Circulation of Blood

Arthropod blood, known as **hemolymph**, may make up 20%–40% of the arthropod's total weight. Most of it is composed of watery plasma that contains various salts, sugars, lipids, hormones, and other materials needed for development.

Additionally, there are various types of blood cells (**hemocytes**) associated with the hemolymph. Depending on the cell type, hemocytes have several different functions: ingestion of foreign particles (phagocytosis), formation of a protective layer of tissue around large foreign particles (encapsulation), coagulation at wounds, and transfer of nutrients, waste products, and hormones around the body.

The hemolymph also functions as an internal lubricant for organs, muscles, and the exoskeleton.

Arthropod blood has many of the same functions as the blood of vertebrates. However, one major difference between arthropod blood and that of vertebrates is that, with few exceptions, arthropod blood is not used in the transport of oxygen (see earlier discussion of the tracheal system). Arthropod blood lacks the red hemoglobin used by vertebrates to transport oxygen and is thus normally nearly colorless. (A few insects found in submerged, oxygen-poor environments do have hemoglobin, which allows them to store oxygen when it becomes available.)

The way the blood is transported throughout the body is also different from the closed circulatory

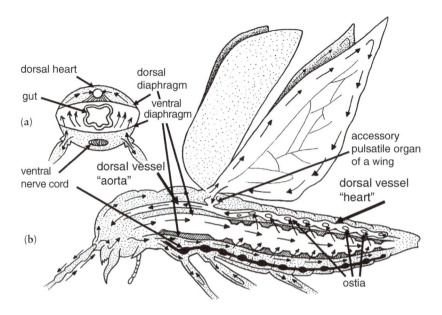

FIGURE 3-8

A generalized illustration of an insect's circulatory system (a, b). Insects, like all arthropods, have an open circulatory system in which blood bathes the tissues and is not confined to vessels. The circulation of the blood is largely produced by the action of a dorsal heart, a tubelike organ that moves the blood, which enters the heart through openings known as ostia, from the back of the body. Alary muscles attached to the heart (not shown) pump the blood forward through the aorta to the head. At some points of the body (e.g., base of wings, legs, antennae), accessory pulsatile organs may help direct blood. Also, diaphragms cross the body, which helps direct the flow of blood. Drawings by Matt Leatherman.

network of arteries, veins, and capillaries found in vertebrates. Instead, the blood freely bathes the internal organs and tissues. A simple pumping system maintains a current of hemolymph moving throughout the arthropod's body.

The principal organ responsible for circulating the hemolymph is the **dorsal vessel**. It is essentially a tube that lies along the dorsal midline and is held in place by connective tissues and the **alary muscles**. Underneath the dorsal vessel is a sheet known as the **diaphragm,** which largely isolates the dorsal vessel from the rest of the body cavity, allowing directed movement of the blood. New blood enters through one-way valves (**ostia**) at the abdominal end of the dorsal vessel, and waves of peristaltic contractions force the blood forward where it ultimately exits near the head.

Fat Body

Insects possess a unique organ known by the poorly descriptive name "fat body." It occurs in the form of sheets or lobes of specialized cells found in many places within the hemocoel, often surrounding other internal organs. Fat body cells are held together and are connected to the body wall by a membranous sheath.

The fat body serves a great many important functions. It is capable of storing fats, glycogen (a form of carbohydrate), and proteins. The fat body assists in the regulation of blood sugar levels and assists in the metabolism of sugars, lipids, and proteins. In some insects, the fat body is used to store metabolic wastes until further processing removes them from the body. Given all of the functions that the fat body plays in insect physiology and metabolism, it is perhaps most analogous to the human liver when comparing insect and human organ systems.

Digestive System

Every living cell in an insect's body requires energy. In order to provide energy to the cells, food must first be ingested, then digested, and finally absorbed into the body. Ingestion, digestion, and absorption take place in the insect's alimentary canal or gut.

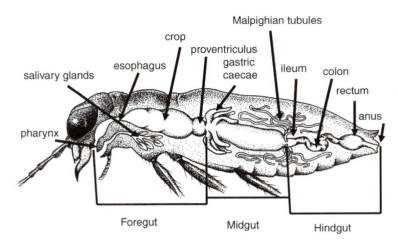

salivary glands — esophagus — crop — proventriculus — gastric caecae — Malpighian tubules — ileum — colon — rectum — anus

pharynx

Foregut   Midgut   Hindgut

FIGURE 3-9

A generalized illustration of an insect's digestive system. The hindgut and the long Malpighian tubules that float in the hemolymph compose the main organs of the excretory system. Drawing by Matt Leatherman.

The alimentary canal is divided into three general regions: the **foregut**, the **midgut**, and the **hindgut**. The foregut is the region where food enters the gut and is initially processed for digestion, later moving to the midgut. The mouth, or oral cavity, is where food is chewed, salivary enzymes are added, and digestion begins. From the mouth, the food moves through the **pharynx**, which functions as a pump that moves the food through the foregut. In insects that feed on a liquid diet, the pharynx is particularly critical in allowing these insects to suck up their food.

Once through the pharynx, food passes through the tubelike **esophagus** and is deposited in the crop where it may be temporarily stored. At the end of the foregut is the **proventriculus**, a structure that can function as a valve controlling entry into the midgut, a filter, and, in some insects, a grinding organ to further fragment food. The entire foregut is lined with exoskeleton.

The midgut is the primary site of food digestion and absorption, and it is the only area of the digestive system that is not lined with exoskeleton. Within the midgut, digestive enzymes are secreted. Although the midgut is largely in the form of a tube, there are often saclike projections at the front end (**gastric caeca**), which increase the surface area.

Separating the living cells of the midgut from the food is the **peritrophic membrane**. This membrane is continuously produced and helps to protect the gut wall from abrasive food particles and to filter and compartmentalize the food as it is processed. Nutrients pass through the peritrophic membrane and are taken in by the cells lining the midgut, where they are then passed into the hemolymph.

After the food has been digested and nutrients have been extracted, it passes to the hindgut. The primary structures include the **colon**, the **rectum**, and the **anus**. Special structures known as **Malpighian tubules** float in the hemolymph and extract nitrogenous wastes, emptying into the hindgut. Although the most obvious function of the hindgut is to process undigested food and waste for ultimate excretion, it has many other important functions. The colon, to some extent, and, more importantly, the rectum are responsible for the reabsorption of water from digested material as it passes down the gut. Much of the ability of an insect to maintain water balance occurs at these points. In addition to reabsorption of water, the rectum may reabsorb various salts, amino acids, and sugars, as needed.

Associated with the digestive system of many insects are beneficial microorganisms (e.g., bacteria, protozoa). Known as **symbionts,** these microorganisms may produce enzymes that allow the insect to digest certain types of foods; for example, gut protozoa in termites allow these insects to digest cellulose. Other types of these symbiotic microorganisms may produce critical vitamins or lipids that the insect cannot make itself. The symbionts may be found in the midgut of some insects and the hindgut of others.

● Insect Symbionts

A great many insects—probably most—harbor in their bodies various types of microorganisms. Many of these provide benefit to the insect and are known as **symbionts**, and often both the insect and the microbe derive benefit from their association, a **mutualistic relationship**. Insect symbionts involve various species of bacteria or protozoa that are usually found nowhere else but within the body of their host insect species.

(a)

(b)

FIGURE 3-10

A great proportion of insects harbor microorganisms known as symbionts in their bodies. The symbionts can produce many positive effects for their insect host, sometimes providing essential functions. For example, symbionts in bed bugs (a) provide B vitamins that are lacking in their all-blood diet. Many termites (b) harbor symbionts in their hindgut that help degrade the otherwise indigestible lignocellulose in plant tissues. Photograph of a feeding bed bug courtesy of Gary Alpert/Bugwood.org. Photograph of subterranean termite workers and soldier by Tom Murray.

Most widespread are bacteria, and these have been found in locations throughout the body of insects. Typically they occur within specialized cells known as **mycetocytes**. Often they are found associated with the cells lining the midgut, and sometimes they group together into large masses, known as **mycetomes**. Mycetocytes may also occur in the fat body or other tissues. The symbionts can also be transmitted to offspring of the insect. During egg development, mycetocytes move to the ovaries, and the bacteria can then migrate into the eggs. This allows them to be passed on to the next generation through what is known as **transovarial transmission** (passage via the egg).

Symbiotic bacteria can perform many functions that are critical to the survival of the insect. Some symbionts help degrade food, forming acetates and other simple molecules. Sterols and certain critical vitamins that insects are unable to synthesize may be provided by symbiotic bacteria. The use of symbionts is particularly common among insects that feed on a liquid diet deficient in proteins (e.g., plant-feeding Hemiptera) or critical vitamins (e.g., blood-feeding lice and bed bugs).

Some symbionts have also been shown to affect immune responses. For example, the symbiotic bacterium *Buchnera aphidicola* has been demonstrated to help protect its aphid host from the attack of parasitic wasps. Parasitoid larvae attempting to develop within a *B. aphidicola*–infected aphid show substantially higher mortality than those in noninfected aphids.

Other symbionts live within the digestive system, particularly in the hindgut. In primitive termites enormous numbers of protozoa are found; these help to degrade and utilize the otherwise indigestible lignocellulose found in wood and other materials. In addition, protozoa and bacteria within the hindgut may utilize the waste nitrogen products that enter the hindgut, helping to recycle these materials into amino acids that are often deficient in the termite's diet. (In addition, microorganisms associated with insects may themselves support their own symbionts; some protozoa that live within the hindgut of termites harbor nitrogen-fixing bacteria.) Most of these products produced by these hindgut-dwelling symbionts cannot be directly absorbed by the insect but are acquired when they consume the microbe-rich feces. Such feeding (**coprophagy**) is also necessary for insects to reacquire symbionts that live in the hindgut following molting, when the linings of the foregut and hindgut are both replaced.

Excretory System

The function of an animal's excretory system is the maintenance of a constant internal environment. Concentrations of water, salts, and other solutes must be maintained at critical concentrations while removing wastes produced by normal metabolism. In humans, the kidneys are largely responsible for this critical task.

The excretory system of insects and many other arthropods primarily involves the Malpighian tubules, which are located at the anterior end of the hindgut.

The Malpighian tubules are typically in the form of long slender convoluted tubes that float in the hemolymph. They remove water, nitrogenous waste products, and excess salts as necessary, shunting them to the hindgut. There the collected waste is further processed along with the undigested food passed from the midgut. Most of the water and any needed salts may later be reabsorbed back into the hemolymph through the rectum. This causes waste products to precipitate out of solution and produce more solid materials for excretion through the anus. (Solid waste excreted by insects is known as **frass**.)

(a)

(b)

(c)

FIGURE 3-11
Frass is the term for the solid waste produced by insects and may be characteristic of different insects. The three illustrated examples are frass from (a) a hornworm caterpillar, (b) a grasshopper, and (c) a drywood termite. Photographs by Whitney Cranshaw/Colorado State University.

Waste products of animals usually contain high concentration of nitrogen-rich materials produced when proteins are metabolized. Typically these waste products are first converted to ammonium, a process that does not require expending much energy. However, ammonium is exceptionally toxic to the animal, and it must be quickly flushed out from the body using copious amounts of water or must be immediately converted to a less toxic form. In aquatic insects, water is readily available for waste excretion and ammonium is often the end product of excretion.

In terrestrial insects, water must be conserved. Lacking access to unlimited amounts of water for the dilution of ammonium, these insects must convert waste products to forms that are less toxic to cells. In some insects (as well as humans), this involves the production of urea as a primary form of excreted nitrogenous waste. Urea is much less toxic to cells than ammonia, and although it is soluble in water, urea requires much less water to excrete. However, the production of urea from ammonium requires expending more energy and some water still must be lost to produce urine.

Conservation of water is particularly critical to insects living in dry environments and most of these species eliminate nitrogenous wastes in the form of uric acid. As water leaves the base of the Malpighian tubules, solid uric acid (urate) is formed and is excreted out of the hindgut. Very little water is lost in the process, although more energy is needed to convert the waste to uric acid. Other animals that must severely restrict their water loss (e.g., certain birds, lizards, and snakes) also excrete nitrogenous wastes in the form of uric acid.

Since Malpighian tubules excel at recycling water, they may play an additional role in water metabolism and conservation. In many beetles that develop on a dry diet with little access to free water, the tips of the Malpighian tubules are in direct contact with the rectum, an arrangement known as a **cryptonephric system**. This allows water absorption directly out of the rectum, achieving a more efficient recycling of water; some species can even extract the moisture from the humid air of the rectum. A similar arrangement of the Malpighian tubules is used by many caterpillars and sawfly larvae to help maintain ionic balance in the hemolymph.

Among insects that feed on plant fluids, the Malpighian tubules may be associated with the foregut. In this way they can help form a **filter chamber** that can efficiently remove excess water

(and sometimes other solutes occurring in excess, such as sugars acquired by sap feeding) from the food before it enters the midgut.

The Malpighian tubules may also have functions not directly related to the excretory system. For example, light produced by the "glowworms" of Australia and New Zealand (larval fungus gnats of the genus *Arachnocampa*) occurs through specialized waste compounds (luciferin) used to generate light. Silk fibers are produced in the Malpighian tubules of Neuroptera larvae and some larvae in the orders Ephemeroptera and Coleoptera.

Reproductive System

Most insects reproduce sexually, requiring eggs produced by females to be fertilized with sperm of the males. In both sexes, the external reproductive system involves paired gonads. In males these are in the form of **testes**, which are responsible for producing sperm. As the sperm travels through the tubes of the reproductive tract, it is mixed with fluids produced by accessory glands, producing a packet of sperm known as a **spermatophore**. In some insects, large amounts of nutritive material are incorporated into the spermatophore, which, when acquired by the female, may help to nourish the fertilized eggs. Most insects have some form of direct fertilization that involves direct coupling of the sexes, and in males the external structure is a penis or **aedeagus**. The shape of the aedeagus is typically unique for each species so that it precisely interlocks with the sex organs of the females of that species, and only with that species.

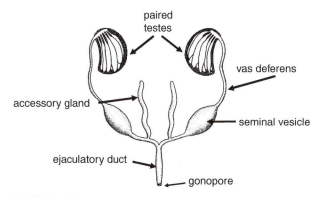

FIGURE 3-12
Illustration of the generalized reproductive system of a male insect. Drawing by Matt Leatherman.

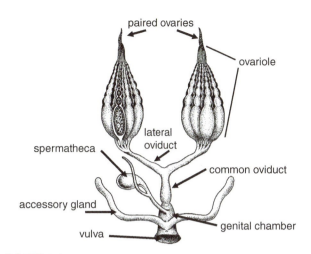

FIGURE 3-14

Illustration of the generalized reproductive system of a female insect. Drawing by Matt Leatherman.

FIGURE 3-13

This female Mormon cricket is feeding on the spermatophore that was passed by the male during mating. This is an example of how male insects may pass on nutrients and other compounds during the course of mating. The spermatophore is produced by the male accessory glands. Photograph courtesy of Nathan W. Bailey.

In females, the eggs are produced within a pair of **ovaries**. Each ovary contains numerous **ovarioles** where the eggs develop before they are fertilized. A unique structure of the female reproductive system is the **spermatheca**. This is used to store sperm after mating, and it releases sperm when the mature egg passes by it en route to being laid (oviposited). Sperm may be kept for a very long time within an insect, and egg laying (**oviposition**) may follow mating by days, weeks, or even longer. For example, honey bee queens mate only during 1 day of their early adult life. They subsequently may lay hundreds of thousands of eggs over the course of 2–3 years, fertilizing them with sperm stored in the spermatheca.

Accessory glands are found in the female reproductive system and these produce secretions that are usually added to the surface of the egg upon deposition. In many insects, the accessory gland produces adhesives that hold the eggs together in masses or simply glues the eggs to an appropriate substrate. In some species, like the mantids and roaches, fluids from the accessory glands form a protective structure called an **ootheca** for the eggs. In stinging insects in the order Hymenoptera (e.g., bees, ants, wasps), the female accessory gland has a very different function, producing venom for the sting.

The egg consists of the **chorion** or shell, the yolk, and the nuclei. Sperm enter the egg through minute holes (**micropyle**) in the chorion. Once fertilized, the nucleus becomes a zygote, which ultimately develops into a juvenile insect that hatches from the egg. During development, the zygote is sustained by the nutritive yolk comprised of proteins, carbohydrates, lipids, and necessary salts.

Most female insects produce eggs that they deposit externally out of their bodies either singly or in groups. These **oviparous** insects have eggs with a tough chorion and enough yolk to provide the developing zygote all the nourishment it needs. Other insects are **viviparous**, retaining the egg within their bodies and giving birth to a live juvenile. Viviparous insects produce an egg that may lack a chorion and have little yolk so that the developing zygote receives almost all of its nourishment from the mother. Somewhat intermediate are the **ovoviviparous** insects. In these species, the zygote receives all its nourishment from the egg yolk, but the egg remains protected within the mother's body until it hatches into a free-living immature insect.

Although a great majority of insects and other arthropods reproduce sexually, there are notable exceptions. Some insects can produce young without egg fertilization through parthenogenesis. Most aphids normally reproduce through parthenogenesis; in this case, all the offspring are female, have the full complement of chromosome pairs (diploid), and are essential clones of their mother. Other insects, such as wasps and ants, produce males from unfertilized eggs. In these cases, males are **haploid**, with only a single set of chromosomes; females in these species are the product of fertilized eggs with both pairs of chromosomes and are **diploid**.

● Do Insects Feel Pain?

For several reasons it is unlikely that insects and other arthropods feel pain, at least as we describe it. Within the definition of the International Association for the Study of Pain (IASP), pain is "an unpleasant sensory and emotional experience associated with actual or potential tissue damage," and the key concept is whether noxious stimuli can induce an emotional response. In humans, and at least some other vertebrates, pain also can induce conditions of fear, anxiety, and stress. These function as a sort of learning tool that teaches the animal to avoid its source in the future.

Among the arguments against insects feeling pain is their lack of pain receptors (nociceptors). Arthropods do have sensory receptors that can cause them to avoid touch, certain chemicals, adverse temperatures, or other stimuli, and these allow them to escape predators or certain other sources of harm in their environment. However, this does not require painful experience, and even bacteria and protozoa without a nervous system will similarly move away from sources of potential harm.

Painful experience has its essential value if it educates the animal to avoid the source in the future. However, the brief life of insects, the tiny number of neurons found within insect brains needed to process and store information, and their very limited memory and learning abilities argue against the existence of any evolutionary value for pain perception in insects.

Finally, observations of arthropod behavior suggest that pain is likely not felt, although some behaviors may appear to us as if they should be painful. For example, an insect treated with certain insecticides (e.g., organophosphates, chlorinated hydrocarbons, pyrethroids) may produce writhing, jerky movements. However, these

FIGURE 3-15
Many animals clearly can feel pain. This sensation sends a powerful message that helps the animal learn to avoid future harm, and many animals have special receptors (nociceptors) of the nervous system, which signal pain. Insects lack specific pain receptors and also have limited ability to learn. It is unlikely that they perceive pain, at least in the manner that pain is perceived by organisms with larger and more complex nervous systems. Photograph of a mantid feeding on a grasshopper by Whitney Cranshaw/Colorado State University.

reactions are a direct response to the mode of action of the insecticides, which short-circuit nerve cell firing, and do not indicate associated pain. On the other hand, grasshoppers may continue to feed—and male mantids continue to mate—while being devoured by a female mantid and insects will continue to put normal pressure on an appendage that has been severed or crushed.

Nervous System

The basic functional unit of the nervous system of arthropods is very similar to that of humans and many other animals: a nerve cell or **neuron**. A neuron consists of a cell body, the **axon** or the long fiber portion of the neuron, and the **dendrites**, which terminate as tiny branches on the opposite side from the axon. Between each nerve cell is a tiny gap, known as the **synapse**, separating the tip of the axon of one nerve cell from the dendrite of the adjacent nerve cell.

The way that nervous impulses pass between two nerve cells, or from a nerve cell to a muscle, in insects is also achieved in generally the same way as it occurs in the human nervous system. An electrical impulse passing through the cell is converted to a chemical reaction at the synapse. At the tip of the axon, a compound known as a **neurotransmitter** is instantaneously released. Passing across the synapse, it is then perceived by receptors on the adjacent dendrite, which converts it back to an electrical signal that then passes through the next nerve cell. Several kinds of chemicals serve as neurotransmitters in insects, each of which can produce different effects on the receiving nerve cell.

There are many different kinds of neurons. **Sensory neurons** detect chemicals, light, touch, vibration, or other things important to the insect's environment. **Motor neurons** receive information from other neurons and lead to muscles that they can then stimulate to move.

Nerve cells that coordinate messages between sensory neurons and motor neurons are known as **interneurons.** Specialized neurons found throughout the body are **neurosecretory cells** that produce hormones.

Although most nerve cells of insects basically function in a manner similar to those found in humans, their arrangement within the body does differ. Most insect neurons do not occur as isolated cells or as simple strings of cells, and the regulating role of a central brain is diminished. Instead, insect nerve cells are aggregated into interconnected masses of neurons known as **ganglia** (singular **ganglion**). At each body segment, there is a pair of ganglia that allows connection to all the muscles and other organs within the segment. Such a decentralized arrangement allows each pair of ganglia to control all the primary activities of its segment. As a result, if an insect is beheaded or even carefully sectioned, the activities of the body segment may continue to function so long as water balance is maintained.

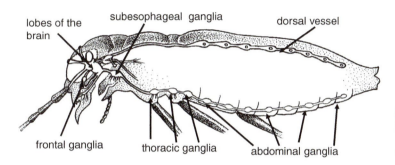

FIGURE 3-16
Illustration of the structures in a generalized central nervous system of an insect. Drawing by Matt Leatherman.

The ganglia occur on the lower (ventral) side of the insect and are connected to each other, forming a double nerve cord. Within the head, the ganglia are fused into two centers, the **brain** and the **subesophageal ganglion**. Together, the nerve cord, brain, and subesophageal ganglion compose the central nervous system of the insect.

The insect brain is formed by the fusion of three smaller ganglia known as the **protocerebrum, deutocerebrum,** and **tritocerebrum.** Each of these ganglia forms a brain lobe responsible for different functions, most of which are concerned with incoming sensory information. The protocerebrum is the most anterior lobe and is usually the largest. It is connected directly to the compound eyes and the ocelli and, consequently, receives all of the visual sensory information. Also within the protocerebrum are small organs known as **mushroom bodies** that help to process and coordinate the information coming into this portion of the brain, providing the very limited learning ability of which insects are capable.

The deutocerebrum is the central lobe of the brain and receives sensory information from the antennae. The tritocerebrum is the hind lobe of the brain that coordinates information received from the rest of the body, while the function of the

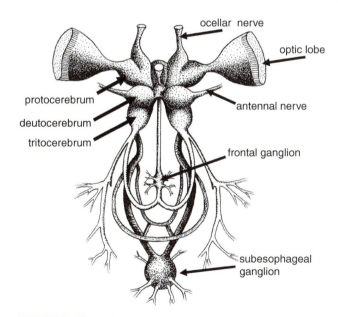

FIGURE 3-17
Front view of the primary structures of the insect nervous system found in the head. Drawing by Matt Leatherman.

mouthparts is controlled by the **subesophageal ganglion.** Found within the brain and the subesophageal ganglion are numerous **neurosecretory cells** that produce hormones.

● Neurotransmitters and Insecticides

Chemicals known as neurotransmitters help regulate the signaling that occurs between cells of the nervous system. Some of the neurotransmitters that insects use (e.g., acetylcholine, dopamine, gamma-butyric acid) are also important among vertebrates, including humans. Conversely, some chemicals may have more limited use among different animal groups, such as octopamine, a significant neurotransmitter in many insects that has far more limited effects in vertebrates. Perhaps the best known neurotransmitter shared by insects and humans is **acetylcholine,** which allows motor neurons to stimulate muscles.

Functioning with acetylcholine are **cholinesterase** enzymes. These enzymes break down the acetylcholine molecule after it has been received by the muscles. Once it is enzymatically degraded (into acetic acid and choline), acetylcholine is no longer available to continue muscle stimulation.

Many insecticides work by interfering with this process. For example, insecticides of the organophosphate (OP) chemical class (e.g., parathion, chlorpyrifos, malathion, diazinon) are best known for their **acetylcholinesterase (AChE)-inhibiting** properties. These organophosphate insecticides cause an irreversible blocking of acetylcholinesterase, producing disruption of nerve function. Most dramatic effects from cholinesterase-inhibiting compounds result from the uncontrolled stimulation of the nerves activating muscles—both in insects and humans—which can cause death following high exposure. Low levels of exposure to organophosphate have been demonstrated to be capable of producing subtle neurotoxic effects, which can be particularly injurious to the nervous system of young children or developing fetus.

That the organophosphates can have serious poisoning potential to humans is not surprising, particularly considering their history of development. The first cholinesterase inhibitors were developed as potential chemical warfare agents, the G-series weapons of the Nazis, such as sarin gas. After World War II, organophosphates were intensively investigated as potential insecticides and several compounds were developed so that organophosphates became the dominant class of insecticides used in agriculture for over 30 years, beginning in the mid-1950s.

Increased awareness of the hazards of organophosphate insecticides—and the development of newer insecticide classes that have substantially lesser hazards as human toxins—has resulted in a steady decline of the use of organophosphate insecticides. Furthermore, regulatory restrictions on their use have greatly increased over the past few decades, and most former uses (particularly those in and around homes) have been eliminated in the United States.

Insect Senses

The primary way that adult insects perceive light is through the **compound eyes** in their heads. These are made of individual light-sensing organs known as **ommatidia,** and the number of these determines visual acuity, such as the ability to distinguish differences in shape. Externally, each ommatidium appears hexagonal, allowing them to pack together, and they are covered with a crystalline lens. Within the ommatidium are 6–10 nerve cells that perceive light intensity and, often, color. No details can be detected by an individual ommatidium, but visual acuity increases with their number. For example, dragonflies have extremely large eyes that may have more than 10,000 ommatidia and may detect good detail at a distance of several meters. Insects with low numbers of ommatidia can detect little more than general differences in light intensity.

Many insects perceive some color but are often particularly sensitive only to certain wavelengths. Those that feed on plants may see yellow or yellow-green colors particularly well, but lower wavelengths that are visibly red to the human eye are not well seen by any insects. On the other hand, some insects can see colors in the ultraviolet range, wavelengths that are invisible to the human eye. Such colors are common in flowers, providing **nectar guides** to pollinating insects. Furthermore, many insects can see **polarized light,** allowing them to detect the position of the sun from a small patch of clear sky, an ability that is critical to insect navigation.

Insects may also be able to process light images much faster than the human eye and brain, giving them exceptional ability to detect motion. The "flicker-fusion" frequency of the human eye is around 45–53 cycles/s so that more rapidly received images are not individually distinguished. (Because of this, television images and movie film that produce numerous images per second are seen by humans as an uninterrupted continuously moving image.) Some insects have a flicker-fusion frequency of 250 cycles/s or more. This gives them

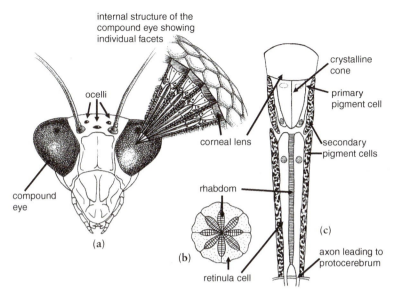

internal structure of the compound eye showing individual facets

ocelli

compound eye

corneal lens

crystalline cone

primary pigment cell

secondary pigment cells

rhabdom

retinula cell

axon leading to protocerebrum

(a)

(b)

(c)

FIGURE 3-18

Illustration of the structures associated with a compound eye. Each facet of the eye (ommatidium) has an exterior lens. This covers a crystalline cone that directs the light to the receiving retinula cells. Each of the retinula cells within the ommatidium continuously transmits signals of the received light to the rhabdom, which links to the nerve carrying the signals to the optic center in the protocerebrum. Pigment cells help confine the light within the individual ommatidium. (a) front view of mantid showing exploded view of compound eye; (b) cross-sectional view of an ommatidium at midsection; (c) longitudinal view of an ommatidium. Drawings by Matt Leatherman.

FIGURE 3-19

Close-up of the compound eye of a moth. The individual facets of the eye are visible, each one associated with a single ommatidium. Photograph courtesy of the Centers for Disease Control Images Library.

(a)

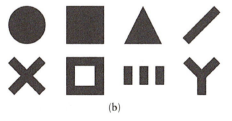

(b)

FIGURE 3-20

(a) A honey bee has fairly large compound eyes and is able to distinguish some detail. Honey bees can distinguish (b) figures in the bottom line as different from those on the upper line. However, they cannot distinguish differences among the shapes within a line. Photograph of the honey bee courtesy of Brian Valentine. Drawing by Matt Leatherman based on original studies conducted by Karl von Frisch.

the ability to clearly see and respond to movement very quickly—one of the reasons why a fly may be so difficult to surprise and swat.

Besides compound eyes, there are other types of light-receiving organs in insects. Typically, adult insects possess three small ocelli on the top of their heads that can perceive light intensity (i.e., light vs. dark) and may track changes in daylight and day length, which guide insect activities. Some immature insects, such as caterpillars, have a different type of simple eye (stemmata), which is light sensing and can provide rough images.

Aside from the relatively few insects that rely on large compound eyes, the ability to detect chemicals is the most important sense for determining what is important in their environment. Some **chemoreceptors** are designed to detect chemicals in the air (sense of smell) and are usually concentrated on the antennae. Chemoreceptors to detect chemicals upon contact (sense of taste) may be concentrated around the mouthparts on the tarsi at the tip of the legs. These various chemical-sensing receptors can be of many forms and structures including pits, hairs, pegs, and plates. Although concentrated on the head, they may occur in many other areas of the body.

Other sensory neurons are used for mechanoreception and can detect touch, body and appendage positions, vibration, air movement, gravity, and pressure differences (sound). Mechanoreceptors are often simple "hairs" that, when moved or touched, will trigger a nervous impulse in sensory neurons located beneath the hair within the exoskeleton. Other mechanoreceptors may function as touch plates, pegs, or stretch receptors.

Mechanoreceptors that are used to detect sound are usually organized into special **tympanal organs**. These are often shaped like a tympanum or drumhead and are usually arranged as a simple membrane with an attached sensory neuron. When sound hits the membrane, it vibrates and triggers an electrical impulse that flows down the sensory neuron. Tympanal organs are located either on the legs or on the abdomen of insects.

One highly specialized mechanoreceptor is the **Johnston's organ**, which is located in the pedicel of the antennae. It detects vibration impacting the flagellum of the antennae and can be particularly sensitive to certain air movements. The Johnston's organ can help flying insects manage their flight speed or detect vibrations that may indicate the presence of other insects. Water striders use the Johnston's organ to detect vibrations on the water surface that may indicate another insect (food) on the surface. Male mosquitoes can detect the specific vibrations produced by the wingbeat of the female of the species.

Other types of receptors are found among insects. Most insects can detect differences in temperature and humidity using specific sensory neurons dedicated to these tasks. Some insects can also detect magnetic fields and changes in atmospheric pressure.

● The Fantastic Pheromones

Several different modes are used among insects to communicate, including sound, light, and color. However, in most insects the best developed senses are those used to detect chemicals, such as odors in the air. Odors can also be used to communicate in the form of **pheromones**.

By definition, pheromones are chemicals used to communicate between individuals of the *same* species. (This is in contrast with the function of **hormones**. Hormones are chemicals released by specific tissues within an organism that elicit an internal physiological response in other tissues—chemical messengers *within* an organism.) Among insects a variety of pheromones are produced to help direct different activities.

Sex pheromones are produced by adults of one sex to attract members of the opposite sex. In some species,

it is the male that produces a sex pheromone, but more often, it is the female. Among the sex pheromones, those produced by some moths are among the most biologically active of all chemicals as they can be perceived in very low concentrations by male moths and direct them toward the female over very long distances. Chemical communication over half a kilometer or more is not unusual with moth pheromones. In some insects, a mixture of long-distance and short-distance pheromones is produced. Short-distance pheromones are used by both sexes to fully identify and locate potential mates at close range (less than few centimeters).

Aggregation pheromones cause insects to come together in some activity. Among the most important of these are those used by bark beetles to mass attack individual trees. Individual beetles are usually killed by

(continued)

FIGURE 3-21
A male lilac/ash borer moth is being drawn to a trap. The trap contains lure-emitting chemicals that simulate the female lilac/ash borer sex pheromone, which attracts mates. Photograph by Whitney Cranshaw/Colorado State University.

tree defenses (e.g., sap flow), but collectively, bark beetles coordinated by aggregation pheromones can overcome the tree's defense.

Spacing or **deterrent pheromones** have the reverse effect, causing other insects to move away. These can be important if there are only limited resources available. For example, female fruit flies will deposit a spacing pheromone (oviposition-deterrent pheromone) to let other females know that eggs have already been laid in the fruit. Bark beetles produce pheromones that inhibit further recruitment when their densities are high.

Alarm pheromones can elicit a variety of behaviors. Aphids will produce alarm pheromones if disturbed, say, by a predator (e.g., lady beetle). This will cause nearby aphids (often twin sisters) to move away or drop from the plant. Social bees and wasps use alarm pheromones to warn their colony of intruders. This can allow them to initiate a coordinated attack. In honey bees alarm pheromones are also associated with the stinger and poison sac left behind after stinging, stimulating other honey bees to attack nearby. In this case, these pheromones also serve as an aggregation pheromone.

Primer pheromones affect the development or physiological state of individuals. Social insects (some bees and wasps, all termites and ants) use several kinds of pheromones to maintain the cohesion of the colony. The reproductive female queen is the source of these pheromones, which suppress the production of other queens or help the workers direct their activities. Other pheromones trigger the development of various types of individuals found within a colony (e.g., workers vs.

FIGURE 3-22
This pine tree has numerous "hits" on the trunk where adult mountain pine beetles have attempted to enter the trunk to lay their eggs. (The resultant pitch at the wound is known as a "pitch tube.") In order to successfully attack a living tree, these bark beetles must make coordinated mass attacks that overwhelm the tree's defenses. Aggregation pheromones are used to produce this behavior. Photograph courtesy of Steven Katovich/Bugwood.org.

soldiers). Primer pheromones released by locusts will trigger individuals to develop into the damaging gregarious phase from the solitary phase.

The effects of **trail-marking pheromones** are most often seen in the behaviors of ants. Foraging ants often lay down a chemical trail that allows them to return to the colony and directs other ants to food sources. The narrow trails of ants seen at picnics are the result of individuals following a chemical scent on the ground. Bees and wasps that forage with winged workers also mark food sources, allowing them to be better located during later visits.

Humans have taken advantage of this type of chemical communication within insects and have developed chemicals that mimic various insects' pheromones in order to manage pest populations.

(*continued*)

FIGURE 3-25
Pheromone traps that capture codling moths are widely used in orchards to determine the best time to spray an insecticide to control this important pest of apples and pears. In this trap the bottom board is sticky, and the pheromone is emitted from the red rubber septa. Photograph by Whitney Cranshaw/ Colorado State University.

FIGURE 3-23
A Mediterranean fruit fly is ovipositing in a coffee berry. Fruit flies typically will also lay down a pheromone when laying eggs, which is a deterrent to other fruit flies considering the fruit to lay their eggs. These pheromones space out egg laying and prevent excessive numbers of eggs from being laid in a single fruit. Photograph courtesy of Scott Bauer/USDA ARS.

FIGURE 3-24
Pheromones are used for many purposes in the management of insects. One of the most common uses is to put the pheromones in a trap that will capture the attracted insects. This can be useful in detecting insects or determining the periods when adults are active and laying eggs. Photograph courtesy of John Ghent/USDA Forest Service/Bugwood.org.

FIGURE 3-26
The white twist tie on this apple tree contains a high amount of the sex pheromone of the codling moth, an important orchard pest. These pheromone sources are placed throughout the orchard and permeate the air with the pheromone so that the male codling moth cannot locate a female codling moth for mating. This method of insect control using pheromones is known as mating disruption. Photograph courtesy of Rick Zimmerman.

(continued)

Sex pheromones are used to bait survey traps to enhance the ability of pest control specialists to detect the presence of damaging insects in an area. In other uses, pheromones can be used in traps that tell when adult stages of an insect are present, allowing insecticides to be appropriately timed.

Perhaps the most elegant use of pheromones in insect pest management is for **mating disruption**. This involves placing large numbers of pheromone-emitting lures at a site so that the air is thoroughly permeated. When properly employed, this can prevent the males from successfully locating and mating with females, resulting in infertile eggs being laid. Most use of mating disruption has involved certain moths that develop as crop pests, such as the codling moth (*Cydia pomonella*) in apples or the pink bollworm (*Pectinophora gossypiella*) in cotton.

4 ● Growth and Metamorphosis

The manner that insects and other arthropods grow can differ greatly from vertebrates. Some of these dissimilarities are related to the very different body design of arthropods, notably a relatively inflexible exoskeleton that encases their bodies, restricting growth. Moreover, arthropods have a limited ability to control their internal temperature; because of this, their growth and activity is largely dependent on the surrounding environmental temperature. Furthermore, as arthropods grow, they not only increase in size but also may take on new forms, a process known as **metamorphosis** ("change in form"). Many of the functions of growth and development are controlled by hormones secreted by neuroendocrine cells associated with the **neuroendocrine system**.

Molting or Ecdysis

The exoskeleton that provides arthropods so many advantages also comes with a serious limitation. Because the sclerotized areas of exoskeleton cannot grow, it has a very limited ability to stretch once it has formed and hardened. (Cell growth can continue in the epidermis of immature insects, which can accommodate some change in shape.) As a result, arthropods grow in stages, each stage terminating with the shedding of the old exoskeleton and the construction of a new one, a process known as **molting**. The physical stage between molts is known as an **instar**, and the duration of the instar period is called the **stadium** (plural, **stadia**). Typical insects may have to molt three to seven times, sometimes more, before reaching the ultimate instar, the full-grown adult.

FIGURE 4-1
As they grow, insects go through a series of distinct stages (stadia), each punctuated by a molting event. The ultimate stage is the adult, which is sexually mature and, if winged, has functional wings. Illustrated are the various life stages of a chinch bug, *Blissus* sp. Photograph courtesy of Jim Kalisch/ University of Nebraska.

The process of molting is triggered by hormones. The most important of these is **ecdysone,** which is produced by several glands but is concentrated in the **prothoracic gland.** Hormones produced by neurosecretory cells of the brain start the sequence that stimulates the production of ecdysone. Once released into the hemolymph, ecdysone initiates a whole cascade of events that initiate and drive the molting process, also referred to as **ecdysis:**

1. The epidermal cells that detect ecdysone contract from the old cuticle, separating from it through **apolysis.** During this time the epidermal cells become active and begin to divide.
2. The new active epidermal cells produce a **molting fluid** that infiltrates the softer areas of the inner cuticle (endocuticle). At the same time, a new layer of cuticle for the new exoskeleton begins to

FIGURE 4-2

Madagascar hissing cockroaches, *Gromphadorhina portentosa,* at various stages in the molting process. The cockroach on the right is at the last stage of shedding the exocuticle of the previous exoskeleton, which the one on the lower left has completed. Both are pale colored and soft as the new exoskeleton has not fully expanded and hardened through sclerotization. The remaining cockroaches in the picture have the normal fully darkened and hardened exoskeleton as the sclerotization process has been completed. Photograph by Whitney Cranshaw/ Colorado State University.

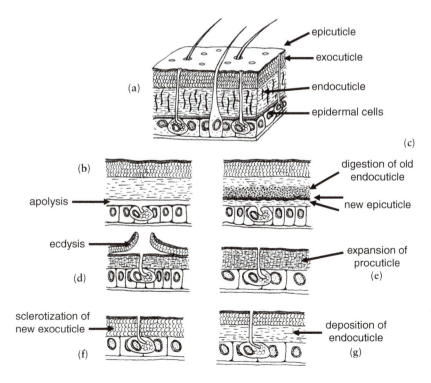

FIGURE 4-3

(a) The sequence of steps involved in the molting of the exoskeleton. (b) During apolysis, the old cuticle separates from the epidermal cells. (c) A new epicuticle is produced, and then enzymes that digest the old endocuticle are introduced above the epidermal cells. (d) The remaining parts of the exoskeleton, the former exocuticle and epicuticle, are then shed during ecdysis. (e) The new procuticle is expanded, (f) which then hardens and darkens during sclerotization to form the new exocuticle. (g) A new endocuticle is then produced. Drawings by Matt Leatherman.

be constructed above the epidermal cells and below the old cuticle.

3. When a new layer of cuticle is in place, the epidermal cells release a material that activates the molting fluid, causing it to digest the soft areas of the old endocuticle.

4. Much of the old endocuticle is then further digested, reabsorbed into the epidermis, and used to help build the new procuticle of the insect. The sclerotized layer of the old exocuticle and the thin surface layer of the old epicuticle cannot be digested.

5. The old exocuticle and epicuticle break apart along the lines of thinner reinforcement and are shed in the form of a "cast skin" or **exuviae**. The new procuticle, still soft, is expanded by increasing the internal pressure by the hemolymph or by filling the air sacs.

6. When the new exoskeleton has expanded to its new maximum size, the process of sclerotization begins, hardening and darkening the layer of procuticle that will become the new exocuticle. When sclerotization is complete, additional procuticle will be produced, producing the underlying endocuticle of the new exoskeleton.

(a)

(b)

(c)

(d)

FIGURE 4-4

A time sequence in the emergence of a convergent lady beetle, *Hippodamia convergens*, and the subsequent hardening of the exoskeleton during sclerotization. In the first picture (a), the adult has just emerged from the pupal skin on which it is resting. The subsequent two pictures (b, c) show the beetle at approximately 1 and 2 h after the first in which the progressive darkening of the elytra is evident. A fully sclerotized adult with normal coloration is shown in the final picture (d). Photographs by Whitney Cranshaw/Colorado State University.

The entire course of molting takes hours and sometimes days. During this period the arthropod may not be able to move, feed, or defend itself well. It is a period of particular vulnerability, and much mortality occurs during molting.

Metamorphosis

Molting almost invariably results in a larger body size, but it may also result in a change in body form. Each type of insect follows its own pattern of metamorphosis, and three general patterns predominate.

FIGURE 4-5

The fourlined silverfish, *Ctenolepisma lineata*, is a representative of the insect order Thysanura. Thysanurans lack wings and they go through ametabolous development, where there are no visible changes in appearance, aside from an increase in size, as they develop. Photograph courtesy of Tom Murray.

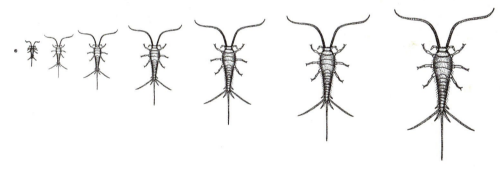

FIGURE 4-6

Generalized life cycle of an insect (silverfish) that undergoes ametabolous development. Drawing by Matt Leatherman.

The simplest type of metamorphosis is known as **ametabolous development,** roughly translated as "no metamorphosis." Other than an increase in size, insects with ametabolous development have very little change in external appearance as they develop through the immature stages and into the adult stage. Furthermore, insects with ametabolous development may continue to molt even after they are sexually active adults. This pattern of metamorphosis is found among the more ancient and primitive wingless insects such as silverfish (order Thysanura) and jumping bristletails (order Microcoryphia). Springtails (order Collembola) and diplurans (order Diplura) also display ametabolous development.

Many other insects (grasshoppers, roaches, aphids, true bugs, etc.) go through **hemimetabolous development,** also known as "incomplete" or "simple" metamorphosis. In these insects, the immature stages are called **nymphs.** They are usually quite similar in appearance to the adult, have the same feeding habits, and are typically found in the same environment. Adults differ primarily by being sexually mature and, if winged, possess wings that are fully developed and functional. Among the winged insects, the nymphs develop increasingly visible **wing pads** following each molt and increase in size.

● Major Evolutionary Changes in Insects

The extraordinary success of insects as life forms is based on many different features. These have accrued over hundreds of millions of years with a few key evolutionary developments considered to be most critical.

Advancing to land. With recovered fossils of ancestral millipedes dating back to the late Silurian period, arthropods are known to be the first animals to move from lakes and seas to land. Among the arthropods, this occurred in several independent events, with the ancestors of the myriapods (e.g., millipedes, centipedes), arachnids (e.g., scorpions), hexapods (e.g., insects), and crustaceans (e.g., isopods) each emerging on land separately. That the earliest fossils are those of predatory animals is suggestive that these first land animals moved only temporarily to land, perhaps to find refuge during vulnerable stages of development or to forage. The exoskeleton provided the ability to conserve water, allowing them to ultimately move away from water. The abundance of land plants provided new opportunities for those that developed phytophagous or scavenging feeding habits.

Development of wings. Probably the most important feature leading to the success of insects was the development of wings. Most importantly, wings allow insects to disperse more efficiently so that they are better able to find new sources of food, mates, or habitats. Wings provide an excellent means of escape from predators. The development of wings also led to the development of a more complex nervous system as the fast-moving, three-dimensional world of flight brought new sensory challenges.

The wings of insects are outgrowths of the thorax, totally new structures unlike the wings of bats and birds, which are modified appendages (legs). Wings likely originated as projections from the exoskeleton that allowed gliding, providing these arthropods with the ability to cover long distances should they drop from their host plant. (A rudimentary form of this is found in some current-day silverfish.) The exact time when wings first appeared is still unknown and will

FIGURE 4-7

Mayflies (Ephemeroptera) were among the first kinds of insects that developed wings. Their wings are held above their bodies because they lack the later-evolved structures that allowed leaf folding. Photograph courtesy of Tom Murray.

remain a subject of considerable debate until the fossil record becomes more complete. Currently, the first winged insects found are from the late Carboniferous period. At that time several kinds of winged insects could be found, many of which showed relatively advanced wing features. It is suspected that the "missing link" of the first winged insects is to be found in some still undiscovered fossils from the early Carboniferous period, or perhaps even earlier.

Later, some groups lost wings as they evolved and adapted to new habitats. For example, all fleas, all lice, and some flies that live as parasites on the bodies of birds or mammals have lost their wings. Wingless species or species that have lost the ability to fly can be found in almost all orders of winged insects.

Wing folding. Among present-day insects, there are two orders where the wings are always held out to the side or above the body: Ephemeroptera (mayflies) and Odonata (dragonflies and damselflies). All other insects have the ability to fold their wings atop their

(continued)

bodies when they are not in flight. This wing-folding mechanism involves the use of a few tiny muscles attached to one of the thoracic sclerites. Some of the first fossil winged insects discovered (order Protoptera, family Paoliidae) had evolved this adaptation by the late Carboniferous period. This minor evolutionary adaptation had tremendous consequences. With the ability to collapse and fold their wings, insects can better move to cover and protect their wings. Further protection of the wings occurred as some insect groups modified the forewing. Thickening of the front wings (**tegmina**) found in many insects improved the protection of the hindwings used for flight. Beetles took this a step further so that their very hard forewings (**elytra**) provide such protection that they can burrow into plants or soil without losing the ability to reemerge and fly away. Rubbing structures on the folded wings are also used by many insects to produce sound for communication (**stridulatory organs**).

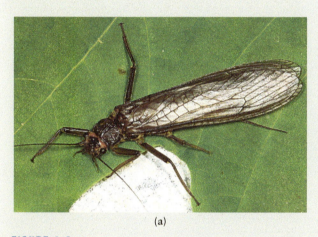

(a)

(b)

FIGURE 4-8

Stoneflies (Plecoptera) are insects that have evolved the ability to fold their wings (a). In other insect orders, the forewings have been further modified so that they help cover and protect the hindwings, such as in grasshoppers (Orthoptera) (b). Photograph of the stonefly courtesy of Tom Murray. Photograph of the grasshopper by Whitney Cranshaw/Colorado State University.

Complete metamorphosis. Among the present-day insects, at least 85% of all species have a form of development involving what is known as complete metamorphosis (chapter 4). These insects have a soft-bodied immature form (a larva), usually with very different appearance and habits from the adult (compare a caterpillar to a butterfly). Wings develop entirely internally and only during the transition stage of the pupa do adult features begin to emerge.

Complete metamorphosis provides several advantages. Most obviously, it allows insects in immature and adult stages to specialize for different functions. The larvae of insects with complete metamorphosis often feed on different foods from the adults and spend much of their life in different environments. This prevents competition for food between adults and juveniles. Insect larvae also have better control of their development and may go into a controlled state of dormancy (diapause) or accelerate to maturation when conditions are unfavorable. Adult insects can specialize in egg production and dispersal. Complete metamorphosis was first found in the order Neuroptera, which evolved in the late Permian period. Since then, insects as diverse as beetles, wasps, fleas, and flies have evolved with this developmental pathway.

(*continued*)

(a)   (b)

FIGURE 4-9
The majority of the insects now present on the planet have evolved to use complete metamorphosis in their development, beetles, flies, wasps, bees, moths, and butterflies being common examples. Insects that have complete metamorphosis, such as the cabbage white butterfly, *Pieris rapae*, can have an adult form (a) that is specialized for dispersal, host finding, and mating and a larval form (b) that is specialized for feeding. Photograph of the adult courtesy of David Cappaert/Michigan State University/Bugwood.org. Photograph of the larvae by Whitney Cranshaw/Colorado State University.

FIGURE 4-10
A boxelder bug, *Boisea trivittata*, is an example of an insect that undergoes simple metamorphosis. Illustrated in this picture are nymphs in different stages showing wing pads and fully winged adults. Some of the adults are a lighter red color, indicating they have recently molted and have not fully sclerotized the new exoskeleton. Photograph by Whitney Cranshaw/Colorado State University.

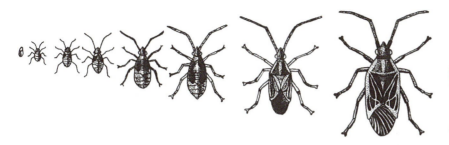

FIGURE 4-11
Generalized life cycle of an insect (boxelder bug) that undergoes hemimetabolous development (simple metamorphosis). Drawing by Matt Leatherman.

There are a few variations upon this development pattern among hemimetabolous insects. These are most obvious in two orders of aquatic insects: Ephemeroptera (mayflies) and Odonata (dragonflies and damselflies). The nymphs of these insects occur in water and differ considerably in appearance from the winged adults. Some insects may also have a stage during their development when they do not feed, as in the whiteflies (order Hemiptera) and thrips (order Thysanoptera).

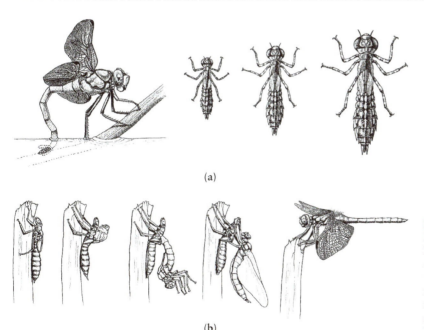

(a)

(b)

FIGURE 4-12
Dragonflies undergo simple metamorphosis but have aquatic immature forms that have a fairly different appearance from the winged adults. (a) Eggs are laid in water and the developing nymphs are aquatic. (b) At the end of the last nymphal stage, they emerge from the water and molt to the adult stage. Drawing by Matt Leatherman.

The majority of insect species undergo **holometabolous development**, often known as "complete" metamorphosis. In these insects, the immature forms look very different from the adults and typically have very different life histories. The juvenile stages are called **larvae** (singular, **larva**) and are often soft bodied, with reduced eyes and antennae. Special names are sometimes given to certain types of larvae such as "caterpillar" to describe the larvae of butterflies, moths, and sawflies; "grubs," to larvae of beetles; and "maggots," to larvae of some types of flies. More common kinds of insects with complete metamorphosis belong to the orders Coleoptera (beetles), Diptera (flies), Lepidoptera (butterflies and moths), Hymenoptera (bees, wasps, ants, sawflies), and Siphonaptera (fleas).

(a)   (b)

FIGURE 4-13
Beetles, such as this ground beetle (Carabidae family), undergo holometabolous development (complete metamorphosis). Insects that undergo holometabolous development have larvae (a) that have very different appearance and very different habits than the adult (b). Photographs by Whitney Cranshaw/Colorado State University.

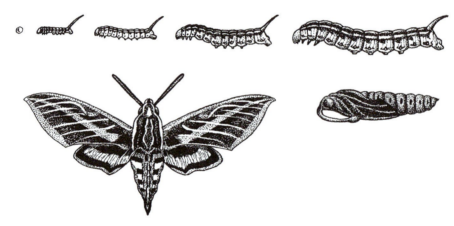

FIGURE 4-14
Generalized life cycle of an insect (sphinx moth/hornworm) that undergoes holometabolous development (complete metamorphosis). Drawing by Matt Leatherman.

Following the larval stage, holometabolous insects have a unique stage, the **pupa** (plural, **pupae**). Often the pupal stage appears to be inactive since the insect typically moves little, if at all, as a pupa. The pupa may be further obscured within a cocoon made of silk. As a result, the pupa is sometimes referred to as a "resting stage." This is a serious misnomer; few stages during the life of any animal are physiologically more active than the insect pupa. During the pupal period, all of the tissues of the larva break down and reform to produce wholly new organs and adult features. In some insects, such as house flies, the pupal period can be extraordinarily short, just a few days. The pupal stage of most insects typically lasts 1–2 weeks, sometimes much longer.

(a)

(b)

(c)

(d)

FIGURE 4-15

The monarch butterfly, *Danaus plexippus*, produces a pupal stage known as a chrysalis. When originally formed (a), it is emerald green with golden spotting. Within the chrysalis, changes take place that allow the insect to transform into the ultimate adult stage, which becomes visible (b) at the end of the pupal stage. The adult then emerges (c) and subsequently pumps blood to extend the wings (d), which soon harden as they sclerotize. Photographs of the chrysalids courtesy of Susan Ellis/Bugwood.org. Photographs of the emerging adults by Whitney Cranshaw/Colorado State University.

Regardless of whether they have hemimetabolous or holometabolous development, the end product of the developmental process is an adult, and there is no further development after that point. Any insect that has adult features, such as fully developed wings (most insects) or adult-form legs (e.g., ants, fleas), will not grow any larger. Therefore, a small fly is not a "baby fly" nor is a small ant a "baby ant." Instead they are small species of fly or ant, or perhaps one that was malnourished when young. Regardless, they will never molt again or increase in size.

Each of the other arthropod classes (millipedes, centipedes, arachnids, crustaceans) have their own patterns of metamorphosis. Most of these species have development rather similar to that of ametabolous insects. Some may increase the number of body segments with each molt or the number of legs or acquire different forms as they develop, a process known as **anamorphic development**.

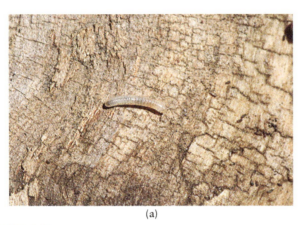

(a)   (b)

FIGURE 4-16

Some other arthropods undergo metamorphosis patterns different from what is seen in insects. For example, millipedes undergo anamorphic development where they progressively add segments, and usually progressively darken, as they change from immature to adult forms. Illustrated here is a (a) young stage and (b) adults of *Cylindroiulus caeruleocinctus*, a European millipede that is now widespread in North America. Photographs by Whitney Cranshaw/Colorado State University.

Different hormones control metamorphosis. The most important are **juvenile hormones** produced in a gland called the **corpora allata**. High concentrations of juvenile hormones during molting cause the insect to remain in the juvenile form (e.g., nymph, larva) from one instar to the next. As they get older, lower concentrations are released so that the next molt produces another form, such as a pupa or the adult.

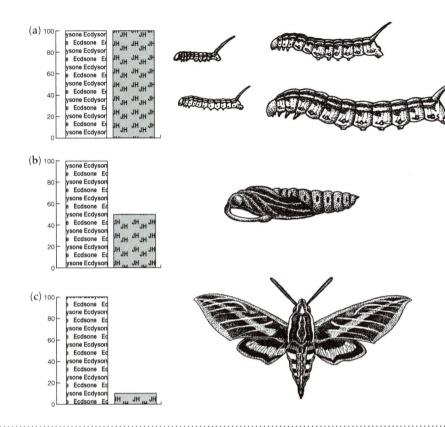

FIGURE 4-17

Hormones control the molting process in insects and regulate the progression of metamorphosis. Ecdysone is the primary hormone that initiates the sequence of events that cause an arthropod to molt. Juvenile hormone (JH) regulates what stage will be produced during the molt. (a) High concentration levels of juvenile hormone cause an insect to molt to an immature form (nymph, larva) at the next molt. When less juvenile hormone is present, the insect will molt to a late stage of development, (b) such as a pupa or, ultimately, (c) an adult. Drawing by Matt Leatherman.

● Insect Growth Regulators—The Holy Grail of Insecticides?

Efforts to find ways to kill pest insects have a long history. Homer, of Classical Greece, extolled the virtues of "pest-averting sulfur," perhaps the earliest insecticide ever identified. By the late 1800s a handful of other insecticides had been identified, some derived from plants (e.g., nicotine, pyrethrum, rotenone) and some from inorganic or heavy metal sources (e.g., arsenic, lead). Later, synthetic organic insecticides were discovered, led by DDT, which had an extremely wide use in the decades following World War II. Other synthetic insecticides, such as the organophosphates, also subsequently become important.

Almost all of these early insecticides had some serious problems associated with their use. Lead and arsenic are now known and widely recognized as persistent environmental poisons. DDT also was highly persistent and concentrated in fat tissues, causing adverse effects on many kinds of wildlife primarily by hormone disruption. Organophosphates were a source of human poisonings, since their mode of action (inhibition of acetylcholinesterase) also affects the nerves of mammals in a manner similar to those of insects.

The recognition of insect hormones and their activities began to be better understood by the early 1960s. With this understanding of these compounds was the promise of ultimately acquiring new kinds of safer and more selective insecticides. Humans and other vertebrates may share a nervous system that is quite similar to insects, but only the arthropods have an exoskeleton and a need to periodically molt. Insecticides that disrupt hormones involved in insect growth and exoskeleton development were seen early on as potentially useful.

In recent years, this promise is becoming realized. Currently, there are several types of insecticides that act on different aspects of arthropod growth, and collectively these are known as insect growth regulators. Some of those are presented in table 4-1.

(continued)

(a)   (b)

FIGURE 4-18
Some insecticides now in use, known as insect growth regulators, act by disrupting the growth of insects. This can occur by disrupting chitin production or by creating an imbalance of hormones regulating molting. (a) Mosquito larvae are often controlled with insecticides that disrupt juvenile hormone levels, preventing the larvae from successfully becoming adult insects. (b) The successful development of a cabbage looper can be prevented by insect growth regulators so that they cannot transform into a normal pupa. Photographs courtesy of Ken Gray/Oregon State University.

TABLE 4-1 Some presently used insect growth regulators that are used for insect control.

ACTIVE INGREDIENT	MODE OF ACTION
Methoprene, kinoprene, fenoxycarb, pyriproxyfen	Juvenile hormone mimics
Diflubenzuron, novaluron	Chitin biosynthesis inhibitors
Tebufenozide, halofenozide*	Ecdysone agonists that disrupt molting
Cyromazine	Molting disruptor of dipterans
Azadirachtin	Uncertain mode of action but effects include molting disruption. A natural product derived from the seeds of neem tree (*Azadirachta indica*)

*Agonists are compounds that bind to and stimulate receptors. For example, ecdysone agonists overstimulate ecdysone production, which results in precocious molting without complete cuticle elaboration, sclerotization, and ecdysis. This is lethal to affected insects.

Development Rate and Diapause

Arthropods are **poikilothermic** ("cold-blooded") and have only a limited ability to regulate their temperature. Basking in sunlight or on sun-warmed surfaces can increase body temperature. Some heat can also be generated by metabolism as in muscle contraction, particularly the use of the large flight muscles. Nonetheless, the temperature of their surroundings greatly influences the activity and development of arthropods.

The rate at which an arthropod develops is very closely tied to temperature. Growth occurs more rapidly at warmer temperatures, slows down when it cools, and completely cease beyond a lower **base temperature**. For many insects this base temperature is around 10°C (about 48°F), but it can be higher or lower in different species.

Because of the close association of temperature with growth, it is often possible to predict how fast an arthropod will complete development or predict when a given stage (e.g., the adult) will appear in a given environment. For each growth stage, a certain period of warmth must occur above the base temperature of the arthropod. For example, a growing cabbage looper (*Trichoplusia ni*) caterpillar has a base temperature of 10.6°C (49°F). The larval stage of the cabbage looper, between egg hatch and pupation, will typically require about 10.6 days to complete at 33.1°C (85°F), 23.8 days at 20.6°C (65°F), and 62.3 days at 14.4°C (55°F).

The amount of heat needed to complete development is sometimes quantified as **degree-days**. A degree-day is one degree of average temperature above the base temperature over 1 day. For example, on a single day with an average temperature of 25°C, an insect with a base temperature of 10°C will accumulate 15 degree-day heat units. After 2 days of this temperature, 30 degree-days will have accumulated, and so on. By rearing insects at various temperatures, one can determine how many degree-days are needed to complete each stage. In the cabbage looper example, about 238 degree-days (C) are needed to complete the larval stage from egg hatch to pupation.

(a)

(b)

FIGURE 4-19

Typical of all moths, the cabbage looper, *Trichoplusia ni*, undergoes complete metamorphosis. Stages include the (a) adult, (b) egg, (c) larva (multiple instars), (d) end-stage larva preparing to pupate (prepupa), and (e) pupa. As an example of how development rate is correlated to temperature, the average time between egg hatch and onset of pupation is 10.6 days at 33.1°C (85°F), 23.8 days at 20.6°C (65°F), and 62.3 days at 14.4°C (55°F). Development of the cabbage looper ceases below 10.6°C (49°F). Photographs courtesy of Ken Gray/Oregon State University.

(continued)

(c)

(d)

(e)

FIGURE 4.19
(*continued*)

Such information can be useful in predicting the appearance of different life stages. One such use of this information is in **forensic entomology**, where the presence of insect life stages at a murder scene may be used to help predict the onset of death. More commonly, degree-day models are used to predict the annual occurrence of pest species in agricultural crops. Using this approach, pest management practices can be better timed and applied.

At some times, activity and growth may slow or cease and become independent of temperature. Insects that are night active will usually be inactive during the daytime, and the reverse occurs with day-active insects. Very hot periods may cause insects to avoid activity, as can periods of drought. When the conditions (light, temperature, humidity) again become favorable, insects soon resume activity.

More extended dormancy occurs during **diapause**. Hormones originating from the brain control diapause,

which produces a semipermanent cessation of development that cannot be rapidly reversed. Conditions that trigger the onset of the diapause condition most often involve changes in day length (e.g., shorter day lengths of late summer); in these cases, diapause is often broken when a different environmental trigger is present, such as the lengthening daylight associated with spring. Other insects require that a certain period of chilling pass before they "break" diapause and resume activity. Wetting, such as with seasonal rainfall or flooding, can be involved in ending a diapause period with other insects.

Diapause can be very important for the survival of insects that live in areas where the environment periodically becomes unfavorable. For example, in areas of cold winters, insects could die of exposure, if not starvation, during winter. (Insects that enter diapause to survive such periods also may undergo prior internal changes, such as increasing the solute concentrations in their blood, allowing them to better survive freezing temperatures.) A few warm days in January will be insufficient to awaken the diapausing insect, which requires other environmental cues that more reliably indicate that conditions are again favorable. Insects living in areas of periodic drought may go into diapause during the dry periods when food plants become unavailable.

Insects in the diapause condition often are also inactive. But inactivity is not a condition of diapause and some insects, such as the monarch butterfly, can undergo long migrations while being in a diapause condition.

Because of diapause, insects often have only a limited number of generations that they will produce each year. **Voltinism** refers to the number of generations an organism produces each year, and insects that have only a single generation have a **univoltine** life cycle. As an example, most tent caterpillars (*Malacosoma* species) hatch from eggs in spring, and these caterpillars develop rapidly (and produce a silken tent) over the course of a few weeks. They then pupate within a silken cocoon and a few weeks later the adults emerge as moths that produce new masses of eggs by early summer. Despite the continued presence of leaves, the newly laid eggs do not hatch in summer and instead remain in a diapause condition until the passage of a chill period and change of day length signal the return of spring the following year.

(a)   (b)

FIGURE 4-20

Tent caterpillars (a), *Malacosoma* spp., are an example of insects that are univoltine and will have only a single generation per year. Eggs hatch in synchrony with the new growth of host food plants in spring, and when the next generation of eggs are laid in early summer, they go into diapause. Greenhouse whitefly (b), *Trialeurodes vaporariorum*, is an example of a multivoltine insect that will reproduce continuously as long as temperature allows activity. Photograph of the tent caterpillars courtesy of David Leatherman. Photograph of the greenhouse whitefly courtesy of Clemson University/Bugwood.org.

Other insects produce two generations a year before entering diapause and thus have **bivoltine** life cycles. Insects that usually produce more than two generations a year are described as being **multivoltine**. Multivoltine insects typically occur in areas where favorable environmental conditions occur year round, and some insects may never enter a diapause period, reproducing continuously. Multivoltine habits also exist among some of the insects that occur as pests in homes, such as the German cockroach.

● Do Insects Sleep?

The role of sleep in humans is complex and still incompletely understood. It is often thought to have a role in consolidating certain types of memory and allowing the brain to process experiences of the day. It may also have growth and restorative functions involving the immune systems. Regardless, it is essential; in humans, sleep deprivation leads to serious behavioral and health problems.

FIGURE 4-21
Insects can have periods of inactivity for many reasons, including unfavorable temperature or light conditions and hormone-induced diapause. However, other than reduced movement, these periods in insects have little, if anything, in common with the biochemical changes that are central to the sleep humans require for normal brain function. This photograph of a bumble bee resting overnight on a flower is by Whitney Cranshaw/Colorado State University.

Although the body remains largely immobile, sleep is a very active process that involves alternations of periods of rapid eye movement (REM) sleep with dreaming, and non-REM dreamless periods when brain waves are slow and are of high voltage. Such changes in brain activity apparently are not present in insects, nor are they present in fish and amphibians.

Insects can remain immobile for long periods. During periods of the year, insects may go into an extended period of inactivity, diapause, that cannot be broken until some key environmental cue has occurred, such as transition to a critical day length or a sufficient period of exposure to cold (chill period). At any time, temperatures below the base temperature of insects can prevent them from moving until it warms sufficiently. For many species this is often around 6°C–10°C, although some alpine species, notably the rock crawlers, thrive at temperatures barely above freezing. Day-active insects become inactive as light fades, and night-active insects hide with daylight. However, these behaviors appear more equivalent to a state called "torpor" where the animal can be roused quickly if the stimulus is strong enough. (On the other hand, some insects feign death when disturbed and may subsequently remain inactive for hours.)

Although insects apparently do not sleep in the manner of humans, other mammals, and birds, some insects do show behaviors that are similar to sleep. Experiments involving fruit flies, cockroaches, and bees show that they may move to a resting area and take on a "sleep pose." Over time they show reduced responsiveness to stimuli but can be awakened with a strong stimulus. The ability to rouse them to activity may increase through time, and "sleep-deprived" insects may have an extended period of immobile rest the next day, much like humans.

The Arachnids—Spiders, Scorpions, Mites, and Other Eight-Legged Wonders

Class Arachnida

According to a classical Greek myth, Arachne was a mortal with a legendary ability to spin and weave. Her boastful pride in her abilities attracted the attention of Athena, the goddess of wisdom, arts and crafts, and strategy, prompting a weaving challenge match between the two. The tapestry Arachne wove was of superior workmanship but illustrated the foibles of the gods—a foolish choice of subject. Attempting to escape the wrath of Athena, she attempted to hang herself but was transformed into a spider by the vengeful goddess, forever doomed to spin. From this story comes the name "Arachnida," given to a class of arthropods that include the spiders.

Arachnida is a huge class of arthropods comprised of over 100,000 described species within 11 orders, second in number only to the hexapods. These include such familiar animals as spiders, opilionids (daddy longlegs), scorpions, mites, and ticks (figure 5-1). Arachnids have six pairs of appendages the first pair of which (**chelicerae**) function as mouthparts. In some arachnids, such as spiders, the chelicerae have two segments: a base and a pointed fang. In others, such as windscorpions and pseudoscorpions, the chelicerae are developed into the form of pincers. The mouth opening of arachnids is usually narrow, restricting most to feed primarily on liquids and small pieces of solid material.

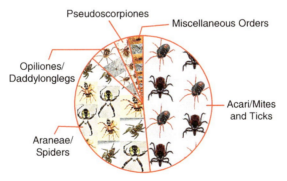

FIGURE 5-1
The relative number of presently described species among the arachnid orders. Just over 100,000 arachnid species have been described, with Araneae (spiders) and Acari (mites and ticks) being particularly species rich. However, many arachnid groups are still very poorly described, most notably the mites. Figures based on *Numbers of Living Species in Australia and the World*, 2nd ed. (2009). Photographs courtesy of Tom Murray.

The closest arachnid relatives are the horseshoe crabs (class Merostomata) and sea spiders (class Pycnogonida), two classes of marine arthropods that also possess chelicerae. This mouthpart feature is used by taxonomists to place these organisms together within the subphylum (or clade) Chelicerata. Insects, crustaceans, millipedes, and centipedes are among the great majority of arthropods that have

FIGURE 5-2
Chelicerae and pedipalps of a male spider. The chelicerae are tipped with fangs and can inject venom. The pedipalps on male spiders are enlarged and involved with sperm transfer. Photograph courtesy of Brian Valentine.

FIGURE 5-4
Windscorpions have immense chelicerae, which are used to tear prey, and leglike pedipalps. Photograph by Jack Kelly Clark and provided courtesy of the University of California IPM Program.

FIGURE 5-3
Chelicerae and pedipalps of a tailless whipscorpion (amblypygid). The pedipalps help to grasp prey, which are then torn with the chelicerae. Photograph courtesy of Jillian Cowles.

FIGURE 5-5
Two body regions are easily seen on most arachnids such as this nursery web spider—the cephalothorax, which contains the head, legs, and the abdomen. Photograph courtesy of Jim Kalisch/University of Nebraska.

FIGURE 5-6
The body regions of some arachnids, such as the daddy longlegs, are not distinct. Daddy longlegs also have small eyes, typical of most arachnids. Photograph courtesy of Brian Valentine.

mouthparts with mandibles and are thus classified together into the separate subphylum Mandibulata.

The second pair of appendages is the **pedipalps**, which may have various functions including capturing prey or manipulating prey, transferring sperm, or sensory abilities. The additional four pairs of appendages are the legs, used primarily for locomotion.

The arachnid body has two primary divisions. The head and leg-bearing thorax are fused to form the **cephalothorax** (or **prosoma**). On the hind end is the **abdomen** (or **opisthosoma**), which may be connected by only a thin area (pedicel) in some groups (e.g., spiders).

Arachnid eyes are of a simple type in the form of ocelli found either on the sides and/or the middle of the head. Only a few species are able to detect much more than general light intensity. (Some spiders, notably the jumping spiders, are an exception and have very good visual acuity.) Antennae are absent in the arachnids.

Oxygen exchange may occur with a tracheal system similar to that found in insects or may involve special structures known as **book lungs**. These are infoldings of the body that open internally similar to the leaves of a book, providing a large surface area for the diffusion of gases into and out of the body.

FIGURE 5-7
Spiders that actively hunt prey during the day have relatively large eyes. They are particularly big in jumping spiders. Photograph courtesy of Brian Valentine.

● Arachnophobia

Phobias are anxieties or fears that are out of proportion to their danger—and the fear of spiders (arachnophobia) ranks very high among animal phobias. Arachnophobic individuals may merely feel uneasy in the presence of spiders or they may be terrified. Intense fears may cause people to alter their lifestyle or habits to avoid even remote possibilities of contact with spiders. Arachnophobia may totally disrupt life, and in extreme cases, people may not be able to stand looking at a picture of a spider or even the word.

In a recent survey of Colorado State University students taking a freshman entomology class, almost one in six (15.7%) responded affirmatively "I fear spiders and cannot stand being in their presence" to the question "What are your attitudes toward spiders?" from among the five answers provided to the responder by the first part of the survey, suggesting strong arachnophobic tendencies. There was a sex bias in this study, with females being three times more likely to admit to strong spider fears than males. Overall, fear levels of spiders in this group were substantially higher than those reported from European populations, suggesting Americans may be more exposed to images and behaviors that stimulate unusually high level of undue spider concerns and fears.

There can be many sources of spider fears, and these vary greatly. For example, some people particularly dislike large spiders, while others are more worried about smaller ones. In general, the following attributes of spiders, described and commented by Colorado State University students, are those that may provoke concerns:

Socialization aspects. Spider fears developed from experiences with other humans (e.g., parents, siblings, teachers, classmates) or from disturbing media (e.g., watching the movie *Arachnophobia*)
Medical/health concerns. Excessive worries about dangers presented by spiders, even where medically important species are not present
Spider appearance issues. Concerns related to spider legs, hairiness, eyes, and size (big or little)
Spider movements. Rapid movements and/or unrealistic ideas about spider jumping ability
Body contact issues. Dread of the touch of spiders may be expressed even among those that do not have concerns about spider appearance
Omnipresence/unpredictability. The omnipresence of spiders and their unpredictable appearance
Presence in the home. Disgust of the idea of spiders within the home, often associated with this being indicative of "dirtiness"
General unspecified feelings. Spiders indefinably disliked and often described by such terms as being "creepy" or "gross"

(continued)

The best known arachnophobe of all time is Little Miss Muffet, a central figure in the nursery rhyme:

> Little Miss Muffet
> Sat on a tuffet,
> Eating her curds and whey.
> There came a big spider,
> And sat down beside her,
> And frightened Miss Muffet away.

A very real person likely inspired this nursery rhyme—Miss Patience Mouffet. She was the daughter of Rev. Dr. Thomas Mouffet (1553–1604), who had a particular interest in developing the use of insects and spiders in medicine. Apparently his daughter was often the recipient of many spider-based treatments and his activities were well known at the time.

Order Scorpiones— Scorpions

Scorpiones is one of the larger arachnid orders with over 1,500 species worldwide and approximately 100 species within the United States. With an abdomen narrowed to a long tail tipped with a stinger, scorpions are readily recognizable. These arachnids also tend to be relatively large, ranging up to 20 cm in length. Scorpions possess several other defining features. Most notably, the pedipalps are greatly enlarged pincers (**chelae**) modified for grasping. The chelicerae are similarly claw-like, although much smaller, and used to rip prey. Also on the head is a pair of simple eyes at the midline and zero to five pairs of eyes along each side.

FIGURE 5-8
A stripebacked scorpion feeding on a house cricket. Photograph courtesy of Jim Kalisch/University of Nebraska.

The cephalothorax and abdomen are broadly joined together with no narrowing that clearly separates these body regions. The first seven segments of the abdomen are broad with the genital opening on the first segment and comb-like structures (**pectines**) on the second segment. The last five abdominal segments narrow to form the stinger-bearing tail (**telson**).

FIGURE 5-9
Close-up of the chelicerae of a *Hadrurus* sp. scorpion. Scorpions feed by tearing apart their prey. Photograph courtesy of David Walter/ Royal Alberta Museum.

Scorpions are also an ancient group. Ancestral, extinct forms of scorpions (protoscorpions) are well represented in fossils from the Silurian period, although these apparently were restricted to shallow waters. Species with modern features that had made the transition to land were present by the early Devonian period, where they joined the millipedes as the first terrestrial arthropods.

During mating, scorpions partake in a ritual dance. The male grabs the pincers of the female, and the two may move together in a complicated "promenade à deux," moving back and forth or in circles. Periodically, their bodies may vibrate ("juddering"), and occasionally they may even come together for a "cheliceral kiss." This behavior may last for over an hour as the male tries to find a suitable spot to lay down his spermatophore. Once he has done so, the male will guide the female over the spermatophore, which she will then pull into her genital opening where the sperm are released to fertilize her eggs.

Scorpions produce an average litter size of about two dozen, although it can range from 6 to almost 100. The young are born alive, dropping into a "birth basket" made of the front two pairs of legs. They then move to the back of the mother where they will remain for at least a couple of weeks until the next molt. The juveniles develop relatively slowly, taking from 6 to 83 months to reach adulthood. Scorpions can be especially long-lived arthropods; at least one species in captivity can live up to 25 years.

FIGURE 5-10
Scorpions give birth to live young, which then climb on the back of the mother and are carried for the first few weeks of life. Photograph courtesy of Gerald Lenhard/Louisiana State University/Bugwood.org.

Scorpions are nocturnal predators that feed primarily on insects, although larger species may capture young lizards or rodents. As scorpions search for prey, their pincers are held forward and are used to grab their victims. Prey is then drawn into the chelicerae, which tear the food and mix it with digestive fluids. The stinger is not generally used for prey capture, except to help subdue particularly large and active prey.

The stinger is used primarily for defense as scorpions are favored foods of snakes, lizards, rodents such as grasshopper mice, and some birds. Several toxins occur in scorpion venom. For example, at least 15 compounds have been found in the venom of various *Centruroides* species, and these may have different functions. A North African desert scorpion *Androctonus australis* produces at least three classes of toxins, each being specifically active on either mammals, crustaceans, or insects. Scorpion toxins are

currently of interest in the area of active medical research and have particularly promising applications for use in cancer treatment.

Only a small fraction (<2%) of all scorpions produce toxins that pose serious medical threats to humans and all of these 20–25 species occur in the family Buthidae. Most scorpions produce a sting that is immediately painful, but the pain is transitory with no lingering effects. In most cases, other than some local redness, the effects of a scorpion sting dissipate within hours.

However, some scorpions are dangerous to humans. The notorious yellow fattailed scorpion of North Africa (*Androctonus australis*) is considered responsible for the greatest number of human deaths caused by any venomous animal, locally exceeding fatalities caused by venomous snakes. The only dangerous species found anywhere in the United States is *Centruroides sculpturatus*. Sometimes known as the "Arizona bark scorpion" because of its association with wooded areas, it ranges from extreme southern Arizona into northern Mexico. It occurs along waterways and is the only scorpion in southern Arizona commonly seen climbing trees or rough walls.

FIGURE 5-11
Fattailed scorpions, *Androctonus* spp., found in parts of northern Africa and the Middle East are the most dangerous of all species, causing most human fatalities from scorpion stings. Photograph courtesy of HTO.

The sting of *Centruroides sculpturatus* causes immediate burning pain that has been described as "electrical." Often there is little swelling or inflammation, but a light finger tap of the injection site causes extreme pain. Symptoms usually subside within 48 h but more serious systemic reactions can develop. These include a general sense of restlessness and may progress to include a sensation of a thickened tongue, slurred speech, drooling, roving eyes, staggering, and convulsions. Deaths from this species have been rare in the United States (two reported in the past 40 years), but many more regularly occur in Mexico.

Scorpions typically hide most of the day and actively hunt and mate at night. Although they are active at night, they can be among the easiest arthropods to locate. All scorpions will fluoresce a soft greenish yellow when exposed to ultraviolet light. This ability has no known function but is due to characteristics found in a layer of the exoskeleton (hyaline layer). Newly molted scorpions cannot fluoresce until the new exoskeleton has completely hardened, but old, discarded exuviae permanently retain this ability.

(a)

(b)

FIGURE 5-12
Emperor scorpion, *Pandinus imperator*, under (a) daylight and (b) fluorescent light. Photographs by Whitney Cranshaw/Colorado State University.

Order Palpigradi—Micro Whipscorpions

Micro whipscorpions are tiny, pale-colored arachnids found under rocks or other high-humidity sites (such as caves). Their bodies are broad and they possess a long, segmented tail, or flagellum, that is covered with sensory setae. They are presumed to be predators of tiny arthropods, although some authors have suggested that they feed on arthropod eggs. Their pedipalps are used for walking and lack obvious features for prey capture found in some other arachnids, such as the much larger whipscorpions (Thelyphonida order).

There are three species known to occur in the United States, particularly in Texas and California. Worldwide, 79 species in two different families have been described, but a great many more undoubtedly

FIGURE 5-13
A micro whipscorpion. Photograph courtesy of Lynn McCutcheon/Kilgore College.

occur as these are overlooked and poorly studied animals. Very little is known about any aspect of their biology.

Order Thelyphonida (Uropygi)—Whipscorpions (Vinegaroons)

Whipscorpions are elongated, slightly flattened arachnids with a slender whiplike tail appendage. The former order name, meaning "tail rump," refers to this feature, which is presumed to be used to detect air vibrations or odors. Whipscorpions have long legs and large, pincer-like pedipalps designed to capture prey. Three pairs of eyes occur along the sides of the head and a single pair on the front. Whipscorpions can be large, with a maximum body length of 80 mm and a tail that may extend it to 150 mm. Among the largest is *Mastigoproctus giganteus*, a species found in Florida and the southwestern United States, where it is sometimes referred to as the "grampus" or "mule killer." The majority of the 106 known species of whipscorpions occur in tropical areas.

Whipscorpions are predators that hunt at night. The narrow, long first pair of legs is sensory in function,

FIGURE 5-15
A whipscorpion in defensive posture. Photograph courtesy of Lynn McCutcheon/Kilgore College.

not involved in walking, and is used to find the insects or other small animals on which they feed. They grab their prey with their powerful pincers and then draw it to their chelicerae where it is crushed and torn.

Whipscorpions display mating rituals that involve caressing the potential mate with their long first pair of legs. A spermatophore is transferred from the male to the female, and after fertilization, the female retreats to a burrow and ceases to feed. A mucous sac is produced within which the eggs are held, and the mother stays with the eggs until they hatch. The young then move and climb onto her back and attach themselves there with special suckers. After molting one more time, they leave and the mother dies shortly afterward. Whipscorpions grow slowly, molting three times over the course of about 3 years.

Their most conspicuous defense is their ability to spray defensive fluid, which can be expelled to a

FIGURE 5-14
A large, tropical whipscorpion, also known as a vinegaroon. Photograph by Whitney Cranshaw/Colorado State University.

FIGURE 5-16
Close-up of the chelicerae of a vinegaroon. Photograph courtesy of Lynn McCutcheon/Kilgore College.

distance of more than 0.5 m. This is produced in a pair of large glands at the base of the tail and may contain a mixture of octanoic acid and acetic acid. The latter provides the vinegar smell that has given these arachnids the common name vinegaroons (or vinegarones). Other whipscorpions spray formic acid or chlorine.

Order Schizomida—Shorttailed Whipscorpions

Shorttailed whipscorpions are tiny arachnids, never exceeding 1 cm. In appearance they most closely resemble the whipscorpions (Thelyphonida) but have a more slender body and a short terminal appendage (flagellum). The latter is knobbed in the males, unique to each species, and is grabbed by the female during mating. The first pair of legs serves as sensory organs and is not used for walking. The hind pair of legs is usually most developed and modified for short jumps, giving these animals an appearance somewhat of a tiny cricket. The name of the order refers to a feature of their thorax, which is split longitudinally into two plates.

FIGURE 5-17
A shorttailed whipscorpion. Photograph courtesy of Lynn McCutcheon/Kilgore College.

Shorttailed whipscorpions are predators of other small arthropods. They have well-developed pedipalps modified to grasp prey and which move vertically, allowing them to pull in the prey, which they dismember with their large chelicerae. They are nocturnal hunters and most species lack eyes. Although they lack venom glands, they can defend themselves by spraying an acidic shot from the tip of their abdomens, much like whipscorpions of the order Thelyphonida.

There are about 236 species known worldwide, almost all in tropical regions. Eight species are found in the United States, where they occur in Florida and southern California.

Order Amblypygi—Tailless Whipscorpions (Amblypygids, Whipspiders)

Tailless whipscorpions superficially mostly resemble whipscorpions (order Thelyphonida) and similarly possess prominent, enlarged pedipalps and long, narrow legs. Unlike true whipscorpions, tailless whipscorpions lack the slender tail, a feature that is referred to in the order name meaning "blunt rump." Their body is flattened, the abdomen elongated and attached narrowly to the cephalothorax. They are of moderate size, ranging from 5 to 55 mm in length. Worldwide, 159 species have been described.

FIGURE 5-18
A tailless whipscorpion. Photograph courtesy of Jillian Cowles.

The first pair of legs is extremely long and can extend many times the length of the body. This pair of legs is described as antenniform and is used for sensing the environment. As they walk using the remaining three pairs of legs, in a sideways crablike crawl, one of their sensory whips extends in the direction of travel while the other continuously probes the area around them. Tailless whipscorpions are excellent climbers and capable of rapid sideways movement. When at rest, they crawl under rocks, into abandoned rodent burrows, or into similar protected sites.

The large, powerful, and spiny pedipalps help to impale and crush their prey. Most often they feed on various insects, but there is a report of a Caribbean species observed eating a hummingbird. Their chelicerae work up and down, back and forth thoroughly crushing their food and mixing it with digestive fluids before it is ingested.

During mating, the males tenderly touch the female with the tip of the antenniform first pair of legs in a ritual that may last an hour. He then produces a stalked spermatophore that the female picks up. Eggs are carried by the female until hatching after which the young move onto her back where they live for another 4–6 days. After leaving the mother, the young may molt five to eight times during the next year as they mature. Adults can be fairly long lived and survive for several years.

Being tropical and subtropical in distribution, representatives of Amblypygi are not common in the United States and are largely restricted to Florida, the Gulf States, and Arizona. Their bizarre appearance—extremely long legs, prominent spined pedipalps, and an unsettling crablike walk—have made them a favorite poster arachnid for horror imagery. Among the more recent amblypygid appearances was in the movie *Harry Potter and the Goblet of Fire*, where it was used to demonstrate an "unforgivable" curse, and they are a periodic staple of various *Fear Factor*–type shows.

Order Araneae—Spiders

Some spider etymology. The order name Araneae is derived from the Latin word *aranea*, meaning spider. The common words for spider in many European languages are also based on this Latin root: French, *araignee*; Italian, *ragno*; and Spanish, *arana*. The English term "spider" is derived from the Anglo-Saxon word *spinnan* ("to spin"/"one who spins"). In modern German, the name for spider is *spinne* and in Swedish *spindel*.

Spiders are a large and diverse order, including over 40,000 described species with about 3,700 found in North America. Their four pairs of legs are approximately similar in size, although some may have larger forelegs to help grasp prey. The pedipalps on males may be grossly enlarged and are used in sperm transfer, while those of females are much smaller and somewhat resemble a small leg. The chelicerae are tipped with sharp fangs connected by a duct to poison glands. The movement of the chelicerae is used to separate two of the three main divisions of spiders (Mygalomorphae, Araneomorphae, and Mesothelae). In the mygalomorphs (suborder Mygalomorphae: tarantulas, trapdoor spiders, purseweb spiders), the

FIGURE 5-19
Many, but not all, spiders produce webs for prey capture. Orbweaver spiders (Araneidae family) produce intricate, geometrically patterned webs and also swath their prey in binding silk. Photograph courtesy of Jim Kalisch/University of Nebraska.

fangs move up and down, parallel to the body. Most spiders are araneomorphs (suborder Araneomorphae: true spiders) with fangs that open sideways and move laterally. Mesothelae includes only one Asian family and has characteristics that are considered primitive among spiders, including segmentation of the abdomen.

FIGURE 5-20
"True spiders" of suborder Araneomorphae have chelicerae that move laterally. Photograph courtesy of Tom Murray.

FIGURE 5-21
Spiders of suborder Mygalomorphae, which includes the tarantulas and funnelweb spiders, have chelicerae that move vertically such that the fangs are inserted in a downward movement. Photograph by Whitney Cranshaw/Colorado State University.

Spiders lack antennae but can be very sensitive to vibration through specialized receptor setae on their body. Chemical cues are also detected by spiders via open pores among the tarsi at the tips of the legs. Most spiders have four pairs of eyes, and the arrangement of the eyes is often useful in distinguishing different families. The size of the eyes is variable; those species that actively hunt possess much larger eyes than those that passively capture prey in webs. To improve light capture, particularly by those that are night hunters, the receptor cells of the eyes of some spiders reflect light through a layer of tissue that functions as a biological reflector, known as a **tapetum**. Those that wish to collect spiders may easily do so at night as the tapetum will readily reflect the beams of light from a flashlight, making spiders with brightly shining eyes an easy target.

Spiders breathe through a combination of spiracles with branching tracheal tubes (similar to insects) and/or book lungs. The latter are leaflike infoldings of the body into which air moves. Oxygen then diffuses through the thin membranous wall of the book lung and is

FIGURE 5-22
The eyes of spiders have a layer of reflective tissue (a tapetum) that helps to capture light. As a result, spider eyes will reflect the light from a flashlight at night, which makes them easier to find. (The spider shown is a jumping spider, a day-hunting species with unusually large eyes.) Photograph courtesy of Brian Valentine.

carried throughout the body via the hemolymph, attached to the respiratory pigment hemocyanin—the spider version of human hemoglobin.

SPIDER SILK

All spiders produce silk that is drawn out of spinnerets located near the tip of their abdomens. Most spiders have three pairs of spinnerets. An additional silk-producing structure (**cribellum**) is present in some families of spiders, known as the cribellate spiders, which include most of the more familiar web-producing species. The cribellum consists of hundreds, if not thousands, of tiny individual spigots through which the silk is extruded. Cribellate spiders also have an elongated comb-like structure on the hind leg (**calamistrum**) that helps to create a fuzzy-surfaced silk that is particularly well adapted to snag insect legs.

FIGURE 5-23
At the end of the abdomen are the spinnerets, from which silk is withdrawn. This orbweaver spider also has a cribellum above the other spinnerets. Photograph courtesy of Jim Kalisch/University of Nebraska.

● Spider Silk and Human Uses

The unusual properties of spider silk—a remarkable combination of strength and elasticity—have long attracted the attention of many. Spider silk has a breaking point far greater than steel and may stretch its full length before breaking: a 0.01 cm thread can support 80 g. Some types of silk, such as those used for draglines, are particularly strong. Spider silk is also very durable and resists breakdown by fungi and bacteria as well as solvents, a feature well illustrated by the persistence of cobwebs.

Throughout history, humans have adapted spider silk for many purposes. For a long period, thick webs such as those produced by funnel weavers (Agelenidae family) were employed in Europe as bandages. There have also been several uses of spider silk for fishing. Twisted webs can be fashioned into fishing lines, with a frayed, baited end that entangles the teeth of small fish without a hook. A very large New Guinea spider can be induced to spin on a bamboo frame to produce a net that is invisible underwater. The net is then used to simply flip the fish out of the water. Spider webs have also been incorporated into fish traps used in the Solomon Islands. Spider silk was also used widely to produce the crosshairs of optical and survey instruments and entered World War II on bomb sights.

Spider silk however has never been successfully utilized for fabrics, although not for the lack of attempts. There were several investigations during the 1700s and 1800s on the use of spider silk, beginning with François Xavier Bon de Saint Hilaire in 1709 who used spider egg sacs to produce a few pairs of silk stockings and gloves that were reported to have a "pleasing" gray color. The French Academy of Sciences in Paris then commissioned René Antoine Ferchault de Réaumur to perform further studies. Unfortunately he concluded that it was not a feasible endeavor, largely because of the problems inherent in rearing large numbers of predators with cannibalistic tendencies. Furthermore, only the egg sac was considered to be a usable material and that it would take 663,522 spiders to produce a pound of silk.

B. G. Wilder took up the challenge a century later and attempted to draw spider silk directly from the spinnerets of a large orb weaver (*Nephila* sp.). He held them in a special "stock" that did not interfere with the flow of silk and reeled it off for 75 min, producing a total of about 45–50 m of silk fiber—without exhausting the silk glands.

Wilder concluded that this process was not feasible for silk production due to the high labor involved, but the challenge was taken up again in recent years by two British textile artists, namely, Simon Peers and Nicholas Godley. In September 2009, they unveiled an elegant golden tapestry at the American Museum of Natural History produced solely from the silk of a Madagascar orbweaver, *Nephila madagascariensis*. The silk used in the project was withdrawn from over 1,000,000 spiders during the course of 3 years, the spiders being collected daily from the wild, a project that involved over 70 villagers. Some 25 m (80 ft.) of dragline silk could be extracted from each spider using a modification of the method used by Wilder. Between 96 and 960 individual threads of spider silk were then combined to create the threads for weaving the finished product, an elaborately patterned 1.3 × 3.7 m (4 ft. × 11 ft.) fringed tapestry of natural golden color.

Interest in spider silk fabrics was renewed when transgenic technologies allowed new approaches. A particularly unique feature of spider silk is its ability to stretch without breaking. This has long interested the Department of Defense where such properties help protective equipment better absorb energy from a projectile; spider silk outperforms Kevlar fibers (used in body armor) in this regard. The production of spider silklike materials has long been considered a goal of material science, and recently there has been substantial progress in its understanding and production. By 2002, genes to produce the proteins found in dragline silk were successfully incorporated into goats that then expressed the proteins in their milk where it could be recovered. Spider silk protein has also been produced in transgenic tobacco, demonstrating the potential of agricultural crops to be a source of these raw materials.

Even if the proteins of spider silk are made available in sufficient quantity, there still remains the big hurdle of how to spin it into usable fiber. A great deal of the strength and physical properties of the silk in a spider's web are a function of how the spider extrudes it through the spinnerets. It is now understood that dragline silk has two components: hard crystalline segments intermixed with tangled areas that together produce the desired qualities. Techniques that can recreate the proper balance of these two components may be able, one day, to produce high-quality spider silk. This research has also stimulated the development of artificial polymers based on the structure of silk, an approach that may also ultimately lead to new materials of similar function.

Spiders can produce many different types of silk, and these can be used for a wide variety of functions:

Dispersal. Young spiders often disperse on wind, a behavior known as **ballooning**. They will play out a strand of silk, and this helps to catch a draft that can carry them aloft.

FIGURE 5-24
A young spider producing a silk strand. This strand may allow the spider to catch wind drafts and disperse from plants. Silk strands are also cast to form attachments with nearby surfaces during web construction. Photograph courtesy of Brian Valentine.

Dragline. Spiders almost invariably lay down silk threads wherever they travel. The dragline allows individuals to safely drop from a plant or catch themselves from an accidental fall. The draglines also allow spiders to find their way back to their retreat.

Structures for stopping and ensnaring prey. The production of webs to intercept prey is the best known use of spider silk. Sticky webs may snare the prey, while other webs serve more as platforms that signal to a waiting spider the vibrations from a potential prey. In bolas spiders, a single thread of silk, tipped with a sticky end, is tossed at a potential prey that is then reeled in like a fishing line.

Bands for binding prey. Web-building species may wrap their prey with silk, particularly those that are large and difficult to subdue (e.g., wasps).

Protection and retreat. Most spiders will produce a retreat where they may rest. In web-building species, this will be at the edge of the web.

FIGURE 5-25
A banded garden spider binding a grasshopper that was caught in its web. Photograph by Whitney Cranshaw/Colorado State University.

FIGURE 5-26
A tarantula at the opening of its retreat. The retreat is a burrow lined with silk. Photograph by Whitney Cranshaw/Colorado State University.

Others may create a silk-lined burrow or line an area under a rock or other protected spot.

Web structures associated with mating. Males produce small webs during the process of transferring sperm from their genital openings to their pedipalps. Webbing may also be used by males during mating to loosely bind the female, presumably reducing the chances of subsequently being eaten.

Protection of the eggs and spiderlings. Tough egg sacs are produced to cover and protect the eggs and newly hatched spiderlings.

FIGURE 5-27

Pedipalps of a male house spider, *Tegenaria domestica*. The pedipalps in male spiders are enlarged and used in sperm transfer during mating. The sperm is first deposited via the genitals onto a small silk web then drawn into the pedipalps. Photograph courtesy of Brian Valentine.

FIGURE 5-28

Spiders protect eggs by covering them in tough silk to produce egg sacs. Illustrated is a western widow spider, *Latrodectus hesperus*. Photograph by Whitney Cranshaw/Colorado State University.

● The Making of an Orb Web

Orbweaver spiders make geometric webs of incredible organization—although not always as organized as those produced by "Charlotte A. Cavaticus," the lead role in the children's story *Charlotte's Web*. (Charlotte is loosely based on the common barn-infesting spider of the Midwest, *Araneus cavaticus*.)

To begin the construction, the orbweaver spider constructs a **bridge line** that runs along the top of the web. This is usually done with ballooning silk filaments that are used to snag a nearby object. The bridge line is pulled tight and subsequently reinforced with several spinning trips back and forth across the main bridge line. Sometimes the bridge line is established by attaching a dragline to a point, crawling down and then up an adjacent surface until about the same level, then pulling the dragline tight.

The spider will then produce a **V-line** by first making a loose loop of silk descending from the bridge. From the center of the dangling loop, the spider then drops to a lower surface and secures another supporting line, producing a "Y" form.

With the fragile original framework now in place, it is then reinforced. A series of **radii** are established; each is attached to the center area of the "Y" and then to the supporting surfaces surrounding the web. Next a **dry scaffolding spiral** of silk is produced that further strengthens the structure. Only after all this is completed can the finishing work be carried out. Here, **sticky coils** are laid down within the web and the dry threads removed.

Each species of orb weaver will customize the web in some manner. Various zones in the web will be free of sticky silk, enabling the arachnid to cross over the web. Parts of the web may be strengthened; other species may cut out the central hub of the web. Some larger orb weavers, including the common "garden spiders" (*Argiope* species), produce dense areas of zigzag silk through the middle, a structure known as a **stabilimentum**.

The construction of the orb web has long been the subject of interest and wonder. Or, as eloquently stated by the French naturalist J. Henri Fabre, "What refinement of art for a mess of Flies!"

(continued)

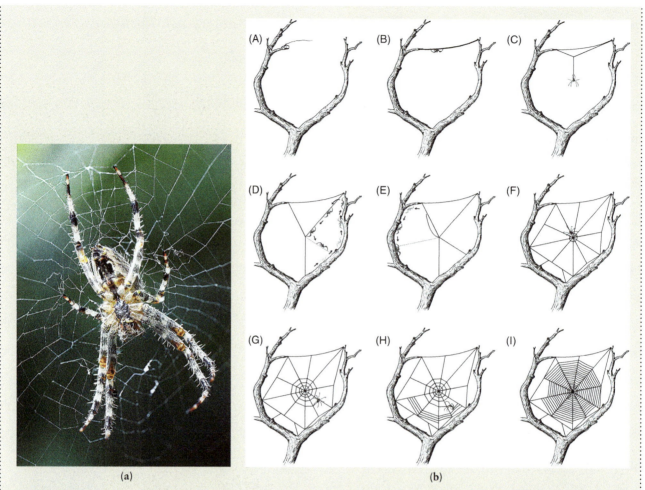

FIGURE 5-29
(a) An orbweaver spider. Photograph courtesy of Jim Kalisch/University of Nebraska. (b) Sequence of steps involved in the production of an orb web. The spider lets out strands of silk (A) that can form an initial bridge (B) if they catch on another point for attachment. This is then reinforced by laying silk back and forth across the bridge line. From the center of one of these strands, the spider then drops (C) to a lower point for attachment, forming a Y brace. At the hub of this is attached a new strand that is carried up to the bridge, then brought down to a lower point, pulled tight, and attached to form a radiating spoke (D). This is then repeated on the other side (E) and then at other points (F), producing a series of spokes that may be reinforced in the center. A dry, scaffolding spiral of silk is then placed down for temporary reinforcement (G). This is then replaced with sticky silk put down closely spaced in a spiral form (H) that ultimately comes to the hub (I). Drawing by Matt Leatherman adapted from *Spiders and Their Kin*.

FEEDING

All spiders are predators and they use venom injected through their fangs to subdue their prey. Although spiders use webs and other forms of traps with silk to catch prey, others seize and physically overcome their prey with their legs and chelicerae. The crushing chelicerae rip and break the prey apart while digestive fluids from the spider's maxillary glands are released. The prey's softer body parts are predigested and sucked up as spiders can only feed on fluids.

As feeding continues, the prey may be rolled and chewed so that all but the indigestible harder parts of the prey body may remain at the end of the meal and are then discarded.

The ability of spiders to kill insects is well known, but their capacity for consumption is less well known. For example, a single female black widow spider was observed to consume 250 house flies, 33 fruit flies, 2 crickets, and another spider. Although admittedly widows are a fairly large and a rather long-lived genus of spiders, they are also only one of many scores, if not hundreds, of species that may be found in any grassy field or woodlot. Estimates of the numbers of spiders in such sites around the Washington, D.C., area were over 27,000/ha (11,000/A) in woodland areas and over 158,000/ha (64,000/A) in grasslands. In England and Wales, it is estimated that throughout the region, spider populations average about 123,000/ha (50,000/A).

MATING AND DEVELOPMENT

The internal sex organs of spiders occur on the abdomen, but male spiders transfer sperm through their pedipalps. A droplet of sperm is first deposited on a specially constructed tiny silk web and then drawn into the pedipalps. The male then wanders in search of a receptive female. When a female is located, the male typically will go through elaborate courtship rituals to coax a female to accept him as a mate. For those spiders with good vision (e.g., jumping spiders, wolf spiders), this can involve vigorous displays of the forelegs or palps and dancelike motions. Others make more subtle approaches that may include a patterned tapping of a web. Regardless of the method employed by the male, it is necessary to announce his presence in a manner that convincingly distinguishes him from prey. Spiders are generalist hunters, not beyond a bit of cannibalism. Males have particularly serious problems with becoming a meal for the female in some of the web-building species where males are often much smaller than the female.

FIGURE 5-31
Male (right) and female (left) wolf spiders. Photograph courtesy of Brian Valentine.

FIGURE 5-32
Male (upper right) and female (center) banded garden spiders, *Argiope trifasciata*. This species shows strong sexual dimorphism, with males being much smaller than females. Photograph by Whitney Cranshaw/Colorado State University.

FIGURE 5-30
Male (right) and female (left) jumping spiders. Photograph courtesy of Brian Valentine.

FIGURE 5-33
Female nursery web spiders (Pisauridae family) carry their egg sacs by means of their chelicerae. Photograph courtesy of Brian Valentine.

FIGURE 5-34
Female wolf spiders (Lycosidae family) carry their egg sacs on the spinnerets. When the eggs are ready to hatch, she tears open the egg sac and the young crawl on her back, where they are carried for the first week or so of life. Photograph courtesy of Brian Valentine.

FIGURE 5-35
Spiderlings clustered together and beginning to disperse shortly after egg hatch. Photograph courtesy of Brian Valentine.

If he is successful (and not eaten for dinner), the male will insert his pedipalps into the genital opening of the female on the underside of her abdomen and inject the sperm where it is stored for later egg fertilization. When the female is later ready to lay eggs, they are fertilized as they emerge and are placed into a silken egg sac. The spider egg sac is a very tough and resilient material constructed from special silk used only for this purpose. Additionally, females of many spider species will stay and guard the egg sac against other predators. As an extreme example of this behavior, a wolf spider female will carry the egg sac attached to her spinnerets. When the young are ready to emerge, she rips it open to release the spiderlings, which then climb on her back where they remain until they are ready to hunt on their own.

Spiderlings may disperse from the area of the egg sac by walking. Many species undergo longer-range dispersal by means of ballooning, by which they are carried to potentially long distances on wind currents. Spiderlings initiate this behavior by crawling to the tip of a twig, leaf, or other prominent point and then tilt their abdomen upward, letting threads from the spinnerets trail outward to catch the breeze. When the pull is sufficiently strong, they let go and are launched into flight.

FIGURE 5-36
A spiderling splaying out a silk strand as it prepares to disperse by ballooning. Drawing by Matt Leatherman.

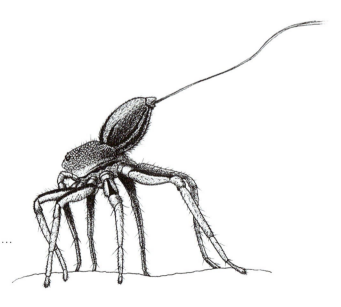

FIGURE 5-37
Spiderlings preparing to disperse through ballooning. Photograph courtesy of Mark Chappell/University of California, Riverside.

Often flights abruptly halt a short distance away if the threads snag a nearby plant. When a large number of spiderlings attempt to balloon from an area, surrounding surfaces may be covered with a shower of shimmering silk threads (**gossamer**). Others, however, may engage in truly epic dispersal. Spiderlings have been found in aerial surveys at altitudes of over 3,000 m and have been known to reach the top of high mountains. (Jumping spiders have been recovered from the slopes of Mount Everest at 6,700 m.) They have also been found on ships in the middle of the ocean, thousands of miles from land. Spiderlings are often the first arthropods to colonize new islands and arrive shortly after violent volcanic eruptions have eliminated other forms of life.

As the spiders grow and develop, they molt repeatedly and grow in size. For most species, the adult stage is reached after 5–10 molts. Males tend to be shorter lived than females and often die shortly after mating. Where hard frosts are present, most spiders live for only 1–2 years. There are notable exceptions among mygalomorph spiders. For example, female tarantulas are known to live more than 20 years; males rarely survive half that long. Also unusual among spiders is that these long-lived females will continue to molt after they are sexually mature. The sperm-storage structure (**spermatheca**) is lost during the molt so female tarantulas must again mate after molting to produce fertile eggs.

MEDICALLY IMPORTANT SPIDERS

Although all spiders are venomous, using their toxins to kill prey, very few species can or will bite humans. Among those that do bite, a spider bite may feel like a sharp pinprick. The mild pain usually dissipates rapidly, unless the wound subsequently becomes infected by bacteria, and there is no further discomfort. The venom produced by the overwhelming number of spider species has no significant effects on humans. However, a small number of spider species that can bite humans possess venom that can cause medical problems. Some spiders produce toxins in their venom that can affect the nervous system (**neurotoxins**), while other spiders have venom that damages tissues (**cytotoxins**).

The best known among medically important spiders are the **widow spiders** (*Latrodectus* species), which worldwide include 31 species. In the United States, probably four are natives, including the southern black widow ("arana capulina" in Mexico), *L. mactans*, which broadly ranges through the southern and eastern states; the northern widow (*L. variolus*) found in the northeast; the western widow (*L. hesperus*) common in areas of the High Plains and west to the Pacific; and *L. bishopi*, which is restricted to Florida where it is considered a threatened species. The brown widow (*L. geometricus*) is now established in southern United States areas, including southern California, but is thought to be native to southern Africa, where it occurs as one of six species of *Latrodectus* collectively known as "button spiders." Other famous representatives of this genus include the European widow ("malmignatte" or "karakurt"), *L. tredecimguttatus*, found throughout the northern Mediterranean and extending into Asia, and the redback spider of Australia, *L. hasselti*.

FIGURE 5-38
A female black widow, *Latrodectus mactans*. Photograph courtesy of Lynn McCutcheon/Kilgore College.

FIGURE 5-39
A male (lower) and female (upper right) western widow spider, *Latrodectus hesperus*. Photograph by Whitney Cranshaw/ Colorado State University.

The venom used by widow spiders is a neurotoxin that commonly produces a set of characteristic symptoms. Often there is a general sense of discomfort shortly after the bite, and acute symptoms increase in severity during the first day after the bite. Muscle and chest pain or tightness is common, with the pain spreading to the abdomen, producing stomach cramping. Sweating, and breathing and speech difficulty may occur if symptoms progress. Symptoms usually decline after 2–3 days but some mild symptoms may continue to be felt for several weeks after recovery. Deaths from widow bites have occurred; however, widow bites resulting in life-threatening effects never involve more than a very small percentage of those bitten as the amount of venom injected is so small. The development of effective medical treatments, including antivenin, has helped greatly so that death from a widow bite is now extremely rare.

The Sydney funnelweb mygalomorph (*Atrax robustus*) of eastern Australia also produces a neurotoxin and is considered by some to be the most dangerous spider in the world. Much of this reputation is based on its aggressive behavior as it may bite with little provocation. Unusual among spiders, it is the males of this species that are involved in almost all bites, and incidents are most common late in the season when the males migrate in search of mates. Thirteen deaths have been reported to have occurred from the bite of this spider, but none since 1980 when antivenin became available. Still, a couple dozen bites requiring medical attention occur annually.

A group of feared spiders in tropical South America include various *Phoneutria* species, sometimes known as "banana spiders." Bites from these species are extremely painful due to serotonin in the venom. More severe symptoms may develop, including cold sweat, excessive salivation, painful erections, and heart irregularities. Deaths are rare, but have been recorded and have only occurred among young children.

Venoms that are cytotoxic and lead to tissue death in humans are produced by a small number of spiders. The most notorious are the brown spiders of the genus *Loxosceles*. Ten species are native to North America and one of these, the brown recluse spider (*Loxosceles reclusa*), is a medically important species in parts of the central and south central United States. *L. laeta*, a brown spider of South America, is also considered a potentially dangerous species where it occurs.

Brown recluse spider bites may be felt initially as a slight pinprick or not noticed at all. The great majority of bites result in no further effects. Even if they bite, spiders may not introduce any venom, resulting in a "dry bite." In a small percentage of cases, an irregular red area develops around the bite within 2–8 h, and the site becomes painful and itchy. A small blister may develop at the point of the bite; typically this heals normally with no further effects.

In a small fraction of these latter cases, further complications can develop. A bluish sinking patch, with the central blister, may occur within 24–72 h after a bite. This may progress to produce an irregular lesion, an inch or two in diameter,

FIGURE 5-40
A female brown recluse spider, *Loxosceles reclusa*. Photograph courtesy of Jim Kalisch/University of Nebraska.

FIGURE 5-41
Brown recluse spiders captured on a sticky trap. Photograph courtesy of Jim Kalisch/University of Nebraska.

with surrounding redness and sensitivity to touch. In a few cases this will further expand, exposing underlying tissues as the dead cells slough away. This ulcerated area is dry, since capillaries are sealed by the effects of the venom, but may take a couple of months to heal, sometimes with permanent scarring.

A few other spiders are sometimes implicated in producing bites that cause some local tissue death, a condition sometimes described as **necrotic arachnidism**. Among those are the whitetailed spiders

(*Lampona* spp.) of Australia and New Zealand, some yellowlegged sac spiders (*Cheiracanthium* spp.), and some large wolf spiders. Most such reports are quite questionable because either the spiders are poorly identified or alternate diagnoses are neglected. For example, brown recluse bites are commonly misdiagnosed in the United States, even by medical personnel—who should know better but often do not. Among the dozens of medical conditions that can mimic the symptoms of brown recluse bite are bacterial infections, particularly those caused by methicillin-resistant "flesh-eating" *Staphylococcus aureus* (MRSA) strains.

FIGURE 5-42

Yellowlegged sac spiders, *Cheiracanthium* spp., are common residents in buildings. They are thought to be one of the most important species involved in human bites in the United States, although no serious symptoms are associated with their bites. Photograph courtesy of Joseph Berger/Bugwood.org.

FIGURE 5-43

Whitetailed spiders, *Lampona* spp., common in parts of Australia and New Zealand, are sometimes portrayed in media accounts as a dangerous species, but repeated studies have shown this to be false. They are harmless to humans. Photograph by Whitney Cranshaw/Colorado State University.

Similarly a poorly designed study in an obscure journal once implicated that a recently introduced (in the United States) European funnel weaver spider, *Tegenaria agrestis* (aka, "hobo spider"), was the cause of necrotic lesions similar to those described for brown recluse. These claims were quickly taken up by the popular press and, later, the Internet,

FIGURE 5-44

Probably no spider in North America has such an ill-deserved reputation as does *Tegenaria agrestis*, the common house spider of Europe known often as the "hobo spider." Repeated studies have clearly shown that this is a species harmless to humans. Early reports that it could produce necrotizing wounds likely confused secondary bacterial infections, which can occur following any wound. Photograph by Whitney Cranshaw/Colorado State University.

producing considerable public concern. (Ultimately these claims even penetrated some medical teaching texts.) Subsequent research has clearly showed the errors of the original study, which could never be replicated. (Again, MRSA or other bacterial infections are thought to be the actual cause of necrotic skin lesions in the initially published study.) It is now recognized that *T. agrestis*, an extremely common species in Europe that has never been of medical importance, is similarly of no significant medical importance in the United States.

TARANTULAS

One group of spiders that are not considered dangerous is the tarantulas. Very often these spiders have a fearsome reputation, but this apparently is based more on their large size and hairiness than on any real threat that they pose. The tarantulas found in North and South America do not have venom that causes any substantial effects on humans, although the fangs of larger tarantulas may break the skin. A few species found in Africa and Asia possess venom that can produce a more painful bite, similar to a wasp sting, and some tree-dwelling tarantulas of Southeast Asia do have more potent venom that can produce cramping, localized paralysis, and severe pain. But no tarantulas have venom that is life threatening to humans.

FIGURE 5-45

An Oklahoma brown tarantula, *Aphonopelma hentzi*. Like other tarantulas native to North America, they do not have venom that could cause any serious injury and they rarely bite. Their main defense is stinging hairs that they can dislodge from the back of their abdomens. Photograph by Whitney Cranshaw/Colorado State University.

FIGURE 5-46

An orange baboon tarantula, *Pterinochilus murinus*. This African species is commonly sold in the pet trade, but it is an aggressive species that can produce a painful bite. Photograph courtesy of Jim Kalisch/University of Nebraska.

About 900 species of tarantulas occur worldwide and they range widely in size. The largest include some of the "birdeater" tarantulas in the genera *Theraphosa* and *Lasiodora* that are found in South American tropical forests. These can display a leg span of 33 cm. In the United States, about 30 species occur, almost exclusively in the southwestern states. The largest may have a leg span of 15 cm, but some tarantulas are only a fraction of that size. None of them have a dangerous bite, and most rarely bite at all unless roughly handled. Instead their primary defense is the use of **stinging (urticating) hairs** that they can brush off their body or dab onto an attacker. The hairs are barbed to lodge into and penetrate thin areas of the skin, producing painful itching.

● Tarantism—A Standard for Mass Hysteria

Few events of mass hysteria have such a bizarre and persisting history as does one blamed on a spider's bite—tarantism. The alleged arachnid culprit was a large wolf spider, *Lycosa tarantula*, whose bite purportedly produced a violent sickness with unusual symptoms. Victims were seized with a sort of insanity that produced weeping, frantic dancing, and skipping. These dances, which often attracted a musical accompaniment, could last for days until the dancer collapsed, exhausted. Music, dancing, and the profuse perspiration that resulted were also the "cure" for the bite, as it allegedly flushed out the venom.

The first reports of tarantism were in 1370 and, originally, were centered around the city of Taranto in southern Italy. The phenomenon spread rapidly through southern Europe and persisted for over 400 years. Some local legends expanded on the symptoms so that they would reoccur each year at the same date of the original "bite."

There was no medical reason for this behavior as the bite of the wolf spider has been shown incapable of producing anything more than a pinch, with perhaps a small local wound that heals without a problem. It has been speculated that there may be a possible spider angle to explain some incidents. The Mediterranean black widow or malmignatte, *Latrodectus tredecimguttatus*, is a common spider species in fields during midsummer and does possess a neurotoxin similar to that found in other widow spiders.

Other explanations are more likely. One is that tarantism originally was a neurotic response to the widespread fears and insecurity that developed in response to the terrible effects of the plague pandemic (page 380), which had recently devastated southern Europe. Its persistence also may have been a subversive means of entertainment. For example, some of the

(continued)

dancers, known as taranti, dress in white with ribbons while carrying bunches of grapes, exactly copying the priestess of the Roman god of wine Bacchus, whose worship had been suppressed by the Church. The taranti would also attract paying viewers, and musicians visited the fields in hopes of being hired by someone who claimed they had been bitten by the spider. After centuries tarantism died out, but a remnant remains in *La Tarantella*, a dance based on the taranti.

FIGURE 5-47
A woodcut illustration of the dancing mania that surrounded the tarantism hysteria during the Middle Ages in parts of southern Europe. Woodcut by Hendrik Hondius from a drawing by Pietre Breughel der Altere.

Order Ricinulei—Hooded Tickspiders

Hooded tickspiders are considered to be the rarest and least known of all arachnid orders. About 57 species are now described, with one from the United States, *Cryptocellus dorotheae*, known from the Rio Grande Valley. The majority of currently described species are cave dwellers.

Hooded tickspiders are small, typically 5–10 mm, with a heavy body form and thick cuticle. The legs are not as long and prominent as found in most arachnid families. They have pincerlike pedipalps, although these also are relatively small. The most unique physical feature of the tickspiders is a flap on the front that can be raised and lowered.

FIGURE 5-48
A hooded tickspider. Photograph courtesy of Brett Opell.

Very little is known about the habits of hooded tickspiders. Several unique behaviors and characteristics are clear. Sperm transfer by the male involves the use of an elaborate organ on the third leg of the male, in a manner similar to how the pedipalps are used by male spiders of other orders. Only a single egg is produced by the female at a time, which is kept by the female under the hoodlike flap on her head. Newly hatched young have only six legs, acquiring a fourth pair after the next molt.

Order Opiliones— Harvestmen (Opilionids, Daddy Longlegs)

Opilionids are among the most diverse of the arachnid orders with over 6,100 species worldwide, some 235 of which occur in North America. They have a globular body form that lacks distinct regions of the cephalothorax and abdomen. A pair of large eyes directed to the sides arises above the head on a short stalk, a bit like a periscope. The pedipalps of the most common North American species (suborder Palpatores) are rather small; these can be much larger in some other opilionids (suborder Laniatores), which also have relatively short legs.

FIGURE 5-50
Head of a daddy longlegs showing chelicerae and pedipalps. Photograph courtesy of Brian Valentine.

FIGURE 5-51
Daddy longlegs feeding on an insect. Photograph courtesy of Jim Kalisch/University of Nebraska.

FIGURE 5-49
Phalangium opilio, a European daddy longlegs that is now widely distributed in North America. Photograph courtesy of Tom Murray.

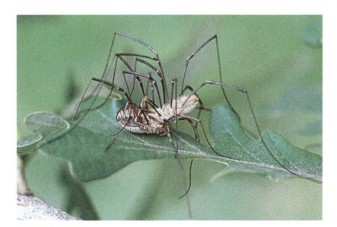

FIGURE 5-52
Mating pair of daddy longlegs. Photograph courtesy of David Leatherman.

The most obvious distinguishing feature is the extremely long, narrow legs possessed by most opilionids. (The largest species, *Trogulus torosus*, has a body length of about 22 mm and, with legs extended, can reach 160 mm.) This feature is the basis for the popular common name "daddy longlegs." (This can be spelled in a variety of acceptable manners: daddy long legs, daddylonglegs, or daddy-longlegs.) Only the hind three pairs of legs are used for walking; the front pair, which is usually held in front of the body, has sensory functions and is used to explore the environment.

These long legs cause some extraordinary challenges during molting, which opilionids undergo about six times during their lifetime. When preparing to molt, they hang upside down, attached at the base of the abdomen, and grab each leg with their chelicerae, pulling out the new leg to free it. Care must further be given to prevent the newly freed legs from sticking to each other until they sclerotize and harden, which may take hours.

A feature observed by almost anyone who has attempted to capture a daddy longlegs is that the legs readily detach when restrained. Furthermore, the detached legs will continue to twitch for as long as an hour. This ability (**appendotomy**) occurs in some other arachnids and apparently is used to distract and confuse predators, allowing the arachnid to escape. Mobility does not seem to be seriously affected by the loss of a leg, or even a couple of legs; however, opilionids will not regenerate the lost appendage.

Among the tremendous number of opilionid species, a range of feeding behaviors occur, and this group of animals includes the most omnivorous of all arachnids. Many are predators of smaller soft-bodied insects. Some scavenge for dead or dying insects and may even feed on the carrion of larger animals. Plant juices, fungi, and tender plant tissues are consumed by others. Their chelicerae help tear apart their food, which is then mixed with digestive fluids. The opening of their mouths is fairly wide and allows them to consume fairly large pieces of solid food.

Opilionids exhibit no special mating behaviors, but males do possess a feature not found among other arachnids, a penis. This allows direct insemination without an externally produced spermatophore. Although their mating is brief, it is often repeated and the male usually then stays with the female, guarding her from later suitors. The female possesses a long, eversible ovipositor that she uses to lay eggs, inserting them into the soil, under tree barks, or in plant stems. In some species, the male prepares and defends the nest site that the female visits and in which she

deposits her eggs. When she finishes, she leaves, and the male is left to guard the eggs. In North America and Europe, most species apparently have a 1-year life cycle, with adult activity occurring in late summer and early fall. This is coincident with harvest time, and thus these are sometimes known as "harvestmen."

Although they lack venom or sharp fangs, opilionids employ a variety of self-defense mechanisms. Some go rigid and feign death, while others will bounce vigorously, blurring their body form. The most important among these defensive mechanisms is the use of chemicals. Special repugnatorial glands occur along the sides of the body at the base of the front two pairs of legs, and these can produce a repellent mixture of phenols, quinones, ketones, and/or alcohols. Often mixed with fluid released from the mouth, these repugnatorial chemicals flow along the grooves on the body, quickly making the opilionid distasteful to potential predators. Some species may use their legs to dab or even flick these chemicals onto a potential predator.

FIGURE 5-53
An aggregation of daddy longlegs. Photograph courtesy of Lynn McCutcheon.

Some opilionids periodically aggregate and rest, usually in dark areas with favorable temperature and humidity conditions. Various hypotheses have been proposed regarding the purpose of these large groupings, which can involve more than one species and hundreds of individuals. One idea is that large groupings reduce the likelihood of any individual being found and eaten. The massed groups may pulsate, disorienting and deterring predators. Furthermore, the effects of combined defensive secretions may intensify their effects.

Opilionids are also recipients of one of the most widespread urban legends involving any arthropod; they are purportedly the world's most poisonous spider, but they can't bite. This is false at many levels;

notably neither are they spiders (different order, Araneae), nor do they even possess poison glands. Despite their complete harmlessness to humans, this story has a long history and can be heard repeatedly in most any area of the world where opilionids occur.

Order Acari—Mites and Ticks

Mites and ticks compose the largest order of arachnids with approximately 50,000 known species. Furthermore, it is widely believed that the order Acari are even less well described than are any of the insect orders. It is estimated that only about 5% of all currently existing mites and ticks have been described, suggesting that this group alone may include close to a million species.

A few general features are used to define the order Acari. Most have an oval body form with little differentiation of body regions. Instead, they have a uniquely developed area on their head, known as the **gnathosome**, which surrounds their mouthparts. The great majority of species are very small, less than

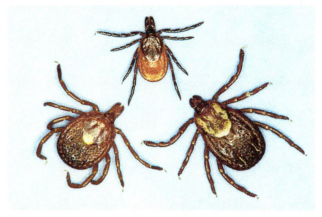

FIGURE 5-54

Three of the more commonly encountered ticks in North America: (top) blacklegged tick, *Ixodes scapularis*; (lower left) lone star tick, *Amblyomma americanum*; and (lower right) American dog tick, *Dermacentor variabilis*. Photograph courtesy of Jim Kalisch/ University of Nebraska.

1 mm, although the ticks and some giant velvet mites may range from 10 to 20 mm. Furthermore, their development is unusual in that the first active stage after egg hatch is a six-legged form known as a larva. This is followed by from one to three eight-legged stages (nymphs) before the ultimate molt to adult.

Top view

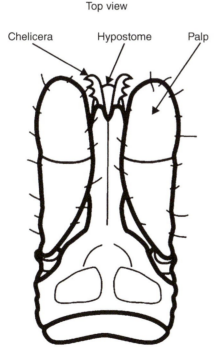

Chelicera   Hypostome   Palp

Base of Capitulum

Bottom view

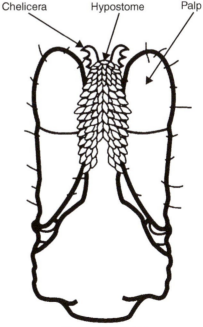

Chelicera   Hypostome   Palp

Base of Capitulum

FIGURE 5-55

Illustration of tick mouthparts. Illustration by Scott Charlesworth and provided courtesy of Purdue University.

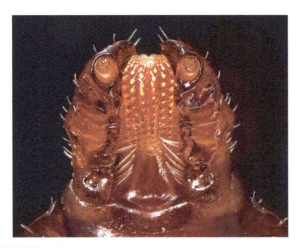

FIGURE 5-56
Head of an American dog tick. Photograph courtesy of Susan Ellis/Bugwood.org.

They are tremendously diverse in their feeding habits and include among them parasites, predators, scavengers, and plant-feeding species. Representatives of the Acari can also be found in almost any terrestrial and freshwater habitat. They are the most abundant arthropods found in soils and are even well represented in oceans, with over 1,000 known marine-dwelling species.

Because of their diversity and abundance, members of the order Acari provide special problems for systematists that try to classify and develop logical evolutionary relationships among different animal groups. For example, in some classification systems, the order Acari is considered a subclass. Within that, there may be multiple superorders, each of which may contain multiple orders. For this book, the Acari will be considered one order that contains three groups (table 5-1).

TABLE 5-1 A systematic arrangement of the mites and ticks to be used in this text. Included under some of the suborders are a few of the families (ending in "idae"), emphasizing those that have the greatest impact on human activities. Recognize that multiple arrangements of this animal group are currently in use. Those families followed by * receive further discussion.

Order Acari
    Group I. Opilioacariformes
    Group II. Parasitiformes
        Suborder Holothyrina
        Suborder Mesostigmata
            Family Dermanyssidae (bird mites, fowl mites)
            Family Macronyssidae (bird mites, rat mites)
            Family Cheyletidae (some mange mites)
            Family Phytoseiidae (predatory mites)
        Suborder Ixodida
            Family Ixodidae (hard ticks)*
            Family Argasidae (soft ticks)*
    Group III Acariformes
        Suborder Prostigmata
            Family Tetranychidae (spider mites)*
            Superfamily Eriophyoidea (three families—eriophyid mites)*
            Family Trombiculidae (chiggers)*
            Family Pyemotidae (straw itch mites)
            Family Demodicidae (follicle mites)*
            Family Tarsonemidae (broad mites, cyclamen mite, tracheal mite)
        Suborder Astigmata
            Family Acaridae (grain mites, cheese mites, mold mites)
            Family Sarcoptidae (scabies, sarcoptic mange mites)*
            Family Pyroglyphidae (dust mites)*
        Suborder Oribatida

FAMILY IXODIDAE—HARD TICKS

Hard ticks are common parasites of vertebrates. The common name of this family is derived from the hard plate behind the tick's head (dorsal shield). Their mouthparts project forward in a structure known as a **gnathosome,** which is largely comprised of a toothed, harpoon-like structure, the **hypostome**. The hypostome (figures 5-55, 5-56) can firmly embed onto the skin of the host, allowing the tick to remain attached and feed. The stylet-like chelicerae then penetrate the skin and the combined mouthparts form a cone through which the blood is sucked.

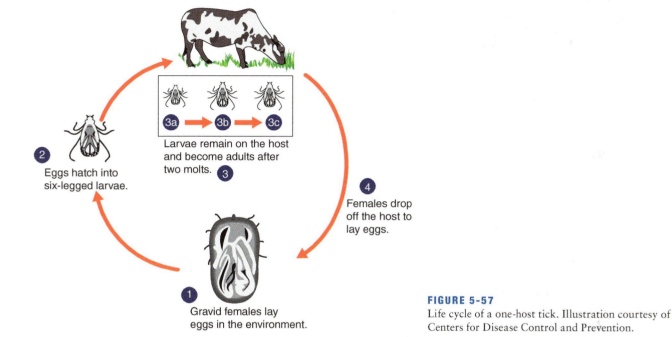

2 Eggs hatch into six-legged larvae.

3a → 3b → 3c Larvae remain on the host and become adults after two molts. **3**

4 Females drop off the host to lay eggs.

1 Gravid females lay eggs in the environment.

FIGURE 5-57
Life cycle of a one-host tick. Illustration courtesy of Centers for Disease Control and Prevention.

The life stages of hard ticks include **eggs** laid as a mass, a minute six-legged **larval stage,** a larger eight-legged **nymphal stage,** and the **adult.** Each of the active stages usually feeds one time, engorging on blood so that they swell to many times their original size. Species of ixodid ticks exhibit different behaviors between meals. Those species known as one-host ticks feed on a single animal after it has been located by the larva. Molting of the larva to the nymph and the nymph to the adult also occurs on the same host animal. After mating, also on the host, the female drops to the ground and lays a mass of eggs. Cattle ticks are an example of a one-host tick.

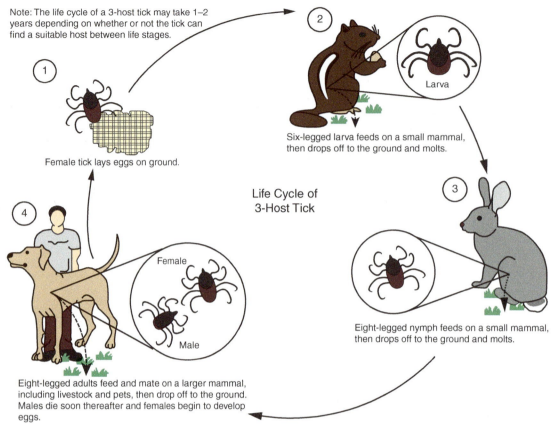

Note: The life cycle of a 3-host tick may take 1–2 years depending on whether or not the tick can find a suitable host between life stages.

1   Female tick lays eggs on ground.

2   Larva   Six-legged larva feeds on a small mammal, then drops off to the ground and molts.

Life Cycle of 3-Host Tick

3   Eight-legged nymph feeds on a small mammal, then drops off to the ground and molts.

4   Female   Male   Eight-legged adults feed and mate on a larger mammal, including livestock and pets, then drop off to the ground. Males die soon thereafter and females begin to develop eggs.

FIGURE 5-58
Life cycle of a three-host tick. Illustration by Scott Charlesworth and provided courtesy of Purdue University.

More commonly encountered are **three-host ticks** that feed on separate animals during each stage. Females lay their eggs on the ground in the form of huge masses containing thousands of eggs. The newly hatched larvae seek small mammals, usually rodents, as the first host. After feeding, the larvae drop to the ground and molt to the nymph stage. Similarly, the nymphs seek other small mammals as hosts. After feeding on this second host, the engorged nymphs again drop to the ground and molt to the adult stage. Adults usually then feed on a large mammalian host such as a dog, deer, or, occasionally, human. After feeding, adults will mate, and the females will produce eggs. Among the most common of the three-host ticks in the United States are the American dog tick (*Dermacentor variabilis*) abundant in eastern United States, the Rocky Mountain wood tick (*Dermacentor andersoni*) found in much of the western United States, and the lone star tick (*Amblyomma americanum*) common in more southern areas.

FIGURE 5-59
A winter tick, *Dermacentor albopictus*, laying eggs. Photograph courtesy of Ken Gray/Oregon State University.

FIGURE 5-60
Larvae of the lone star tick. Photograph courtesy of Tom Murray.

The life cycle of these multihost ticks may take a few months to several years to complete, depending in large part on how successful the tick is with respect to locating a new host. Ticks are remarkably tolerant of starvation and may survive more than a year between meals, usually waiting patiently along pathways used by animals. They are highly sensitive to carbon dioxide, which is utilized as a cue alerting the tick of a nearby potential host animal. Usually poised at the top of vegetation, once they detect a host, these ticks often will wave their legs in a behavior known as **questing**. This activity allows the tick to readily snag and cling to a passing animal.

FIGURE 5-61
A Rocky Mountain wood tick, *Dermacentor andersoni*, in typical questing pose. Ticks wait for a suitable host to brush alongside and then grab the passing host. Photograph courtesy of Ken Gray/Oregon State University.

Once a host has been successfully encountered, ticks often require several hours to settle before they can begin to feed. If left undisturbed on the host, ticks will remain in place for several days before their blood meal is complete and they drop to the ground. Most often tick bites result in minor inflammation at the site of the bite with no serious effects. One can become infected particularly if a tick is improperly removed such that the mouthparts remain embedded in the tissues of the host. The proper way to remove a tick is to carefully pull it out with tweezers, grabbing it as close to the skin as possible and avoiding crushing its body. Disinfecting the area is also recommended.

FIGURE 5-62
A tick nearly fully engorged. Photograph courtesy of Lynn McCutcheon/Kilgore College.

FIGURE 5-63
Demonstration of the proper way to remove a tick. Care should be taken to grasp the head so it can be fully extracted. Photograph courtesy of Jim Kalisch/University of Nebraska.

Ticks are most important to humans as vectors of pathogens. At least nine tick-vectored diseases occur in the United States, with Lyme disease and Colorado tick fever being the most common (table 5-2). Rarely, the Rocky Mountain wood tick has been known to directly cause a disease known as **tick paralysis**. Also known as tick toxicosis, this can produce a gradual paralysis when the tick remains feeding around the head or neck for a long period. Human cases in the United States are only reported every few years, but sheep, cattle, horses, pigs, and dogs are also sometimes affected by tick paralysis. Symptoms are reversed when the tick is removed.

TABLE 5-2 Tick-borne diseases that occur in the United States.

PATHOGEN—DISEASE	PATHOGEN—TYPE OF ORGANISM	SCIENTIFIC NAME	TICK VECTORS
Colorado tick fever	Virus	Coltivirus group	*Dermacentor andersoni*
Rocky Mountain spotted fever	Rickettsia	*Rickettsia rickettsii*	*Dermacentor* spp.
Tularemia	Bacteria	*Francisella tularensis*	Various
Lyme disease	Spirochete	*Borrelia burgdorferi*	*Ixodes scapularis*
Babesiosis	Protozoan	*Babesia microti*	*Ixodes scapularis*
Human ehrlichiosis	Bacteria	*Ehrlichia chaffeensis*	*Amblyomma americanum*
Human granulocytic ehrlichiosis	Bacteria	*Ehrlichia phagocytophila*	*Ixodes scapularis*
Tick-borne relapsing fever	Spirochete	*Borrelia hermsii*	*Ornithodoros hermsi*
Southern tick-associated rash illness	Unknown	Unknown	*Amblyomma americanum*

● Cattle Tick Fever and Arthropods as Vectors of Pathogens

In the mid-1800s cattle drives out of Texas moving on their way to northern markets left a trail of sick and dying animals in their wake suffering from a new disease called cattle fever. Texas cattle, long exposed to cattle fever, were immune to its effects, but northern cattle, grazing in areas through which the southern herds from Texas had passed, would often become sick or die. The cause of this disease was completely unknown at the time, and only later was it determined to be the result of tick-transmitted microorganisms, the protozoans *Babesia bigemina* and *Babesia bovis*.

The idea that microorganisms could cause disease in anything—the "Germ Theory"—had only recently been well demonstrated with the studies by Louis Pasteur, Robert Koch, and others, but it was still not widely accepted.

The first real progress in understanding the epidemiology of cattle fever (or bovine babesiosis) was made in 1889 by Theobald Smith. Smith conclusively proved the *Babesia* spirochetes produced cattle fever, reinforcing the newly emerging awareness of pathogens. However, Smith had only half the story as

(continued)

FIGURE 5-64
A cattle dip for killing cattle ticks. A dilute acaricide is present in the water, which thoroughly drenches the body during immersion. Photograph courtesy of Scott Bauer/USDA ARS.

he thought that these diseases were transmitted from animal to animal through contaminated feces, urine, or saliva. It wasn't until 4 years later that two researchers, F. L. Kilbourne and Cooper Curtice, demonstrated that the cattle ticks *Boophilus annulatus* and *Boophilus microplus* were responsible for transmitting the pathogens from animal to animal. The efforts of Kilbourne and Curtice resulted in the first demonstration that arthropods were involved in the transmission of animal diseases; that is, they were **vectors**. This breakthrough idea stimulated the research that in the next 20 years led to the understanding of some of the most important human diseases that are transmitted by arthropods: from malaria and yellow fever (by mosquitoes) to typhus (by lice) and plague (by fleas).

With this information in hand, the National Cattle Fever Tick Eradication Project was started by the US Department of Agriculture in 1906. This program aimed to eradicate the cattle ticks from a 14-state area of the southern United States and was considered complete in 1943. Currently, a quarantine zone running along a 500-mile swath of the Texas/Mexico border is maintained where all cattle must be thoroughly checked to ensure they are tick-free. Fifty mounted "tick riders" patrol the area for strays, livestock smuggled from infested areas, and evidence of cattle tick infestations.

● Lyme Disease

Among the most important human diseases to emerge within the United States in recent years is Lyme disease, an infection caused by the bacterial spirochete *Borrelia burgdorferi* transmitted by ticks in the genus *Ixodes*, notably the blacklegged tick *Ixodes scapularis*. It first came to attention in 1976 when an unusual cluster of juvenile arthritis cases were investigated in the area around Lyme, Connecticut. It has since exploded, with over 20,000 cases a year occurring over broad areas of the United States and southern Canada. Lyme disease is most commonly reported now from New York, Pennsylvania, New Jersey, Connecticut, Massachusetts, Wisconsin, and Minnesota; however, a few cases have occurred as far west as California.

In humans, an early symptom is often the development of a ringlike rash (erythema migrans) at the bite. Body aches, headache, fatigue, and other

flu-like symptoms are typical of early Lyme disease. What makes the disease so important is that, if left untreated, Lyme disease can develop much more serious effects. This can include numbness and partial paralysis, severe headaches, meningitis, encephalitis, severe fatigue, and other symptoms. Occasionally, the disease progresses to produce arthritis, particularly of the large joints, and symptoms that mimic multiple sclerosis or Alzheimer's disease.

Lyme disease primarily occurs among rodents, particularly white-footed mice and chipmunks in the eastern United States. The larval and nymphal stages of the blacklegged tick feed on these rodents and *Borrelia burgdorferi* survives through winter in the midgut of nymphal ticks. The ticks then transmit it again to rodents in spring, renewing the cycle of infection. Infected nymphs are also the primary stage that may incidentally bite people and

(continued)

FIGURE 5-65
A female blacklegged tick, *Ixodes scapularis*. The blacklegged tick is the vector of the organism that causes Lyme disease. Photograph courtesy of Scott Bauer/USDA ARS.

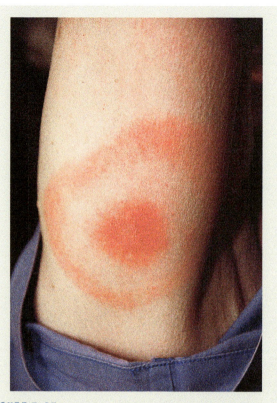

FIGURE 5-67
A ring rash symptom developing around the site of a tick bite can be a characteristic symptom of Lyme disease. Photograph courtesy of Centers for Disease Control and Prevention.

transmit the pathogen. After the nymphs develop to the adult stage, the ticks then seek out a deer or other large animal host, occasionally biting humans at this time as well. Deer, their primary host in the adult stage, are not susceptible to Lyme disease but are critical for the survival and reproduction of the blacklegged tick.

The rapid rise of Lyme disease has transformed the way that many people now look at wild areas, and the prevention of tick biting has become a major concern for millions that live in areas where it has become common. What also contributes to the mystery of this "emerging disease" is that it likely has been present for a very long time. Cases dating to the 1960s from Cape Cod have been recently confirmed as previously unrecognized. Furthermore, ticks in museums that were collected over a century ago contain *Borrelia burgdorferi.* Apparently a series of changes in the ecology, including increasing deer populations, increasing rodent populations, and changes in the landscape that favor tick survival and reproduction have combined in recent decades to produce ideal conditions for the spread of Lyme disease over wide areas of the country.

FIGURE 5-66
A blacklegged tick deutonymph. Photograph courtesy of Scott Bauer/USDA ARS.

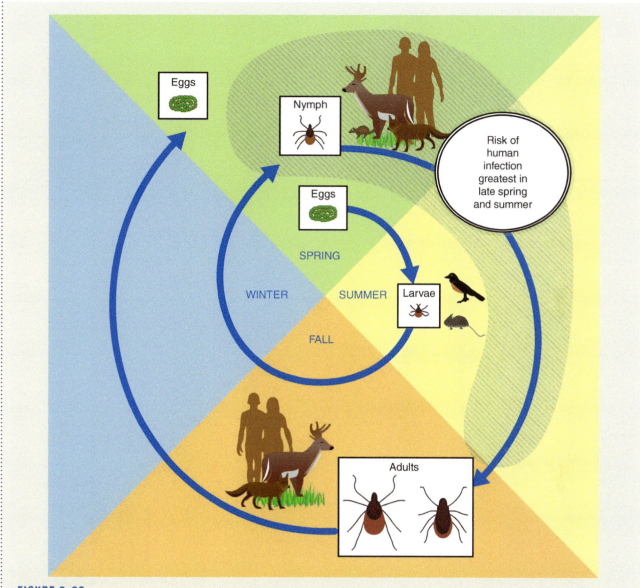

FIGURE 5-68
Typical annual cycle of Lyme disease. The pathogen is primarily maintained over winter in rodent hosts. Blacklegged tick nymphs moving from these hosts produce the infections that occur in humans and other large animal hosts. Illustration courtesy of Centers for Disease Control and Prevention.

FAMILY ARGASIDAE—SOFT TICKS

Several features differentiate soft ticks from the more commonly encountered hard ticks (Ixodidae). Perhaps most conspicuous is that the mouthparts are not directed forward in soft ticks as they are in the other group. Instead they are hidden below the body and are smaller in size. No dorsal shield behind the head is present and the body is less sclerotized and more flexible ("softer"). Viewed from above, the soft ticks display various patterning; some are wrinkled or grooved while others are covered with small bumps or have a granulated surface appearance.

The soft ticks are all blood-feeding parasites of mammals or birds; however, their associations with their host animals differ from hard ticks. Most notably, upon first finding a suitable host, soft ticks will remain with the host, living within its nest or burrow. Later stages (nymphs, adult) may feed often, but for short periods of time—less than an hour. In addition, female adults may repeatedly lay eggs rather than produce one large egg mass following a single final engorgement of host blood. Soft ticks are also highly resistant to starvation since their host may periodically leave the nest when it migrates (host birds flying south for the winter months). Under these conditions soft ticks will go into a state of deep inactivity (torpor) and may survive many years while waiting for a host animal to return.

Soft ticks rarely bite humans and are far less important as vectors of disease than hard ticks are.

FIGURE 5-69

Dorsal and ventral view of a spinose ear tick, *Otobius megnini*, an example of a soft tick (Argasidae family). Photograph courtesy of Jim Kalisch/University of Nebraska.

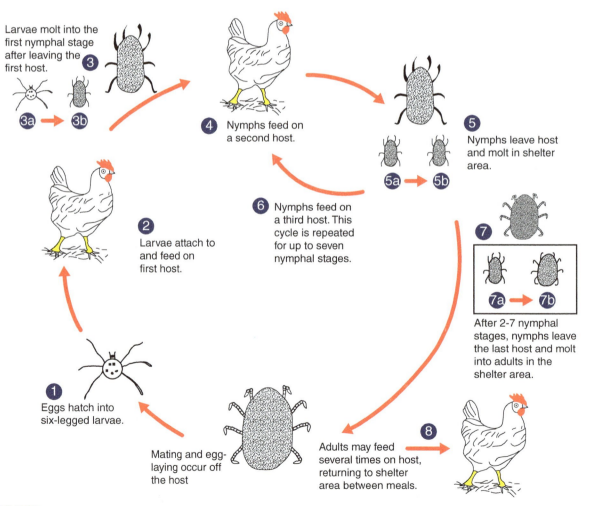

FIGURE 5-70

Life cycle of a typical soft tick (Argasidae family). Soft ticks are associated with the nests of their hosts and intermittently feed throughout their development. Illustration courtesy of Centers for Disease Control and Prevention.

Nonetheless, tick-borne relapsing fever, caused by infections with various *Borrelia* species of the spirochete class, is transmitted by soft ticks and occurs in many areas of the world. A few cases may occur every year in some mountainous areas of the western United States where it exists endemically at low levels in populations of rodents such as chipmunks and pine squirrels. Transmission to humans usually occurs when camping in rustic cabins also inhabited by nesting disease-carrying rodents and soft tick vectors of the genus *Ornithodoros*.

FAMILY PHYTOSEIIDAE—PHYTOSEIID MITES

The predatory mites of the family Phytoseiidae also feed on blood, preying on other mites and small insects such as thrips and aphids. (Occasionally they will also feed on pollen as well.) They are much smaller than ticks, only about 0.5–0.8 mm in length. The phytoseiid mites can be found inhabiting the soil, leaf litter, bark, the surfaces of leaves, or wherever they may find their prey.

FIGURE 5-71
A phytoseiid mite (lower) feeding on a spider mite. Photograph by Jack Kelly Clark and provided courtesy of the University of California IPM Program.

Because of their prodigious appetites, short generation time, and tremendous fecundity, phytoseiid mites have been successfully used as biological control organisms. Several species are also commercially available for purchase and release to control damaging populations of plant-feeding pest mites, thrips, and other small insects.

FAMILY TETRANYCHIDAE—SPIDER MITES

Spider mites are among the most important of all plant pests, and at least some species can occur on almost any crop. They damage plants as they feed, their chelicerae whipping the surface cells of the plant and releasing the cell fluids that they then suck up. Small whitish leaf flecks (stippling) are a typical symptom produced wherever they feed. When abundant, they can cause the leaves to turn yellow and look scorched and often prematurely drop from the plant. Some species of spider mites also produce webbing that may cover leaves.

There are several reasons why spider mites can be so damaging as plant pests. One is their short life cycle (egg, larvae, protonymph, deutonymph, adult) that can be completed in about a week when conditions are favorable. This allows their populations to increase very rapidly, even exceeding reproductive efforts of aphids.

FIGURE 5-72
Twospotted spider mites, *Tetranychus urticae*. This spider mite has an extremely wide host range and is damaging to numerous crops and garden and greenhouse plants. Photograph courtesy of Ken Gray/Oregon State University.

Spider mites are also notoriously difficult to kill with pesticides. Strains of spider mites often rapidly develop resistance to pesticides that originally were effective. Furthermore, the use of ineffective pesticides often is more detrimental to their natural enemies, thus allowing spider mites to thrive unchecked by the predators that would normally help control their numbers. Indeed, on many crops, spider mites are

FIGURE 5-73

Webbing produced by twospotted spider mite. Visible webbing is usually only observed during heavy infestations. Photograph by Whitney Cranshaw/Colorado State University.

FIGURE 5-74

Clover mites, *Bryobia praetiosa*, are cool season spider mites that are most active in spring and fall. They frequently enter homes accidentally during late winter and early spring. Photograph by Whitney Cranshaw/Colorado State University.

considered "secondary pests," species that only become damaging when there has been ineffective or improper insecticide use.

Arguably the single most important spider mite in North America is the twospotted spider mite (*Tetranychus urticae*). This mite damages a wide variety of crops, including fruit trees, roses, cotton, corn, beans, and berry crops, and many kinds of houseplants. Other important species affecting fruit crops and ornamental plants include the Pacific spider mite (*Tetranychus pacificus*), European red mite (*Panonychus ulmi*), spruce spider mite (*Oligonychus ununguis*), Southern red mite

(*Oligonychus ilicis*), and honeylocust spider mite (*Platytetranychus multidigituli*). A cool season species that is most active on lawns in late winter and spring is the clover mite (*Bryobia praetiosa*). During this time they often accidentally enter buildings through cracks on the structure's south and west sides, where they may be described as "little walking dust specks."

SUPERFAMILY ERIOPHYOIDEA—GALL AND RUST MITES/ERIOPHYID MITES

Several families of extremely minute plant-feeding mites are known as gall or rust mites. The largest of these families, Eriophyidae, also lends these animals the commonly used name "eriophyid mites." Over 1,250 species have been described, but because of their very small size (0.1–0.3 mm), they are commonly overlooked and a very high percentage of existing species undoubtedly remain undescribed.

FIGURE 5-75

An eriophyid mite, *Aceria malherbae*, showing their typical elongate body form and two pairs of legs. This species has been purposefully introduced into North America to help suppress the invasive weed known as field bindweed. Photograph courtesy of USDA ARS/Bugwood.org.

Eriophyid mites are unique among arthropods in that they only possess two pairs of legs. Usually their body is elongated but some that live on leaf surfaces may be more flattened. Their life cycles are usually simple including eggs, two immature stages, and the adult, with males and females appearing fairly similar. On species that occur on trees and shrubs in areas of cold winters, the stages that survive winter (**deutogynes**) may differ somewhat in appearance than those found during most of the growing season (**protogynes**).

Eriophyid mite injuries produce very diverse injuries to plants. Many feed on the leaf surface as "leaf vagrants." Often there is little evidence of their activity, although some, known as "rust mites," may produce a slight brownish discoloration of leaf surfaces.

Other eriophyid mites can induce leaves to produce odd growths within which the eriophyid mites develop. These are known as **galls** and

FIGURE 5-76
A fingergall symptom produced by eriophyid mites on a leaf of American plum. Photograph by Whitney Cranshaw/Colorado State University.

FIGURE 5-77
Erineum symptoms on mountain maple produced by eriophyid mites. This type of injury is produced when the mites induce the plant hairs to grow in dense profusion. The mites live within the tangle of plant hairs. Photograph by Whitney Cranshaw/Colorado State University.

many insects (e.g., gall wasps, gall midges) and some plant pathogens can also cause gall formation. Galls produced by eriophyid mites take on various forms, some of which are unique to this group of arthropods. Among the kinds of galls eriophyid mites produce are "pouch" or "finger" projections on leaves, dense patches of leaf hairs that resemble a small bit of felt (erinea), and enlarged and disorganized buds or flower parts.

FAMILY DEMODECIDAE—FOLLICLE MITES

Two species of follicle mites are very common associates of the human face. *Demodex folliculorum* is directly associated with hair follicles, while *Demodex brevis* is associated with the sebaceous glands. Populations are most dense on the upper cheeks, forehead, nose, eyelashes, and eyebrows. They are minute (0.3–0.4 mm), highly elongated mites with a semitransparent body. They are often found at the base of a recently plucked eyelash hair examined under a microscope. They very slowly move about on

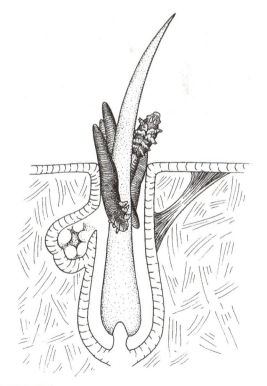

FIGURE 5-78
A diagram of follicle mites, which live at the base of hair follicles on the face of humans. Drawing by Matt Leatherman.

the skin surface and are transferred between people in close physical contact.

Follicle mites lay eggs that hatch in about 3–4 days and the mites may become full grown within a week after eggs hatch. With few exceptions, no symptoms are produced by these ubiquitous parasites that cause little, if any, damage to living cells. They so efficiently feed that they do not produce waste materials and even lack an excretory opening. The percentage of individuals that support follicle mites increases with age and is estimated to be over 95% in older adult populations in North America and Europe. Incidence is usually somewhat greater among males than females, perhaps because of differences in hair follicles and sebum secretion. Populations also can increase on individuals with suppressed immune systems. Rashes may develop when infestations are extremely large.

FAMILY SARCOPTIDAE—SCABIES/HUMAN ITCH MITE

The scabies mite (*Sarcoptes scabiei*) develops in tiny burrows made under the skin of its human host, producing a pimple-like itchy rash. The mite occurs worldwide, and although not very common in the United States, it is regularly encountered by dermatologists. It is moved person-to-person by close and prolonged physical contact and only affects humans.

Scabies mites burrow into the outer layer of the skin. The mite first secretes enzymes that break down the outer skin cells. This process then allows the mite to begin digging into the skin using special spined hooks on the front legs in combination with the chelicerae that are used in a chewing manner. These burrowing activities produce the generalized itchiness that is associated with scabies (human itch mite).

Most digging is produced by the mature, fertilized female. She produces a permanent burrow that is continuously expanded, about 0.5–5.0 mm/day. During the next 4–6 weeks she remains within the burrow and deposits two to three eggs per day.

Minute six-legged larvae hatch from the eggs in 3–4 days. They may remain within the tunnel or move to the skin surface where they find shelter around hair follicles. Wherever they settle, they will then produce a small burrow (known as the molting pocket) just large enough to conceal them. Upon molting, they become known as a first stage nymph, with a full complement of eight legs. This stage is followed by another molt to the second stage nymph. Both of these stages similarly produce small burrows within the skin. After the ultimate molt, adults are produced. The entire period from egg deposition to the emergence of adults requires about 2 weeks. The adult female will then begin constructing her permanent burrow where she soon will be joined by a male and mates.

Symptoms of scabies infestation are slow to develop. Itching symptoms usually occur within 3 months after the initial infestations; however, a longer period of time before symptoms appear is not unusual. Furthermore, the initial infestation of a healthy adult human may involve only a few individual mites, typically less than two dozen. Immune system responses are important in limiting survival of scabies.

There are several distinctive features in the pattern of scabies. Most suggestive is the location of the itching (hand, wrist, elbows, etc.) and that the itching tends to occur at night. The adult female also produces a minute (ca. 5 mm) linear burrow that may be visible and somewhat resembles a faint pencil line. A fine scale may also be present where the mite initiated its burrow. These features are difficult to see and positive diagnosis of scabies requires trained medical personnel using skin scrapings that expose the mites. Effective treatments exist for scabies that can eliminate the mites.

Related species of mites cause **mange** in other animals. Very rarely these mites can be transferred to humans if they are in very close contact with the animal. A transitory irritation may be produced by these mites, arising shortly after contact with the animal, and often will produce itching on the face. Mange-producing mites cannot survive on humans and die quickly in the absence of their animal host.

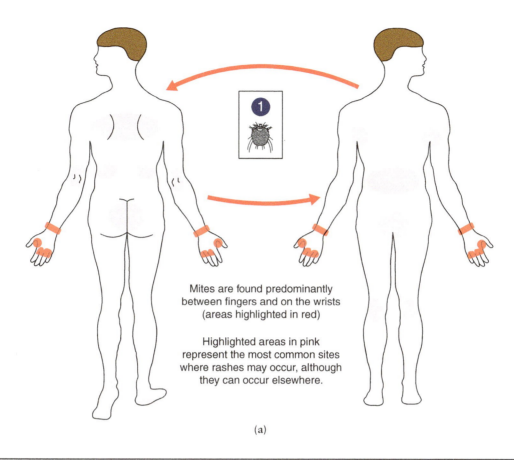

Mites are found predominantly between fingers and on the wrists (areas highlighted in red)

Highlighted areas in pink represent the most common sites where rashes may occur, although they can occur elsewhere.

(a)

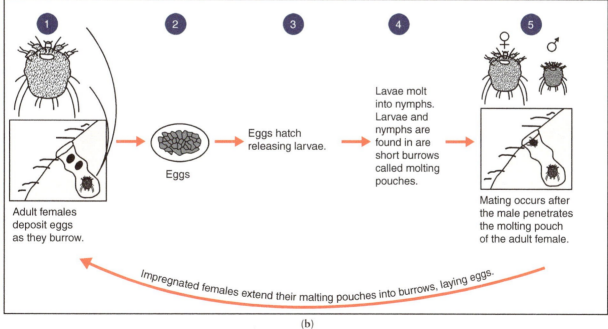

Adult females deposit eggs as they burrow.

Eggs

Eggs hatch releasing larvae.

Lavae molt into nymphs. Larvae and nymphs are found in are short burrows called molting pouches.

Mating occurs after the male penetrates the molting pouch of the adult female.

Impregnated females extend their malting pouches into burrows, laying eggs.

(b)

FIGURE 5-79

(a) Areas of the body most frequently infested by scabies mites. (b) Life cycle of the scabies mite on a human. Illustrations courtesy of Centers for Disease Control and Prevention.

FAMILY TROMBICULIDAE—CHIGGERS

There are many kinds of mites known as chiggers, most of which are found in the family Trombiculidae. All have a life cycle that involves a parasitic first stage larvae associated with larger animals. This parasitic stage is followed by later stages that are predators of other arthropods. Various species can be found worldwide, with the most common chiggers affecting humans in North America being the species *Eutrombicula alfreddugesi*, *E. splendens*, and *E. batatas*. All are found in moist, well-vegetated areas, although habitat preferences vary. For example, some chiggers are most abundant in areas of dense grassy or weedy growth. Others like moist but sunny locations with relatively little leaf shade.

Eggs are laid as clumps in the soil and will subsequently hatch in a week or two. When the minute six-legged larvae emerge, they swarm over the soil or climb low-growing plants and cluster along leaf edges where they wait. Although they can withstand starvation for about a month in this stage, they readily transfer if brushed by a passing animal host: a reptile, small bird, or mammal (including human). Once on the host, the larval chigger seeks out a soft area of skin and settles to feed. On humans they typically attach at a skin pore or hair follicle and concentrate where clothing constricts the skin—around the tops of socks, belt line, or bra straps.

Chiggers do not feed on blood and technically do not bite. Instead they produce digestive fluids that liquefy the cells of the inner skin. To reach this layer, they form a small feeding tube through the outer skin, known as a stylostome, produced from the hardening

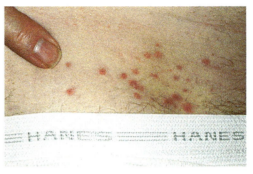

FIGURE 5-81
Chigger "bites" clumped near waistline. Chiggers tend to feed around points where clothing is close to the skin, such as around the tops of socks and waistline. "Bites" are a reaction to the digestive saliva that the chiggers secrete on the skin surface. Photograph courtesy of Jim Kalisch/University of Nebraska.

of fluids resulting from feeding. After the chiggers cease feeding, oozing fluids (digested skin, lymph) will continue to emerge from this tube, forming a tiny cap that is diagnostic of chigger feeding.

Effects from chigger feeding results from the reaction to the digestive enzymes produced during feeding. Symptoms often do not begin to show for about 6–12 h and initially appear as a tiny reddish spot. Within the next 20–30 h these increase in size, and itchiness increases rapidly and subsequently can persist for almost 2 weeks before subsiding.

FIGURE 5-82
A velvet mite. Velvet mites are adults of mites belonging to the family Trombiculidae, which includes the chiggers. Velvet mite adults are predators of other arthropods and do not feed on humans. Photograph courtesy of Jim Kalisch/University of Nebraska.

Despite the discomfort they produce, humans are poor hosts for chigger larvae and most fail to further develop on a human and die. On a more

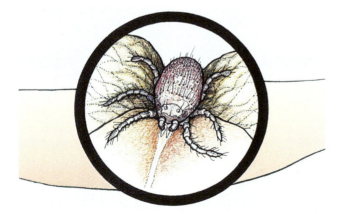

FIGURE 5-80
Illustration of a chigger feeding at a hair follicle. Illustration courtesy of Jim Kalisch/University of Nebraska.

favorable animal host they will feed for several days, ultimately dropping off when fully engorged. They then molt to an eight-legged nymph which feeds as a predator of small arthropods. Upon two more molts the ultimate adult form occurs, a large bright red mite covered with hairs known as a velvet mite. A common name for chiggers, "red bugs" refers to this feature in its life cycle. In Great Britain chiggers are called "harvest mites."

Some chiggers are also involved in transmission of the human pathogen that produces the disease known as scrub typhus. This is produced by a rickettsial bacterium (*Orientia tsutsugamushi*) that is common among rodents in Japan, the Philippines, SE Asia, northern Australia, and many Pacific Islands. It is transmitted by various *Leptotrombidium* species of chiggers and produces a debilitating disease that can be fatal. During World War II over 20,000 American troops fighting in the Pacific were infected with scrub typhus.

HOUSE DUST MITES—FAMILY PYROGLYPHIDAE

Various house dust mites are the single most important source of allergens in house dust. Most frequently encountered is *Dermatophagoides pteronyssinus*, but *Dermatophagoides farinae* and *Euroglyphus maynei* are also common. All are minute, about 0.4 mm in length, pale colored, and globular in body shape. With good lighting and a dark background surface, they are just barely visible.

House dust mites are scavengers that will feed on many foods, particularly materials of animal origin. Skin flakes, which typically comprise about 80% of household dust, are the most common food but other high protein materials such as dried fish food are also eaten. Highest populations of house dust mites are almost always found in and around beds, because skin flakes typically concentrate at these sites. In addition, sleeping humans provide favorable warmth and increase the humidity with their respiration and perspiration.

The primary allergens come from the feces that the mites continuously produce. These are often mixed with highly allergenic enzymes from the mite's digestive tract. Furthermore, the fecal pellets are usually colonized by various fungi (primarily *Aspergillus penicilloides*) that are also allergens. The action of the enzymes and the fungi helps to remove fats and further digest the feces, which then may be eaten again by the mites until all the nutrients are extracted. An individual may produce about 20 of the tiny (0.02–0.5 mm) fecal pellets/day. The shed skins and particle of dead mites also are potential allergens.

Although house dust mites potentially may occur almost anywhere, their populations are limited by low humidity. Relative humidity (at 25°C) above 70%–75% appears to be critical and house dust mites are most common in lower-elevation coastal areas. Populations also may vary seasonally, declining during drier seasons.

Order Pseudoscorpiones— Pseudoscorpions

As the name of the order implies, these animals superficially resemble scorpions. Currently there are more than 3,260 described species of pseudoscorpions worldwide, and approximately 200 species are known in North America. Pseudoscorpions are tiny arachnids (typically less than 5 mm), and they possess conspicuous pedipalps modified into pincers in a manner similar to scorpions. Unlike scorpions, pseudoscorpions have a broad flattened abdomen and lack a stinger.

Pseudoscorpions usually occur under rocks, among fallen leaves, under bark, or on similar moist sites where they hunt mites, springtails, and small insects. Typically they wait in ambush within small crevices and grab passing prey with their pincers. Connected to the movable "finger" of the pincer of some species is a venom gland. The venom rapidly incapacitates their prey, which are then brought to their chelicerae where they are crushed and covered with digestive fluids. The pseudoscorpion drinks the liquids that are extracted from their prey through this form of exodigestion (similar to most other orders of arachnids).

FIGURE 5-83
A pseudoscorpion. Photograph courtesy of Brian Valentine.

Prior to mating, males may fight or challenge each other over territories. The male with the largest pincers is almost always the victor in such territorial battles. Subsequent mating behaviors vary among pseudoscorpions. Males of some species leave packets of sperm (spermatophores) on a silken stalk that females then locate through associated attractant chemicals. Other species have complex mating rituals, at the end of which the male guides the female to his spermatophore, which is then taken into the genital opening by the female. Males of some pseudoscorpion species actively push the spermatophore into the genital opening.

Pseudoscorpions can produce silk that is associated with the glands on the chelicerae. One use of silk is to produce an egg sac that the female uses to hold the eggs next to her abdomen. A dozen or more eggs may be deposited within the sac; hatching will occur sometime later, with juveniles remaining within the sac. The young pseudoscorpions remain within the sac until they again molt and will ultimately leave to hunt on their own. Pseudoscorpions will usually then molt two more times during the course of a year before they are fully grown. They are fairly long lived, about 2–3 years being typical for their life cycle, and will remain dormant within a silken cocoon during winter.

Although their ability to move about is limited by their small size, pseudoscorpions can disperse long distances by catching a ride on a large insect. Hitchhiking pseudoscorpions simply grab onto a leg or other appendage of a passing insect and go along for the ride. Such transport of one species by

another is known as **phoresy** and occurs with some other arthropods, such as mites, bot flies, and blister beetles.

FIGURE 5-84
Pseudoscorpions attached to the hind leg of a longhorned beetle. Long distance dispersal of pseudoscorpions by attachment to more mobile species, known as phoresy, is a behavior common among pseudoscorpions. Photograph courtesy of Sean McCann.

Pseudoscorpions can be quite common but are rarely seen because of their small size and secretive habits. They sometimes even occur in homes, usually near sinks, bathtubs, and other moisture sources. The majority of pseudoscorpions occur in tropical areas and, due to their predatory nature, are considered to be beneficial.

Order Solifugae— Windscorpions (Sunspiders, Solpugids, Camel Spiders)

Grossly enlarged chelicerae often most immediately characterize the appearance of members of the order Solifugae, known variously as windscorpions, sunspiders, and many other names. The chelicerae may take up over half the length of the cephalothorax and are articulated into two sections so that they can act as vertically shearing pincers. The windscorpions are a fairly large order of arachnids with about 1,090 species mostly found in more arid areas. About 75 species occur in the United States, concentrated in the western states.

FIGURE 5-85
Eremochilus bilobatus, an example of a windscorpion or sunspider.
Photograph by Whitney Cranshaw/Colorado State University.

The pedipalps are leglike in appearance but are used for sensing and to help manipulate prey. Many species have an adhesive and eversible terminal organ in the pedipalp that can assist prey capture and helps them climb. They move on all four pairs of legs and can be very fast, some running about 16 km/h for short periods. (The common name windscorpion reflects their speed, as they can "run like the wind.") The body ranges from a couple of millimeters to 10 cm, but the long legs further accentuate size and fearsome appearance.

The order name means "those that flee from the sun," and windscorpions seek shade – despite another common name for these arachnids, "sunspiders." Windscorpions are also competent burrowers, allowing them to conserve moisture and avoid excessive heat during the day. Windscorpions are primarily nocturnal predators that most often appear to locate their prey by random foraging. Prey are detected by touch with the pedipalps and drawn into the chelicerae. The latter rapidly crush the prey to a pulp, which may be further ground by teeth along their surfaces. Insects are their primary prey, but they are also known to kill small lizards. Like most other arachnids, they use exodigestion to break down their prey.

During mating, the male massages the female with his pedipalps, causing the receptive female to go into a state of torpor where she remains immobile. The male of most species then produces a spermatophore, picks it up with his chelicerae, and forces it into the female's genital opening. Upon recovery, the female produces a mass of eggs and may, or may not, stay with them until they hatch. The newly hatched windscorpions, called larvae or post-embryos, cannot move but after a second molt they more resemble the adult form. Most windscorpions have a 1-year life cycle, with the adults dying shortly after mating and egg production.

FIGURE 5-86
Side view of the head of a windscorpion, showing the chelicerae.
Photograph by Whitney Cranshaw/Colorado State University.

FIGURE 5-87
The notorious "camel spider" photograph. This shows two enormous windscorpions attached together. This image is created by a trick of camera angle and the actual animals have a body length of about 9 cm at most.

With the combination of a voracious appetite, two powerful and massive segmented chelicerae, and fast speed, windscorpions combine features that stimulate myriad rumors and legends. Belief in their toxicity is widespread, including the idea that when one drops into a water trough it will poison animals that come to drink. In the Middle East, where some of the largest windspiders occur, they are sometimes known as "camel spiders" because they purportedly can leap and attack camels. Another myth is that they may attack sleeping humans, using an anesthetizing bite so that upon awaking the victim has a massive flesh wound. None of these stories are true, and none of the windscorpions found in the Middle East or North America possess venom glands. Larger species will attempt to bite if handled, and larger individuals can break the skin. Although a bite from a windscorpion would cause some immediate pain, the wound would be small and heal normally unless secondary bacterial infections occur.

6

The "Other" Arthropods

Myriapods and Crustaceans

Among the various kinds of terrestrial arthropods, the insects and arachnids are overwhelmingly dominant. But they are not alone in the ability to have conquered the problems of colonizing land. Several other kinds of arthropods may be commonly encountered in yards, fields, forests, and inland waters. A few crustaceans can be found here, such as the common pillbugs ("roly-polys") and some terrestrial sowbugs. Also present are the myriapods, arthropods marked by having an extreme number of legs, such as the centipedes and millipedes.

Subphylum Crustacea— The Crustaceans

Crustaceans are an exceptionally diverse set of animals with over 44,000 described species and include such familiar animals as crabs, shrimp, lobsters, barnacles, and pillbugs. Despite a wide variety of body forms and habits, all crustaceans share some basic characteristics: two pairs of antennae, some appendages that are branched (biramous), and oxygen acquisition through gills. Most will show substantial physical changes as they grow and molt, undergoing a process known as anamorphic development. Crustaceans are the dominant arthropods found in marine environments, but a few are common in freshwater and a much smaller number have adapted to life on land. (Only the crustaceans found on land or inland waters are emphasized in this book.)

FIGURE 6-1
Side view of an Atlantic lobster illustrating features of a typical crustacean. Photograph by Whitney Cranshaw/ Colorado State University.

Within the Crustacea (a subphylum in most taxonomic arrangements), the class Malacostraca contains those that are most widely known and recognized as crustaceans: lobsters, shrimp, crayfish, crabs, and pillbugs. These animals are characterized by having three body regions: a head (cephalon), an eight-segmented thorax, and a six- to seven-segmented abdomen. In most species the head and thorax are fused to form a single body region, the **cephalothorax**, which may be covered with a shield-like **carapace**. The head supports two pairs of antennae and a set of compound eyes. Each segment of the thorax and most segments of the abdomen typically possess a pair of appendages that serve various functions (e.g., walking, swimming, mating). Most members of the class Malacostraca can be found in marine environments, but there are representatives commonly found within inland waters and on land in North America.

Crustaceans within the class Branchiopoda primarily inhabit fresh or brackish water and are typically much smaller than members of class Malacostraca. In most groups the thorax and abdomen are fused with appendages that are often leaflike in appearance. There are approximately 900 species within the class, separated into four orders.

Two of these orders are present in inland waters of North America—the tadpole shrimp and the fairy or brine shrimp.

ORDER ISOPODA—ISOPODS

Isopods are one of the largest orders of crustaceans with approximately 5,000 species. The majority are marine animals found in shallow coastal waters; however, there are many freshwater species. A handful of species have adapted to life on land, and these are familiar animals to most—the sowbugs and pillbugs (Armadillidiidae and Porcellionidae families, respectively).

The body of an isopod is covered with a series of prominent plates. They have seven pairs of legs, each of approximately equal size and form. (The term "isopod" means "equal foot.") Distinct and flexible plates are associated with each leg-bearing segment but the segments of the abdomen are more or less fused. A pair of appendages used for absorption of water, the **uropods**, project from the end of the body. Two pairs of antennae appear on the head, but one is much longer than the other, with the shorter pair projecting from below the body. The vision of terrestrial isopods is minimal as they possess only a few simple light-detecting ocelli.

(a)

(b)

FIGURE 6-2

Dorsal (a) and ventral (b) views of a sowbug. Photographs courtesy of Tom Murray.

FIGURE 6-3

Front view of a sowbug. Photograph courtesy of Brian Valentine.

Terrestrial isopods go by various common names. In Great Britain, they are known as "wood lice" and are ubiquitous crustaceans in forested areas. In the United States, only a small number of terrestrial species are present, and most are nonnatives incidentally introduced from Europe. The pillbug, *Armadillidium vulgare*, attracts most attention as it is able to roll into a ball for defense, a favorite trick that rarely fails to delight children. Often these are known as "roly-polys" or, less commonly, "potato bugs." The sowbugs (*Porcellio scaber* and *Oniscus asellus* are among the most commonly encountered species) have a generally similar appearance but cannot curl completely.

FIGURE 6-4

Armadillidium vulgare, commonly known as the pillbug or "roly-poly," can curl into a tight ball when disturbed. Photograph by Whitney Cranshaw/Colorado State University.

Pillbugs and sowbugs are **scavengers**, feeding primarily on decaying plant matter and less commonly on animal matter. Sometimes they will chew on tender garden seedlings, but they rarely cause any significant injury and are considered to be minor pests, at worst, of gardens. In overall habit, these terrestrial isopods can be very important as macrodecomposers, accelerating the process of decomposition as they shred and partially digest dead plant material. Pillbugs and sowbugs are sensitive to drying and usually feed at night, spending the day under cover, but rainy and overcast conditions will often draw them out. Following very wet periods they may migrate from lawns and gardens and enter nearby buildings.

Being crustaceans, pillbugs and sowbugs acquire oxygen through **gills** that are present in the form of thin membranous areas on the underside of their bodies. In order to function in these terrestrial species, the gills require continuous wetting. This requirement consequently restricts where and when these animals may be found. The taillike uropods also help them acquire moisture by wicking up water from the substrate, allowing them to drink from both ends of their body.

In areas of cool winter weather, pillbugs and sowbugs in the adult stage spend the winter in protected sites such as under logs or among leaf litter. In spring, they become active, and mating occurs. The male often guards the female to prevent other males from mating with her. In some species, females can reproduce parthenogenetically, producing all-female clones in the absence of mating.

Females will produce a number of eggs that are then carried in a special pouch on the abdomen, known as a **marsupium**. About two dozen eggs are deposited at a time. After the eggs hatch, the young remain within the marsupium for an extended period before ultimately leaving to forage on their own. Females carry the eggs and young for about 6–7 weeks. The general appearance of the isopod, plus the continued care of the young, is reminiscent of a pig with her litter, hence the name "sowbug." Often two broods of eggs are produced annually.

The young pillbugs and sowbugs molt for the first time within 24 h after leaving the marsupium. Their molting process is unusual in that it occurs in two stages. The front half is shed first,

followed 2–3 days later by the rear half. Caught in between this process, they may have a two-toned appearance. This unique molting behavior apparently helps protect them from desiccation.

As they develop, molting occurs fairly regularly at approximately 2-week intervals for the first 4–5 months. After this time, molting becomes an irregular event, with adults continuing to molt (unlike insects). The young stages appear very similar to the adults, the primary difference being smaller size and lighter color in the juveniles. Pillbugs and sowbugs are relatively long lived, often surviving 2 or 3 years.

FIGURE 6-5
An aquatic isopod. Photograph courtesy of Tom Murray.

A far greater number of isopods found in North America (some 130 species, all in the family Asellidae) are aquatic. They are usually found in shallow areas around the edges of ponds and in springs and seeps. The aquatic isopods are **omnivores** that feed on dead plant and animal matter, occasionally consuming tiny animals and plant material.

ORDER DECAPODA—DECAPODS

The decapods include the largest and best recognized of all crustaceans. The great majority of decapods are marine animals, and many are staples of the seafood industry—shrimp, lobsters, and crabs. A few groups have adapted to land, notably certain hermit crabs, and others are common residents of inland waters.

FIGURE 6-6
Certain hermit crabs are the most common decapods to have adapted to life on land. Photograph courtesy of Jim Kalisch/ University of Nebraska.

In a typical decapod two body regions are readily evident: the cephalothorax and the abdomen. The cephalothorax includes the head and the entire thorax, which is fused and covered by a carapace. Decapods have five pairs of walking legs, although the front pair is usually modified into grasping claws.

From the decapod head protrudes a pointed structure known as a **rostrum**. Two large compound eyes protrude as well, and one pair of antennae is always substantially larger than the other. The abdomen often terminates as a broad flipper, and appendages that aid in swimming may occur on the underside. Most decapods have a more or less cylindrical body form, although crabs are an exception having a greatly reduced abdomen.

FIGURE 6-7
Front view of an Atlantic lobster, a species that well illustrates the features found in the decapods—and is considered highly edible. Photograph by Whitney Cranshaw/Colorado State University.

Crayfish are decapods that are best represented in North America with 315 species found in two families (Astacidae and Cambaridae). Known variously as "crawfish" or "crawdads," almost all live in shallow water. One species or another can be found in almost any type of inland water—ponds or lakes, streams, marshes, swamps, and ditches. A few species occur on land, in damp meadows where they tunnel to groundwater and produce conspicuous chimneys of pellets from their excavations. One species that occurs in Texas has adapted to relatively dry sites by tunneling into hillsides. This allows it to catch runoff water from rainfall so that it can have sufficient moisture to allow its gills to function.

FIGURE 6-8
A crayfish, *Orconectes virilis*. Photograph courtesy of Tom Murray.

Crayfish are omnivores that feed on a wide variety of foods. Mostly, they scavenge decaying plant matter or carrion. Some will feed on living plants as well, and crayfish may be predators of snails, aquatic insects, and even small fish. Normally, they crawl slowly on the bottom of bodies of water, but they can make rapid darting movements to escape, propelling backward by rapidly contracting their paddle-like tail.

When preparing to lay eggs, the female crayfish produces a sticky material that cements the eggs to the abdomen. She then carries the eggs until they hatch, and the young remain on the mother for the first couple of molts. Only when the mother again molts are they dislodged and begin life on their own.

Crayfish usually become full grown after 6–10 molts, and they will continue molting even after they are sexually mature. Frequently they may lose an appendage through fights or predator avoidance, a behavior known as appendotomy. The lost limbs will regenerate at the following molt; however, they will be smaller than the original. With each subsequent molt the new limb increases in size and ultimately may regain its normal size. Crayfish can be long-lived animals, with some living for 6–8 years.

Less commonly observed in inland waters are freshwater shrimp. Only 15 species are known to occur in North America, and almost all are in the family Palaemonidae. They are found in still waters, usually in quiet backwater areas of large rivers. Most shrimp feed on algae that they scrape off of surfaces, but they may also engulf small insects. The common freshwater shrimp found in North America have a 1-year life cycle. One species of freshwater shrimp native to China, known as the giant river prawn (*Macrobrachium rosenbergii*), is cultivated in small shrimp farms.

Shrimp farming has increased enormously since the practice was initiated in the 1970s and the total production of farmed shrimp now approximates that of wild shrimp caught by traditional methods. The great majority of farmed shrimp are species in the family Paneidae that require access to saltwater for at least part of their lives, with two species (*Litopenaeus vannamei, Penaeus monodon*) composing over 80% of farmed shrimp. Shrimp farming is thus

FIGURE 6-9
Farm-raised paneid shrimp. Photograph by Whitney Cranshaw/Colorado State University.

concentrated along coasts and production has become so intensive in some areas that there are serious concerns about pollution and habitat destruction to estuaries and mangrove stands. The great majority of farmed shrimp are raised in Southeast Asia (China, Thailand, Vietnam, Indonesia), followed by Latin America (Ecuador, Brazil, Mexico).

● Entomophagy—Why Not Eat Insects?

Consumption of various crustaceans is widespread in much of North America, including crayfish in Louisiana cuisine, Maine lobsters, Chesapeake Bay crabs, and shrimp hauled from the Gulf of Mexico. However, it is very rare to find their close cousins, the insects, as a main course anywhere in the United States or Europe. This situation is not the norm for much of the world where all manner of insects may be consumed, either as primary ingredients or condiments.

The eating of insects is called **entomophagy** and many species are considered delicacies. Certain grasshoppers are prized in parts of Mexico, Korea, and Africa. Large caterpillars and large beetle grubs of various species are prized in many parts of the world, notably Africa and Australia. Cicadas, ants, and wasps are part of traditional Chinese cuisine and giant water bugs are a market staple in Thailand.

(continued)

TABLE 6-1 Comparison of some nutritional characteristics between yellow mealworm (*Tenebrio molitor*) larvae, greater wax moth (*Galleria mellonella*) larvae, adult house crickets (*Acheta domestica*), raw ground beef (20% fat), and raw peeled shrimp.

	MEALWORM LARVA	WAX MOTH LARVA	HOUSE CRICKET	GROUND BEEF (80% LEAN)	SHRIMP
Moisture (%)	61.9	58.5	69.2	61.9	77.9
Basic analysis (corrected for dry weight)					
Crude protein (%)	49.1	34.0	66.6	45.0	84.1
Crude fat (%)	35.0	60.0	22.1	52.4	7.2
Fiber (%)	6.6	8.1	10.2	0	0
Ash (%)	2.4	1.4	3.6	2.2	5.0
Minerals					
Ca (mg/kg)	440	590	1,320	472	2,153
K (mg/kg)	8,950	5,320	11,270	7,074	7,659
Fe (mg/kg)	54	50	63	51	100
Vitamins					
Vitamin E (mg/kg)	20	21	43	11	0
Thiamin (mg/kg)	6.3	5.6	1.2	1.1	1.1
Riboflavin (mg/kg)	21.3	17.6	110.7	3.8	1.4
Niacin (mg/kg)	107	90	125	111	106
Choline (mg/kg)	4,839	3,953	4,932	1,478	3,349

Source: Values for insects from Finke, M.D. (2004). Nutrient Content of Insects in *Encyclopedia of Entomology*. J.L. Capinera, Editor. Values for ground beef and shrimp calculated from figures supplied by USDA National Nutrient Database for Standard Reference http://www.nal.usda.gov/fnic/foodcomp/search/.

FIGURE 6-10
A wide range of marine crustaceans are staple items at fish markets. Photograph by Whitney Cranshaw/Colorado State University.

Several insects were used by Native Americans that lived in the western part of the continent. Pupae of the shore fly (*Hydropyrus hians*) could be collected by the bushels around Mono Lake in California and, once dried, stored well as a nutritious food. During outbreaks of Mormon crickets, these large flightless katydids could be herded so that tremendous numbers could be captured, dried, and ground for food. When the large caterpillars of the pandora moth occurred in periodic abundance, native Americans avidly collected them. The use of these insects by some Native Americans was not due to a dearth of alternative foods. Instead, they were actively collected as a favored item for their diet.

Insects can be quite nutritious and compare favorably to many foods we commonly accept on the dinner table (table 6-1). Correcting to dry weight (55%–85% of the insect's body being typically water), the highest fraction of an insect's body composition is usually protein, ranging from about 21% to 80%. Much of this is of high quality and the amino acids are in favorable balance with regard to human nutrition.

Insects can also contain large amounts of fat, providing a high-energy source, most often a mixture of palmitic, oleic, linoleic, and linolenic acids. Most of

(continued)

FIGURE 6-11
Piled grasshoppers, prepared as chapulines, for sale in a market in Oaxaca, Mexico. Chapulines are a local delicacy that use a local species of grasshopper considered to be particularly flavorful and is prepared with lime and chili. Photograph by Whitney Cranshaw/Colorado State University.

these are polyunsaturated, producing a healthy ratio of polyunsaturated to saturated fats.

Calcium is a bit low in insects, but phosphorous content is high. Most insects can provide good levels of magnesium and potassium along with the essential trace minerals iron, zinc, copper, and manganese. Vitamins A and E also tend to be low, but insects often are a good source of B vitamins. The chitinous exoskeleton can be a good source of fiber.

It is unclear where the selective aversion to arthropod consumption among Americans originated, but it is definitely a minority position in world opinion. Perhaps some prohibitions on entomophagy derived from European religious teachings. Biblical writings in Leviticus declare most insect consumption "unclean" and they are definitely not kosher. There is a notable exception with locusts (grasshoppers), a fortunate loophole for John the Baptist who sustained himself in the wild lands on a diet of "locusts and honey." (The honey was most likely **honeydew** secreted by a common scale insect associated with tamarisk, a sweet material collected as food to the present day.)

(a)

(b)

FIGURE 6-12
Although crustaceans are commonly eaten in the United States, most insects are offered (a) as only novelty food items. Some insects, such as the pupae of silkworms (b), have more widespread use as human foods. Photographs by Rick Redak/University of California, Riverside.

(*continued*)

Although nutritionally insects appear to be generally wholesome food source, there are some limitations. Some insects are too spiny or hard to safely ingest. Insects that sequester toxins from the plants they feed on (e.g., cardiac glycosides of milkweed consumed by monarch caterpillars) could produce illness, as could poisonous insects such as blister beetles. Also important are potential allergic reactions. Allergies to ingested insects are not well documented, but allergies to "shellfish" are fairly common, affecting about 2% of the US population. Since many shellfish are arthropods (e.g., shrimp, lobster, crabs), there are shared proteins and potential for some cross allergenicity between ingested insects and ingested marine crustaceans.

ORDER AMPHIPODA—AMPHIPODS

Amphipods are an extremely diverse set of crustaceans with over 7,000 species described. Most occur in marine environments although many are common species found in fresh or brackish waters. Amphipods somewhat resemble miniature shrimp, although they lack the carapace. Three body regions are evident: a cephalothorax (comprised of the head and the first segment of the thorax), thorax, and abdomen. The body is compressed laterally, and the segments of the abdomen are usually fused, together constituting a small part of the body. The largest body region is the thorax to which seven pairs of appendages are attached, the first two of which are enlarged for grasping. The two pairs of antennae are about the same length.

An amphipod found very commonly in most freshwater ponds is *Hyalella azteca* (Hyalellinae family). They are active swimmers, characteristically swimming on their side, and are sometimes called "sideswimmers" or "scuds." Most often, they occur near the bottom of shallow water, particularly among tangles of plant stems.

Walking along the high water mark of an ocean beach, one is likely to encounter another group of amphipods, the "beach fleas" (Talitridae family). They can become extremely abundant among piled seaweed and other plant material that accumulate above the high-tide level on beaches. They are active animals that readily jump when disturbed, and other names sometimes given to them include "sand fleas" or "sand hoppers." A few species of introduced species can be found in moist yards and gardens in parts of Florida and southern California where they are known as "lawn shrimp." Normally living in mulch, leaf litter, and moist areas under shrubbery, migrations of "lawn shrimp" may be triggered by saturating rains, which can cause them to invade homes and garages.

FIGURE 6-13
The amphipod *Hyalella azteca* is a ubiquitous resident of inland waters in North America. Photograph courtesy of Scott Bauer/ USDA ARS.

FIGURE 6-14
Amphipods that occur along shores are often known as "sand fleas" since they are capable of jumping. Photograph by Whitney Cranshaw/Colorado State University.

Amphipods are omnivores that chew on small particles of decaying organic matter and associated microorganisms that they sweep into their mouth. They may also graze on films of algae and bacteria. Amphipods brood their eggs in a special pouch where the young remain for a week or so after the eggs hatch, and the juveniles are released from the pouch the next time the female molts.

ORDER NOTOSTRACA—TADPOLE SHRIMP

Tadpole shrimp have adapted to life in some of the harshest environments, temporary pools of water that may last only a few weeks (vernal pools). Most are rarely seen, appearing for only a brief period following an unusual flooding event. The name "shrimp" applied to these animals may be a bit misleading, as their unusual body form is less reminiscent of commonly encountered shrimp than it is of a trilobite, the ancient arthropods that ruled the oceans during the Paleozoic era before their ultimate extinction some 250 mya. (The tadpole shrimp are somewhat more recently evolved, first emerging about 300 million years or so.) Only 15 species of tadpole shrimp within two genera, *Triops* and *Lepidurus*, are known from North America.

FIGURE 6-15
Longtail tadpole shrimp, *Triops multicaudatus*, are usually associated with very temporary pools of water. They produce eggs that are highly drought resistant and may survive many years until a wetting event causes them to hatch. Photograph by Whitney Cranshaw/Colorado State University.

The most obvious feature of a tadpole shrimp is its broad shield-like carapace. Their paired antennae protrude from the lower part of the body and there are prominent paired tail filaments. On the underside are pairs of leaflike appendages (phyllopods) that propel food particles into a food groove that leads to the mouth. They are usually olive or gray and blend well with the muddy substrate they inhabit. Tadpole shrimp produce drought-resistant eggs that can remain viable for years. When moistened, as with a flooding event, the eggs hatch within 1–3 days. The young develop rapidly, reaching adulthood within a week to 10 days, but some species will also continue to grow after reaching reproductive maturity and egg production. One of the larger tadpole shrimp in North America can reach a body length of 3–4 cm within a few weeks.

Tadpole shrimp can be found in some of the most arid deserts of the western United States, appearing suddenly when a drenching rainstorm provides the pools of water they require. They may be very restricted in where they occur and one California species, the vernal pool tadpole shrimp (*Lepidurus packardi*), is listed as an endangered species. However, another species (*Triops longicaudatus*), commonly known as the "longtail tadpole shrimp" or "rice tadpole shrimp," is so common in the intermittently flooded California rice fields that it can be an occasional pest, feeding on tender emerging plants. Longtail tadpole shrimp, which occur widely across western North America, are also packaged as a novelty item under trade names such as "Triassic Triops" and "Aquasaurs."

ORDER ANOSTRACA—BRINE SHRIMP/ FAIRY SHRIMP

Brine shrimp occur in highly saline inland waters where fish and most other predators cannot survive but algae are present. All those that occur in North America are in the genus *Artemia* (Artemiidae family). Brine shrimp are tiny crustaceans with an elongated and distinctly segmented body. They have 11 pairs of swimming legs and swim upside down. These features are also shared by several other families within the order Anostraca, all of which are found in either highly saline waters or transient vernal pools and are collectively known as "fairy shrimp."

FIGURE 6-16
Fairy shrimp or "brine shrimp." Photograph courtesy of Tom Murray.

During adverse conditions brine shrimp survive as highly drought-resistant eggs (**cysts**) that can remain dormant for years and resist extreme temperatures. Upon return to proper conditions of water salinity and temperature, eggs hatch within hours, producing tiny first instar stages known as **nauplii** that still retain a yolk sac. As the yolk is consumed, the developing animal feeds on microalgae and grows rapidly, becoming full grown in a few weeks. Adult stages may then produce either live nauplii or egg cysts as many as 300 every 3 days. Multiple generations are often produced annually, with adults living about 3 months.

In some areas, notably the Great Salt Lake of Utah and salt basins around San Francisco Bay, brine shrimp are harvested for commercial purposes. During the mid-1990s, about 15 million pounds of egg cysts were collected annually from the Great Salt Lake alone. Most of these eggs are used to rear brine shrimp for fish food for commercial aquaculture or for the pet trade.

Another unique commercial use of brine shrimp is for purchase as "Sea Monkeys" or "Instant Life." This has been a staple novelty item for over 50 years, formerly advertised widely in comic books and still found in toy stores. They still have an avid following of fans who promote May 16 as National Sea Monkey Day.

FIGURE 6-17
Brine shrimp sold under the trade name "Sea Monkeys" have been a staple novelty since the 1950s. Photograph by Whitney Cranshaw/Colorado State University.

Myriapoda— The Myriapods

Several classes of arthropods are known as myriapods. Most myriapods have an elongate body and a large number of legs, with legs found on almost every body segment behind the head. Myriapods include many different kinds of arthropods with substantial differences in other physical features, behavior, and, likely, ancestry. Most systematists believe the classes of myriapods to be only distantly related and recognize Myriapoda as an animal group of convenience rather than one having closely shared relationships.

Class Chilopoda— Centipedes

Centipedes are a relatively small group of animals, numbering only about 2,800 described species. Their unique body form comprises a head and a multisegmented trunk equipped with a single pair of prominent legs on every segment. Due to their uncommon body form and predatory nature, centipedes are generally more familiar than their many-legged myriapod cousins, the millipedes, symphylans, and paurapods.

FIGURE 6-18
The tiger centipede, *Scolopendra polymorpha*. Photograph by Whitney Cranshaw/Colorado State University.

Most centipedes are fast-moving, agile animals propelled by large legs. Although the common name "centipede" means "100 legs," very few species reach this watershed number. The most commonly encountered centipedes possess only between 15 and 23 pairs of legs, one pair per segment. Only in the infrequently seen soil centipedes (Geophilomorpha order) do the numbers of legs meet or exceed the 100 mark. One Fijian soil-dwelling species (*Gonibregmatus plurimipes*) is reported to possess up to 382 legs (191 pairs).

FIGURE 6-19
A soil centipede. Photograph by Whitney Cranshaw/Colorado State University.

The centipede body is elongate and usually flattened dorsoventrally; this allows them to crawl into tight spaces. Within some groups, notably the common stone centipedes found in gardens (Lithobiomorpha order), alternate body segments are shortened. This feature, known as **tergite heteronomy**, helps to dampen the swaying of the body when the centipede rushes forward at high speed.

FIGURE 6-20
Stone centipedes are common residents in and around many gardens. These centipedes show tergite heteronomy, where body segments alternate in width. Photograph by Whitney Cranshaw/Colorado State University.

Slightly different body plans occur among some centipedes. The order Scutigeromorpha, represented by the house centipede, *Scutigera coleoptrata*, has extremely long legs and is able to hold its body well off the ground. Conversely, the legs of the soil centipedes are tiny, although numerous, and assist these creatures to move slowly through the soil. Indeed possessing long legs would be counter-adaptive for animals living within the soil.

FIGURE 6-21
A house centipede, *Scutigera coleoptrata*. Photograph courtesy of Gary Alpert/Bugwood.org.

The front pair of centipede legs is uniquely modified for prey capture. They form a pair of hard, sharply pointed **maxillipeds** that are fused at the base. The maxillipeds project forward from the head and are used to capture and penetrate prey while injecting paralyzing toxins that flow from a grove found near the tip. All centipedes are predators and most feed on small insects and other arthropods. Some centipedes in the order Scolopendromorpha achieve a very large size (up to 30 cm) and are known to occasionally prey upon small birds, reptiles, amphibians, bats, and rodents. In addition to the maxillipeds, centipedes possess chewing mouthparts that can crush and tear their food.

FIGURE 6-22
Ventral view of the head of a scolopendrid centipede, illustrating the maxillipeds, which are used to subdue prey, and the mouthparts. Photograph by Whitney Cranshaw/Colorado State University.

The simple eyes of centipedes are tiny and function for little more than detection of light. Most centipedes hunt at night, and prey are usually located and identified by the centipede's long antennae. On many centipedes the hind pair of legs is unusually long and trails from the body in a manner similar to the antennae. This produces the appearance of a pseudohead at the tail end that can confuse predators. Centipedes are very agile and, if held incorrectly, can easily twist their body to bite their attacker.

FIGURE 6-23
The largest centipede native to the United States, *Scolopendra heros*. Photograph courtesy of Tom Coleman/USDA Forest Service/Bugwood.org.

Centipedes with jaws large enough to penetrate the skin can produce a painful bite. Such a bite may produce immediate sharp, even fiery, pain. Fortunately, the effects are usually of short duration, largely subsiding within a few hours. Some tenderness may persist for days, but centipede bites fortunately are not life threatening.

Reproduction in these animals is indirect and is somewhat unique. Males deposit packets of sperm (spermatophores) onto silken webs. In some species, these webs are then abandoned and left for the females to encounter. Other species take more direct action and engage the female in a courtship ritual. The pair form a mating ring, palpitating each other with the antennae, and may move in a circle for up to an hour. Ultimately, the male guides the female to his previously deposited spermatophore, which she then picks up and uses to fertilize her eggs.

Maternal care of the young occurs with the large scolopendrid centipedes and the soil centipedes.

The mother will remain with the eggs, protecting them with her body and carefully moistening them with salivary secretions to prevent fungal infection. If the nest is disturbed so that the mother is forced to permanently abandon it, the eggs almost always succumb to a fungal infection.

FIGURE 6-24
Scolopendrid centipedes show a high level of maternal care and protect the eggs until they hatch. Photograph courtesy of Ken Gray/Oregon State University.

Class Pauropoda— Pauropods

Pauropods are minute arthropods (ca. 1.25 mm) that occur in soil. Approximately 500 species have been described, most from the tropics and subtropics. They are poorly studied animals, about which little is known.

FIGURE 6-25
A pauropod. Photograph courtesy of Tom Murray.

Adult pauropods are elongate with paired body segments that are fused, producing **diplosegments**. Although somewhat resembling the millipedes (class Diplopoda) in this fusion of segments, pauropods do not possess paired legs on each segment. Adults have nine pairs of legs but at birth pauropods possess only three pairs of legs. Undergoing a form of anamorphic development, they add legs with each successive molt, possessing five, six, and then eight pairs of legs following the next three molts. Unique three-branched antennae further separate pauropods from other arthropods.

Pauropods are found under stones and leaf litter but most often occur within the soil. They are often most abundant a few inches below the surface. Some are known to feed on molds and hyphae of fungi.

Class Symphyla— Symphylans

Symphylans are tiny arthropods (1–8 mm), rarely reaching more than 6.5 mm. With an elongated body, long antennae, and up to 12 pairs of legs, they somewhat resemble miniature centipedes and are sometimes known as "garden centipedes." This similarity is only superficial as they have very different habits and can be readily distinguished by their pale translucent coloration. Approximately 200 species of symphylans are known worldwide.

Eggs are laid in small masses, and the newly hatched symphylan has six to seven pairs of legs. As they develop, they will add segments with each molt, reaching a full complement of legs after the final molt. They can be fast moving and will readily run away to avoid light. A stiff spine present at the base of the legs probably assists them when they move through soil.

Reproduction is through indirect fertilization but involves an unusual twist. Males produce hundreds of packets of sperm atop small stalks of silk that they place rather randomly throughout their habitat. When a female encounters one, she cuts the packet and ingests it. Some of the sperm is digested and used as food, but a fraction is diverted and kept in special pockets of her mouth. Later, as she lays eggs, she carefully moves each one to her mouth where it is fertilized with the stored sperm.

FIGURE 6-26
A symphylan. Photograph courtesy of Tom Murray.

Symphylans are restricted to the soil, where they move through cracks and channels. Often they move up and down through the soil with changes in soil moisture. Many symphylans feed on decaying plant material, but a few chew on plant hairs.

One species found in North America is an occasional crop pest, the garden symphylan (*Scutigerella immaculata*). It occurs in localized areas, particularly where heavy soils amended with organic matter provide numerous channels for their movements. Root feeding may seriously retard plant growth, and garden symphylans may directly damage root crops, such as potatoes, by chewing small pits in the tubers that expose the plant to fungal and bacterial infections. A simple technique used to sample for the garden symphylan is to place a potato slice on the soil surface, covering it and then checking the underside for the presence of symphylans after a day or two.

Class Diplopoda— Millipedes

Diplopods (millipedes) were among the very first animals to venture onto land. Fossil diplopods date back to the late Silurian period, over 425 mya, apparently preceding the later arriving hexapods by tens of millions of years. Since then they have continued to thrive with over 10,000 known extant species, most of which occur in the tropics.

About 910 different millipede species are known from the United States and Canada.

FIGURE 6-27
Close-up of the head of a millipede. Photograph courtesy of Tom Murray.

Despite their age as a life form, the basic body plan of diplopods has remained the same. Five or six segments are fused to form the head, which includes a pair of relatively short antennae and a pair of mandibles designed to chew. The remainder of the body (trunk) is usually cylindrical or slightly flattened and is made up of a series of segmented plates. Each of these apparent segments is actually a fused pair of segments, known as a **diplosegment**. Associated with each diplosegment (after the first four) are two pairs of legs. The class name Diplopoda means "two foot" and refers to this feature.

FIGURE 6-28
A large African species of millipede, *Archispirostreptus gigas*. Photograph courtesy of Jim Kalisch/University of Nebraska.

Their paired legs and sheer number largely define the most obvious physical characteristic of millipedes. They are clear champions when it comes to legs, easily surpassing the runner-up, centipedes.

Nonetheless, the name millipede, meaning "thousand legs," is a bit of an exaggeration. The apparent record holder is a California species (*Illacme plenipes*) that only musters 750 legs (375 pairs).

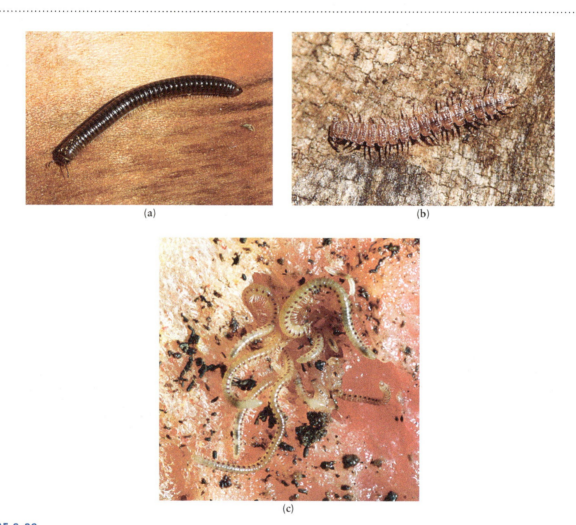

(a)

(b)

(c)

FIGURE 6-29
Millipede representatives of three families: (a) Julidae, (b) Polydesmidae, and (c) Blaniulidae. Photographs by Whitney Cranshaw/ Colorado State University.

At the other end of the spectrum, the "duff millipedes" (Polyxenida order) have only 12 pairs of legs. They are also among the smallest millipedes, typically only 2–4 mm in length. Very large millipedes occur in more tropical areas, with the giant African millipede, *Archispirostreptus gigas*, reaching 30 cm. (This species has been common in the pet trade.) In the distant past, when these animals had the world more to themselves, some species were even larger, with one fossil diplopod exceeding 1.8 m.

As millipedes develop, they increase not only in size but also in both number of body segments legs (anamorphic development). Newly hatched millipedes usually have only three pairs of legs and add more with each molt. Commonly encountered millipedes usually have molted a great many times, over the course of 2–5 years, before they reach the ultimate adult form.

Prior to mating, males produce packets of sperm (spermatophores) that they then transfer to specially modified legs on the seventh segment, known as **gonopods**. During mating, the male grasps the female and presses the spermatophore onto the genital opening of the female, who then draws it in.

Mating takes a bit more involved form among the pill millipedes (Glomerida order). In this group the male constructs a small cup of soil into which sperm is ejaculated. Grasping the female with a pair of clasping hindlegs, he then passes the cup down the legs of the female. Eggs are laid in areas of damp soil, sometimes within small cavity nests constructed by the female.

Almost all millipedes are scavengers, feeding primarily on decaying plant material and associated fungi and bacteria. Soft plant materials, such as seedlings and overripe fruit lying on soil, are also sometimes eaten and some millipedes may be considered minor garden pests on occasion. (The spotted snake millipede, *Blaniulus guttulatus*, is one of the more common garden pest species.) A small number of tropical millipedes are predators, primarily of land snails.

Millipedes are almost always found in moist environments. Despite their hard body covering, they lack the waxy protective layer of insects and are very sensitive to desiccation. In part to compensate, millipedes are formidable tunnelers and can find adequate moisture belowground. The effects of their tunneling can also have substantial ecological impacts. In parts of the tropics where earthworms are scarce, millipedes can be the major animals involved in soil building as they mix and incorporate dead plant matter through their tunneling activities.

Although millipedes are slow moving and lack either a sting or powerful jaws, they are by no means defenseless. When disturbed most will roll or coil, protecting the vulnerable legs and underside with the more heavily armored parts of the body. Alternately, the tiny duff millipedes are covered with specially hooked hairs that readily dislodge to entangle and incapacitate ants or other potential predators.

Millipedes also use a variety of chemical defenses. Many have special **repugnatorial glands** that produce ill-smelling and/or caustic fluids that ooze through openings along the side. These fluids, which may consist of various mixtures of hydrocyanic acid, iodine, and various quinones, can

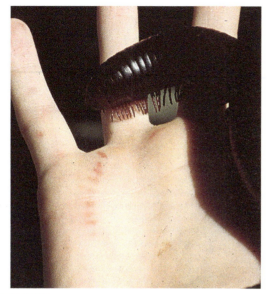

FIGURE 6-30
Skin irritation produced by defensive chemicals secreted by a millipede. Photograph by Whitney Cranshaw/Colorado State University.

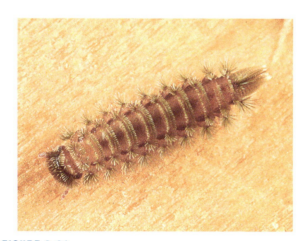

FIGURE 6-31
Tiny polyxenid (duff) millipedes have hooked body hairs that can readily incapacitate ants or many other potential predator. Photograph courtesy of Tom Murray.

stain and irritate the skin and even kill small animals. Furthermore, some tropical millipedes are capable of forcibly discharging these defensive compounds as a spray, temporarily blinding potential predators. At least one species of pill millipede (*Glomeris marginata*) is known to produce quinazolinones, a class of compounds that include the sedative Quaalude®. Wolf spiders confined with this millipede become immobilized and may not recover for days.

● **Oldies but Goodies**

The Entognathous Hexapods and Primitively Wingless Insects

Among the six-legged animals known as hexapods, there are two groups of animals that never had winged ancestors and that demonstrate many features that are considered to be primitive, that is, resembling those of the earliest hexapods. One of these is the **entognathous hexapods**—three orders of six-legged arthropods that are now recognized as being distinct from insects. Two orders of insects, known as **apterygote** insects, are also considered to be derived from ancient lines that arose before the first insects developed wings for flight.

The Entognathous Hexapods

The class Hexapoda is made up of the insects and three odd groups of animals—diplurans, springtails, and proturans. Collectively they are known as the entognathous hexapods because of one of the more notable features that distinguish them from insects—the concealment of mouthparts within a pouch formed by a fusion of the labium and labrum. Within the pouch the mandibles can work horizontally or can be moved up and down.

FIGURE 7-1
A springtail drinking from a water droplet. Photograph courtesy of Brian Valentine.

All of the entognathous hexapods are wingless. Most are small and rarely seen, although the springtails can be extremely abundant in soils. All of the entognathous hexapods use various methods of **indirect sperm transfer** (e.g., production of external spermatophores), and immature stages generally resemble adults except for the changes in size.

The classification of these hexapods is subject to continuing debate. Almost all current taxonomic systems now recognize these as being different from insects. Some classification schemes consider them to be different orders within the class (or subclass) Entognatha. The arrangement used here will be that they are each a separate order within the subclass Entognatha.

Order Protura—Proturans

Entomology etymology. The order name "Protura" means "first" (pro) and "tail" (ura), a name implying that they have primitive physical features. Collectively, these animals are commonly referred to as proturans.

FIGURE 7-2

A proturan. Photograph courtesy of David Shetlar/The Ohio State University.

Proturans are tiny (ca. 1 mm), pale-colored animals without eyes or wings. They also lack antennae, which is unique among hexapods. Instead their front pair of legs is used to sense their surroundings, and they are carried forward so that they function much like antennae. Only about 500 species of proturans are described worldwide, with 19 in North America.

Proturans are found in soil, in decaying vegetation, and under bark, where they primarily feed on fungi, occasionally scavenging other materials. Most are fluid feeders and they use their mouthparts to scrape, not chew; they then mix their food with saliva before it is ingested. Little is known about their habits, but they undergo anamorphic development, increasing the number of body segments after each molt, from 9 at birth to 12 at adulthood, in a manner similar to millipedes and other myriapods.

Order Collembola—Springtails

Entomology etymology. Collembola means "glue" (coll) and "bolt" or "wedge" (embola) in reference to the collophore on the underside of the abdomen, a structure originally thought to allow these hexapods to attach to surfaces. It is now known that the collophore has a different function: to regulate water uptake and excretion. The common name "springtail" is derived from the ability of many to jump by the use of a taillike appendage. Collectively, these animals may also be known as collembolans.

In sheer numbers springtails are—by far—the most abundant of all hexapods, and in areas of moist soil, their populations typically exceed that of all insects combined. (Soil-dwelling mites are usually even more abundant.) However, they are so minute that it has been estimated that it would require about 13,000–15,000 of them to weigh a single gram. They are associated with moist conditions and most springtails are found in soil. There they may feed on decaying plant matter, feces of other arthropods, pollen, fungi, or microorganisms; in these roles they are considered to be very important in decomposition and nutrient cycling. Some feed on algae or plants and a very small number of species (e.g., *Sminthurus viridis*, the "lucerne flea") may damage crop seedlings. Others are predators of nematodes and other small animals.

Small size does have advantages, allowing springtails to absorb oxygen (and excrete carbon dioxide) directly through the body wall, bypassing the need for more elaborate systems of air exchange.

FIGURE 7-3
An entomobryid springtail. Photograph courtesy of Tom Murray.

There are a reduced number of abdominal segments (6), often producing compact body forms. A long tubelike structure, known as the **collophore**, can evert from the abdomen and is thought to help with water balance; springtails are also able to absorb moisture through their bodies. Body coloration is variable but some species have vivid colors and patterns. Antennae are present but are short, and there are no cerci.

FIGURE 7-4
A globular springtail and (smaller) entomobryid springtail. Photograph courtesy of Tom Murray.

The common name is derived from a unique feature found among many in the order—a "spring tail." Technically known as the **furcula**, it is a forked taillike structure attached to the fourth segment of the abdomen. It can tuck underneath the body where it is held in place by a clasping appendage

(**retinaculum**). When released, the furcula swings backward, propelling the springtail upward, with some recorded to jump almost 20 times their body length. This structure is not present on all species and a large number of springtails have a reduced, nonfunctioning furcula, making them unable to jump. (This is particularly common among those species that remain in the soil.) Other features that tend to be present among soil-dwelling springtails are their pale coloration and shorter antennae.

FIGURE 7-5
Underside of a springtail showing the forked furcula and retinaculum. Photograph courtesy of Brian Valentine.

FIGURE 7-6
Springtail with furcula in released position. Photograph courtesy of Tom Murray.

Immature stages undergo five molts before reaching the adult stage, but molting continues for life. As with other entognathous hexapods, mating involves indirect sperm transfer. Nonetheless, mating behaviors can be elaborate and quite unique with these animals. Some springtail males engage in highly energetic dances around the female to help guide her so that she contacts and thus picks up the spermatophore. In the genus *Dicyrtomina*, the male locates a female and then surrounds her with a "picket fence" of spermatophores, virtually ensuring that she will eventually contact it.

This collection of odd features in body design apparently has served springtails well. They are ancient and well represented in fossil records dating back to the early Devonian period; a fossil collembolan dating at least 396 mya is presently the oldest record of any hexapod. At least 6,000 species are known and they occur on all continents, even the Antarctic, where they are common wherever bare soils occur. However, it is their staggering abundance that is particularly distinctive. A 1945 survey of pasture soils in England estimated populations of over 600 million/ha; other estimates have suggested that they can be even more abundant in some sites.

Springtails are usually only noticed where extraordinary numbers are present on the surface. For example, the dark-blue/reddish-brown species *Podura aquatica* is sometimes observed in the form of large floating rafts on the surface of puddled water. Along seashores of the North Atlantic, the slate-blue species *Anurida maritima* may be seen teeming among the debris of the high tide line. Springtails may also occasionally be observed when they migrate into homes from lawns, mulch, and other sites around the building. These indoor migrations occur most often when large

numbers of springtails have developed following an extended favorable period of moist weather that abruptly changes with hot, drying conditions.

Perhaps the oddest aggregations of springtails occur when they appear as "snow fleas." Such events usually occur in forested areas during periods of thaw that allow small channels to appear along the sides of plants, rocks, and other objects projecting through the snow. The springtails follow these channels and may appear in such tremendous numbers that they darken large patches of the snow surface. The aggregations that they form, sometimes described as "jumping dirt," most commonly involve the species *Hypogastrura nivicola*, but other springtail species may be present. However, the purpose of this behavior remains a mystery; few of the migrants are able to ever return to the soil, particularly after a subsequent snowfall.

FIGURE 7-8
"Snow fleas" are springtails that mass on the surface of snow. This phenomenon usually occurs when melting soil provides the springtails access to the surface via cracks that develop around plants and rocks. Photograph courtesy of Linda Corwine.

Regardless, the ability of many springtails to thrive and remain active at such low temperatures has stimulated research into how they protect themselves from the lethal effects of crystal formation during freezing. Proteins isolated from the blood of "snow fleas" are being studied for their potential use in protecting human organs at low temperature prior to transplant.

FIGURE 7-7
Springtails massed on the surface of a compost pile. Photograph courtesy of Brian Valentine.

● The Berlese Funnel—A Simple Method of Extracting Arthropods from Soil

The greatest number of terrestrial arthropods are minute species of soil-dwelling collembolans, mites, and other tiny animals that live unseen lives in soil. These normally are only rarely found when large numbers come to the surface due to flooding or when they happen to be incidentally exposed when turning over a log. Methods that can extract them from soil allow the soil dwellers to be much more easily collected and, for their abundance, to be better appreciated.

A simple device known as a Berlese funnel can effectively accomplish this task. The basic design is a large funnel with a collecting container at its tip. Screen mesh is placed across the funnel and samples of soil, leaf litter, or various plant materials are then placed on the screen. An incandescent light bulb is suspended above the funnel; the light and drying heat of the bulb force insects in the sample to move down away from the source. They ultimately drop through the screen into the collecting container. Alcohol in the collecting container rapidly kills the collected arthropods and also serves as the most effective way to preserve these generally soft-bodied animals.

FIGURE 7-9

A Berlese funnel, a device used to collect arthropods from soil and leaf samples. The sample is placed on a screen in the funnel with a light above it. The light and heat from the light drive the arthropods in the sample down and they can be captured in a collecting container at the bottom. Photograph by Whitney Cranshaw/Colorado State University.

Order Diplura—Diplurans

Entomology etymology. The order name "Diplura" is derived from the Greek words "dipl" (two) and "ura" (tail) in reference to the taillike appendages that these insects possess. Collectively these animals are referred to as diplurans.

The diplurans are a relatively small order (approximately 1,000 species worldwide, 84 in North America) but they are widely distributed and some species are common. Their general body form is elongated, and, while they have long unsegmented antennae, they lack both compound eyes and ocelli. Most are small (less than 7 mm) but some large Australian species (*Heterojapyx* spp.) reach 50 mm. Small eversible vesicles along the sides of the abdomen are present and are thought to help them absorb water. Diplurans tend to occur in moist soils where they are most often found under rocks or fallen leaf debris.

FIGURE 7-10

A dipluran. Photograph courtesy of David Shetlar/The Ohio State University.

Diplurans display an interesting mixture of features and habits parallel to those of a variety of insects. The form of the cerci, from which the order name is derived, varies and reflects the feeding habits of diplurans. Long taillike cerci are present on diplurans that feed primarily on live and decaying plant matter, and these diplurans

somewhat resemble two-tailed thysanurans. Predatory species have stout pincerlike cerci that resemble those of earwigs, which are used to help capture small arthropods. *Heterojapyx* species that hunt by ambush bury their heads in the soil, leaving the cerci above the ground, where they can grab passing insects.

Diplurans lay eggs in small groups and among some species the mother remains with the eggs until the young have undergone several molts, a level of maternal care somewhat similar to earwigs (order Demaptera). After the young begin to forage on their own, they molt repeatedly—as many as 30 times a year—and molting continues even after the adult stage has been reached. They are also able to regenerate lost appendages following molts, a feature nearly unique among hexapods.

The Primitive Wingless Insects (Apterygota)

Two orders of insects (Subclass Insecta)—Microcoryphia and Thysanura—are known as the **apterogyte** insects and possess certain primitive features that separate them from the remaining orders, which are all classified as **pterogyte** insects. Most fundamentally they not only are wingless but

also lack thoracic structures that indicate that their ancestors never possessed wings. (Some pterogyte insects now are wingless, such as fleas and lice, but all these evolved from winged ancestors.) The pattern of metamorphosis among apterogyte insects is also so simplified that it is sometimes referred to as being ametabolous—"without metamorphosis"—the only obvious physical difference between nymphs and adults being size. Molting can continue even after the adult stage is reached, and mating is indirect, with males producing external spermatophores that are collected and picked up by the female.

Order Microcoryphia— Jumping Bristletails

Entomology etymology. The name of this order is derived from the Greek words translating to "small head," a reference to their relatively small head area. An alternate name for this order of insects is **Archaeognatha**, a name that refers to the primitive form of their mouthparts.

Jumping bristletails are small- to medium-sized (up to 15 mm) insects that usually occur in grassy or wooded areas. They have an elongate body, with a humped thorax and three long terminal "tails." Scales, often of elaborate pattern, cover the body. On the head extends a pair of long antennae and two long maxillary palps extend from the sides of the mouth. A pair of large compound eyes meets at the top of the head, and they also have well-developed ocelli.

Jumping bristletails are active insects but are usually nocturnal and infrequently seen. (During nighttime searches, the eyes may reflect light from the beam of a flashlight.) In addition they are also capable of jumping considerable distances, up to 25–30 cm, an effect produced by arching the thorax and sudden flexing the abdomen.

Mating of bristletails involves indirect sperm transfer, but bristletails have developed several methods for fertilizing eggs. Some bristletail males will deposit the spermatophore directly onto the ovipositor of the female, who will then guide it to her genital opening. Other bristletail males use a silken thread that guides droplets of sperm directly to the female, and they remain in close contact during the transfer.

Lichens or algae are the most common foods of jumping bristletails. They pick at these foods using a mandible that has only a single point of attachment and can rotate, rather in the manner of an auger. This is considered to be a very primitive feature. Indeed, jumping bristletails are generally thought to be the most primitive of all living insects; fossil records date them to at least the middle Devonian period (ca. 380 mya). At present there are about 500 known species but they are among the most poorly studied of the insect orders.

Order Thysanura— Silverfish

Entomology etymology. The name of the order is derived from the Greek words "thysan" (bristle or fringe) and "ura" (tail) in reference to the bristle-like hairs on the long terminal filaments. An alternate name for these insects used in some taxonomic arrangements is **Zygentoma**. Collectively, these insects are often referred to as thysanurans.

Superficially silverfish share some features of appearance with jumping bristletails, having a general similar body shape and possessing prominent terminal filaments. Many earlier taxonomic arrangements combined the two orders into the Thysanura, but it is now well recognized that both are distinct insect orders. Furthermore, the silverfish are now thought to be more closely related to some orders of winged insects than they are to the superficially similar jumping bristletails.

The body of a silverfish is elongated much in the manner of the jumping bristletails, but it is much more flattened, allowing them to easily slip into tight crevices. They cannot jump but can run quickly and easily move sideways. The body is covered with scales, often silvery gray, which lend them the common name "silverfish." On the terminal end of the body extend three long, often bristly, filaments. The antennae are very long and thin. There are about 400 species of silverfish worldwide; 18 occur in North America.

FIGURE 7-13
Silverfish. Photograph courtesy of Gary Alpert/Bugwood.org.

For mating, males produce spermatophores that often have attached silken threads. The thread helps to snag the cerci of a passing female; she then guides it to the genital opening, ejects the sperm, and consumes the nutrient-rich remnant of the spermatophore.

In contrast to the jumping bristletails, the mandibles of the silverfish mouthparts have two points of attachment, resulting in the mandibles moving in a single plane. This arrangement allows silverfish to more effectively chew. They are omnivores that may scavenge a wide variety of foods. Most are found in moist environments such as under leaves or bark or in association with insect nests or animal burrows. Some silverfish are well adapted to very dry conditions, for example, the sand dune dwellers found in the Namib Desert.

A few species can occur indoors and are considered significant household pests. The best known is *Lepisma saccharina*, the common

FIGURE 7-14
Silverfish egg mass. Photograph courtesy of Ken Gray/Oregon State University.

silverfish. They have a metallic steel-gray sheen and reach about 12 mm in length when fully grown. The common silverfish will scavenge a wide variety of materials common in homes and prefer proteins to carbohydrates. Stored foods and paper products are most commonly damaged. Their flattened body allows them to hide easily in narrow cracks and complicates efforts to control them.

The common silverfish also has an impressive ability to reproduce. Females may produce over 1,500 eggs and the young become full grown within a few months after undergoing three or four molts. Adults are long lived, up to two and a half years, over which time they may molt over 40 times. Common silverfish adults are also quite resistant to starvation and can survive months without food.

The firebrat (*Thermobia domestica*) is a mottled tan or brown silverfish that frequents warm areas such as around furnaces. They feed primarily on starches and can be damaging to certain paper products and linens. The largest (ca. 18 mm) silverfish common in homes is *Ctenolepisma longicaudata*, known as the gray or longtailed silverfish. It is a particularly long-lived species, taking 2–3 years to become sexually mature and surviving as an adult for up to 5 years. This silverfish is also exceptionally resistant to heat and drying. They do not need access to sources of free water, being able to extract sufficient amounts from their food.

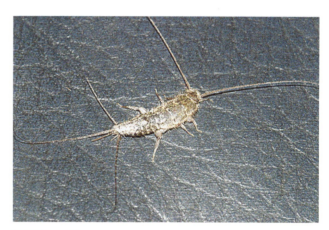

FIGURE 7-15
Firebrat. Photograph courtesy of Tom Murray.

● **Insects Fly!**

Mayflies, Damselflies, and Dragonflies

The ability of insects to develop wings is one of the key evolutionary steps contributing to their success as life forms. Two insect orders continue to have wing features that largely resemble those of the earliest winged insects in fossil records—the mayflies (Ephemeroptera) and the dragonflies and damselflies (Odonata). Both have immature stages that develop in water.

Order Ephemeroptera—Mayflies

Entomology etymology. The name of the order is based on the brief span of the adult stage of these insects: ephemero ("lasting a day" or "short-lived") and pteron ("winged"). The common name "mayflies" is not particularly accurate as many of the more abundant and most often encountered species emerge and are active in late spring and early summer.

Mayflies are considered to be the most ancient of all extant winged insects, being found in early Permian fossils. (Some other insect orders with wings preceding the mayflies are now extinct.) About 2,500 species of mayflies are described worldwide, with 630 occurring in North America.

Adult mayflies have membranous wings that are held vertically when at rest, and the triangular hindwing is substantially smaller than the forewing. (Some species have only a single pair of wings.) They possess an elongate body that is further extended with two or three long tails. Small bristle-like antennae project from the head, but the mouthparts are greatly reduced in size and are nonfunctional; adult mayflies don't feed. The sexes can be distinguished in some species by the longer forelegs and larger eyes of males.

Mayflies go through a modified form of simple metamorphosis. Adult females deposit eggs in or on fresh water, and the nymphs are entirely aquatic. Mayfly nymphs are recognizable by the presence of three (rarely two) caudal filaments, often called "tails." Gills are present along the side of the abdomen. Distinct wing pads become evident on the thorax of later stage nymphs. Other physical features that distinguish different mayfly families may be present. For example, one of the most commonly encountered mayfly families is Ephemeridae (common burrowing mayflies) that live within U-shaped burrows and have tusks that project from the mouthparts of the nymphs. These tusks allow them to burrow into the soft sediment at the bottom of lakes and ponds.

FIGURE 8-1
An adult female mayfly, *Isonychia* sp. Photograph courtesy of Tom Murray.

FIGURE 8-2
An adult male mayfly, *Eurylophella prudentalis*. Note the greatly enlarged eyes characteristic of the males, which are used to locate the females during the mating swarm. Photograph courtesy of Tom Murray.

Most mayfly nymphs feed by chewing upon decayed plant material or grazing the film of algae, diatoms, and cyanobacteria (**periphyton**) that cover underwater surfaces. A few are predators. All crawl and/or swim with undulating movements of the body. Mayfly nymphs molt frequently, a dozen or more times before reaching winged stages. Typically, the life cycle is completed in a year (univoltine). A few species are multivoltine and may produce multiple generations each year.

The winged stages of mayflies involve one of their most unique aspects: there are two developmental stages with functional wings. No other present group of insects develops in this manner, although repeated molting of immature winged stages is thought to have occurred among many long-extinct insect orders. The stage that initially emerges from the water is known as a **subimago** (or "dun" to fly fishermen). It is

(a)

(b)

(c)

FIGURE 8-3
Examples of mayfly nymphs: (a) *Isonychia obscura*, (b) *Baetis tricaudatus*, and (c) *Drunella doddsi*. Photographs courtesy of Tom Murray.

often dull colored and somewhat hairy with incompletely developed genitalia. Within a day or two, the subimago molts again, producing the **imago** (or "spinner" to fly fishermen), the reproductively active adult. This final form is usually smooth and glossy with longer legs, longer tails, and fully developed genitalia. In some mayflies, the legs are lost at this molt and they are subsequently fated to fly for the rest of their brief life.

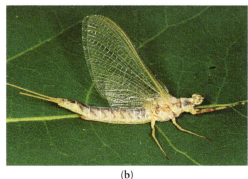

(a)   (b)

FIGURE 8-4

A female (a) subimago and (b) adult of one of the larger species of mayflies, *Heptagenia limbata*. The subimago of mayflies is the only non-adult stage of any insect that is winged. Within days a further molt will occur, producing the adult stage that will usually be less dark colored and hairy, have longer legs and tails, and be sexually mature. Photographs courtesy of Tom Murray.

Adult mayflies are graceful fliers that may make undulating flights above the surface of waters. Males typically emerge first from the water and produce large swarms above the surface while awaiting receptive females. Females that fly into the swarm are grasped by the males and mating takes place during flight. Very shortly after copulation, egg laying (**oviposition**) begins. Some species drop the eggs directly onto the water surface; others lay them directly into the water, and a few attach them to surface structures on the water. The adults die very shortly after mating and oviposition.

Because the adult stage is extraordinarily brief, sometimes lasting only a few hours and never more

FIGURE 8-6

Mayflies massed on a porch screen during a swarming event. Photograph courtesy of Charles Hesselein.

than a couple of days, the emergence of winged mayflies is highly synchronized. Such a synchronized mass emergence often consequently overwhelms predators like birds and fish such that the probability of survival increases for any individual mayfly. In waterways that support large populations of some mayfly species, the emergence of adult forms can be spectacular, producing clouds of insects over the waters. Should this happen near human development, street lights may confuse them and attract large numbers. There have been reports where mayflies have collected in 0.5 m deep piles around lights, and snow plows have been used to clear them from bridges.

FIGURE 8-5

A large mating swarm of mayflies. Photograph courtesy of Suzanne Wainwright.

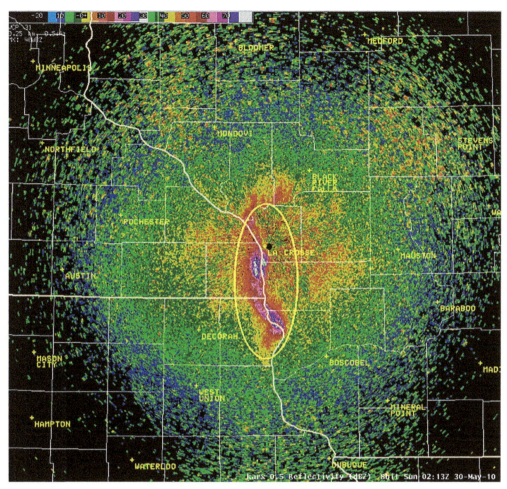

FIGURE 8-7
During a mass emergence of some common mayflies, swarms may be so dense that they are detectable on radar. This image from the National Weather Service captures a mayfly swarm south of La Crosse, Wisconsin, that occurred on May 29, 2010.

Order Odonata— Dragonflies and Damselflies

Entomology etymology. Odonata means "toothed jaw." Although not particularly descriptive of the order's most prominent features, the name remains from an early method of insect classification using features of the head developed by Danish zoologist Johan Christian Fabricius. Collectively, members of this order are sometimes referred to as odonates, although the term is rarely used by the general public. The origins of the common English names "dragonflies" and "damselflies" are unknown but have a long history.

With their large size, colorful patterning, and stunning aerial acrobatics, insects in the order Odonata are some of the most familiar, and also popular, of all insects. There are two primary suborders—Anisoptera (dragonflies) and Zygoptera (damselflies). (The small third suborder, of debatable relationship, Anisozygoptera contains two living Asian species and several species known from fossils.) Worldwide, there are about 4,000 described species, with 450 recorded in North America.

(a)   (b)

FIGURE 8-8
Side view of a typical (a) dragonfly and (b) damselfly showing the general differences in appearance. Photographs courtesy of Brian Valentine.

The body shape of dragonflies is distinctive. The large thorax is packed with flight muscles; the abdomen is long and thin. The thorax is pushed forward so that, while flying, the forelegs form a basket, allowing them to catch insects on the wing. Damselflies are much thinner and more delicate in form than dragonflies. The largest damselfly (*Megaloprepus caerulatus*), found in the forests of Central and South America, and the largest dragonfly, the giant Hawaiian darner (*Anax strenuus*), both have a maximum wing span of about 19 cm.

● "Giant Dragonflies" and the Great Dying

Insects reached their most massive size during the late Carboniferous and Permian periods, a time when oxygen concentrations were much higher than today. At that time several now-extinct orders of insects were common, notably various palaeodictyopterids along with ancestral forms of today's cockroaches and orthopterans. Mayflies were also abundant, as were other arthropods such as scorpions and millipedes.

The most massive of the flying insects appear to have been the griffinflies (sometimes poorly described as the "giant dragonflies") of the order Protodonata. Some of these reached very impressive sizes, with one recovered fossil having a wing span of 71 cm. From the late Carboniferous through the later Permian periods (about 70 million years), griffinflies, along with mayflies and members of several extinct insect orders, ruled the airways of the world. Unfortunately, the massive disasters associated with the Permian–Triassic extinction event (252 mya) wiped out the griffinflies, along with seven or eight other insect orders and their associated families.

Dragonflies are only very distantly related to the griffinflies. Although both groups use large membranous wings and possess a similar body form, the similarities between griffinflies and today's dragonflies and damselflies are only superficial. The more recent developing Odonata first appeared some time during the middle of the Permian period, near the end of the griffinflies' aerial reign, and was one of the few winged orders of insects that survived "the Great Dying."

Adults of all Odonata have large, intricately netted, straight wings that they are incapable of folding and that protrude when the insect is at rest. Wings of dragonflies are generally held in a plane horizontal to the body when perched, while the shape of the damselfly thorax causes its wings to appear swept back over the abdomen at rest. A wide variety of colors and patterning are present among both dragonflies and damselflies; males and females of the same species often are quite different in appearance, with males tending to be of brighter colors.

Both dragonflies and damselflies have well-developed binocular vision, which is among the best of all insects. The very large compound eyes bulge from the sides of the head and each may contain over 10,000 individual light-receiving ommatidia. Furthermore, the head is flexibly attached to the body, allowing the insect to pivot for a full range of view. The eyes are sensitive to colors, and they can distinguish markings of potential mates or rivals from several meters of distance. Large optic nerves pack the brain and are dedicated to rapidly processing the constantly changing images acquired while in flight, allowing the insects to quickly respond to small movements. Dragonflies also possess three large simple eyes (ocelli) that are apparently used for orientation and stability during flight.

(a)   (b)

FIGURE 8-9
Both (a) dragonflies and (b) damselflies possess very large compound eyes, which provide them with high visual acuity and the ability to readily detect motion. The head is also capable of pivoting, further improving these insects' range of vision. Photographs courtesy of Brian Valentine.

The immature stages of these animals are strikingly different in form from that of the adult. Damselfly nymphs are quite elongate, with three platelike gills extending from the end of the body.

Unlike damselflies, dragonfly nymphs lack external gills. They draw water up into their rectum where internal gills extract oxygen. Forceful expulsion of water from the rectum of a

FIGURE 8-10
Damselfly nymphs have a slender body form and a pair of gills that extend from the end of the abdomen. Photograph courtesy of Tom Murray.

FIGURE 8-11
A damselfly adult emerging from the nymphal skin. Photograph courtesy of Brian Valentine.

dragonfly nymph can propel it forward, providing the nymph with jet propulsion bursts that also allow it to escape predators. Dragonflies that sprawl on the bottom of ponds and hunt by ambush tend to have a broad body form. Species that are active hunters have a more elongated body form with well-developed legs that are used for crawling in search of prey.

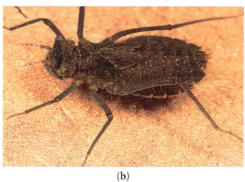

(a)   (b)

FIGURE 8-12

Dragonfly nymphs lack the external gills of damselflies but have internal gills in the rectum. Body forms vary depending on, and often reflecting, their hunting habits. Longer body forms, such as that of (a) *Anax junius*, occur in species that actively crawl and hunt. Stouter body forms, such as that of (b) *Libellula* spp., occur among sprawlers that remain stationary and hunt by ambush. Photographs courtesy of Tom Murray.

In many cultures throughout the world, Odonata have historically played notable roles. The Japanese, perhaps more than any other culture, have long revered dragonflies and damselflies, devoting to them many folktales, poems, and songs. Even the Japanese military honored these insects as symbols of strength among samurai. Dragonflies and damselflies also appear in the art and folklore of Indian tribes from the southwestern United States, Europe, and various island cultures scattered across the South Pacific. Presently in the United States they are receiving a great surge of attention by naturalists (including many present or former bird watchers), in large part due to improved equipment for viewing (e.g., digital cameras) and the recent production of some outstanding identification guides.

FEEDING HABITS OF DRAGONFLIES AND DAMSELFLIES

Both the adults and nymphs of dragonflies and damselflies are active predators. Adults are aerial hunters that scoop insect prey from the air with their legs and then quickly dispatch them with their powerful jaws. Adults consume a wide range of flying insects, including midges, small flies, and an occasional day-flying mosquito. Fairly large insects are occasionally taken as prey, including butterflies and even other odonates.

FIGURE 8-13

Side view of the dragonfly *Anax junius*. Their legs allow them to scoop and hold insects caught in flight. Photograph courtesy of Tom Murray.

Immature forms (nymphs), with very few exceptions, occur in freshwater. A unique distinguishing feature of Odonata nymphs is their enlarged, extensible labium (lower jaw), sometimes called the "mask," used to capture prey. When suitable prey is encountered at the right distance, the labium rapidly extends forward and grasps the prey; the action is not unlike a snake strike in rapidity and effect. The prey item is then drawn back into their powerful jaws and consumed. They have a prodigious appetite, as documented by William Calvert of the University of Pennsylvania.

A single dragonfly nymph, in the course of its life in Calvert's laboratory, consumed 3,037 mosquito larvae, 164 mosquito pupae, and miscellaneous other insects, including 17 nymphs of other dragonflies/damselflies. Larger nymphs can and do capture small fish and amphibians.

(a) (b)

FIGURE 8-14

(a and b) A nymph of the dragonfly *Sympetrum obscurum* demonstrating the mobility of its extensible labium. This can be used to snatch prey, which are then drawn into the mandibles. Photographs courtesy of David Leatherman.

MATING AND EGG LAYING

Male dragonflies and damselflies typically establish territories near or over bodies of water. The size of the territory depends on the species, with dragonflies typically defending a much larger area than damselflies. Often the males will rest on a perch with a good view of the area, occasionally darting out for inspection and then returning. Some may cruise for extended periods seeking females within a territory, a behavior known as "hawking." Meanwhile, the females may fly over a much wider area, returning to water only for mating and egg laying.

Mating by dragonflies and damselflies is unique. Just a brief period of watching insects by a pond in midsummer will usually yield the opportunity to view a pair of dragonflies or damselflies joined in flight. The male is in front, grasping the female with claspers at the tip of his abdomen. Dragonflies will clasp the female at the head, damselflies clasp the prothorax. Often the female has curved her abdomen to bring it to the base of the abdomen of the male, completing the "wheel position" Although the male's genital opening is present in the normal location (ninth segment of the abdomen), prior to mating, sperm are transferred to a special structure (the secondary genitalia) on the second or third abdominal segment. After pairing in flight, dragonflies and damselflies will usually move to a nearby perch to consummate the mating. This is achieved by the female who twists the tip of her abdomen underneath to draw the stored sperm from secondary genitalia of the male into her oviduct. They may remain conjoined in this wheel position for several hours. In most species of damselflies, special structures of the male genitalia are used to remove from the female stored sperm from previous matings. This adaptation allows improved chances that her eggs are fertilized with sperm from her latest mate. Similarly, many dragonflies have ways for males to push out sperm from previous matings.

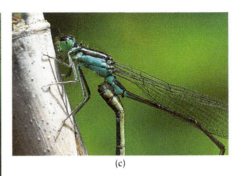

(a)   (b)

(c)

FIGURE 8-15

(a–c) Sequence in the mating of damselflies, which mate in the "wheel position." Males grasp the female behind the head using a special pair of claspers. The female pivots her body to access the sperm that are located in a special structure, the secondary genitalia. Prior to mating, the male transfers the sperm from the genital opening on the ninth segment of the abdomen to the secondary genitalia, which are located on the second or third segment. Photographs courtesy of Brian Valentine.

FIGURE 8-16

A mating pair of dragonflies, *Aeshna verticalis*. Photograph courtesy of Tom Murray.

Female dragonflies lay their eggs on the surface of water, typically dipping the tip of their abdomen into the water in flight. Damselflies lay eggs in a more deliberative manner, attaching or inserting them to plant stems. Many damselflies and some dragonflies may remain attached in tandem after mating as the female lays eggs, an activity that further ensures eggs are fertilized by the last male with whom the female

mated. During egg laying, damselfly pairs may be seen along a plant stem, with the female submerged as eggs are laid several inches below the surface and the male typically remaining joined above the water.

FIGURE 8-17

A pair of damselflies conjoined while the female is laying eggs. Photograph courtesy of Brian Valentine.

Eggs usually hatch within a couple of weeks of being laid. Development of the nymphs is variable, depending on species and environmental conditions such as water temperature. In general, damselflies have shorter life cycles, and multiple generations can be produced in a single season. Some dragonflies have quite long life cycles that may take 2–3 years to complete and involve as many as 20 instars. When full grown, the nymphs crawl out onto rocks or along the edge of the water. The exoskeleton splits and the adult form emerges, leaving behind the "shuck" of the exoskeleton of the last stage nymph.

FLIGHT

The flight mechanisms used by dragonflies and damselflies are considered to be of primitive design with the wings directly attached to the thorax. Each pair of wings is powered separately by individual flight muscles. Each wing beat results from individual contractions of these muscles, known as **direct flight muscles**, which are attached to the base of the wings. This allows a wing beat frequency of about 30 strokes per second, considerably slower than that of insects of some insects that use a more "advanced" design for driving their wings (**indirect flight muscles**). However, the ability to control each wing has advantages in maneuverability that makes dragonflies and damselflies aerial experts that are a marvel to watch. By controlling the beat of individual wing pairs, they can manage to alter forces of thrust or lift to meet the need of the moment. Some damselflies are even capable of asymmetric strokes, with front and hindwings moving in opposite direction to allow very sharp turning. The shape of the wing, its angle of attack, and the length of the stroke can all be altered to produce extremely complex aeronautics.

Many dragonflies are also capable of long-distance flights. At least nine North American species of dragonflies make long annual migrations, much in the same manner as that of migratory birds. Spring migrations from Mexico, the Caribbean, and southern United States progress northward in late March through early June, colonizing the northern United States and southern Canada. Triggered in part by cold weather, a reverse flight occurs in late summer and early autumn as these migrants return to their winter breeding grounds. As with migratory songbirds, topographic features such as shorelines, cliffs, and ridgelines appear to be used to direct dragonfly migrations.

FIGURE 8-18
The green darner, *Anax junius*, is one of several North American dragonflies that make annual north–south migrations. Photograph by Whitney Cranshaw/Colorado State University.

But perhaps the greatest migrations among dragonflies are achieved by the globe skimmer, *Pantala flavescens*. Long recognized as the most widely distributed dragonfly in the world, only recently have its abilities at long-distance flight been fully recognized. One annual migration route carries them from India to East Africa, often with brief

FIGURE 8-19
The global skimmer, *Pantala flavescens*, has the widest distribution of any dragonfly. They are highly migratory and may make annual round trips that cover thousands of kilometers. Photograph courtesy of Mark Chappell/University of California, Riverside.

stopovers on the Maldives and other islands en route. Traveling at heights of over 1,000 m above the oceans, they catch winds that carry them progressively eastward in October through early December with intermittent island stopovers. Their arrival in East Africa coincides with the monsoons that provide the temporary pools of water in which they breed. A reverse migration occurs in spring, returning them to the Indian subcontinent, for an annual round trip of 14,000–18,000 km. Probably three to four generations are produced during the course of this annual migration, on par with that of the better known migrant, the Monarch butterfly. Concurrent with the migrations of the globe skimmer are many migratory birds, which feed on the dragonflies in the course of their own travels.

Jumpers and Strollers—Grasshoppers, Crickets, and Walkingsticks

The wing designs that first emerged in most ancient winged insects (e.g., mayflies) were subsequently modified in a great many kinds of insects that followed. One important development was the ability to fold wings, making the body forms more streamlined with the ability to squeeze into tight spots while allowing better protection of the wings. Another adaptation that evolved was thickened forewings that, when folded, protect the hindwing and abdomen. Several insect orders have benefited from these developments. Among these are two orders that seem to share many common features: the Orthoptera (grasshoppers, crickets, and relatives), and the Phasmatodea (walkingsticks and leaf insects).

Order Orthoptera—Grasshoppers, Crickets, and Katydids

Entomology etymology. The order name is derived from the Greek words *orthox* (straight) and *ptero* (wing) in reference to the leathery, straight-sided forewings of these insects. Collectively, these insects are sometimes referred to as orthopterans.

FIGURE 9-1
In the order Orthoptera, the forewings are thickened and leathery. They cover and help protect the hindwings, which are used for flight. Photograph by Whitney Cranshaw/Colorado State University.

The order Orthoptera includes some of the most familiar insects we see in fields or gardens. It also includes most of the insects that produce the sounds we often associate with spring and summer. In terms of species number, Orthoptera is of medium size, with about 25,000 species. It is divided into two suborders: Ensifera, or long-horned Orthoptera (crickets and katydids), and the Caelifera, or short-horned Orthoptera (grasshoppers).

One of the key distinguishing features of the order is the thickened and leathery front wings, known as **tegmina**. The tegmina are flexible enough to provide a small amount of lift for flight but are tough and strong enough to provide protection for the membranous hindwings that fold underneath when orthopterans are at rest.

Flightless species of Orthoptera either do not possess tegmina or have very reduced and small tegmina.

Most orthopterans also have rear legs designed for jumping (**saltatorial legs**), with greatly enlarged femurs. These enable the insects to leap, often over long distances, and to kick for defense. Winged orthopterans often use their jumping ability to launch themselves into the air before taking flight.

As a group, members of this order possess chewing mouthparts and large compound eyes. Antennae are in the form of thin filaments and, in crickets and katydids, they are quite long, often longer than the body. Other distinct morphological features include the presence of a well-developed, often long and sword-shaped, ovipositor, which is used to deposit eggs within the soil or host plants. Many orthopterans have also evolved various modifications on the legs, wings, or body (**stridulatory organs**) that allow them to produce sound. Body size ranges tremendously within the order Orthoptera, including some rather small members (0.5 cm), but also arguably the heaviest insect (71 g for a pregnant female Little Barrier Island giant weta) and some grasshoppers with wingspans exceeding 20 cm.

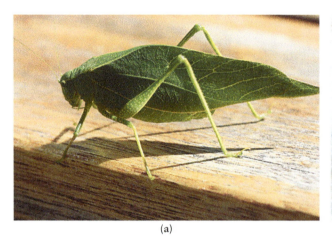

(a)   (b)

FIGURE 9-2

A katydid (Tettigoniidae family) (a) and field cricket (Gryllidae family) (b) demonstrate most of the features typical of the order Orthoptera. These include having hindlegs that are adapted for jumping, long filament antennae, chewing mouthparts, and thickened forewings (tegmina). Photographs by Whitney Cranshaw/Colorado State University.

FAMILY ACRIDIDAE—SHORTHORNED GRASSHOPPERS

"Typical" grasshoppers occur within the family Acrididae, the largest orthopteran family, with some 620 species in the United States and Canada alone. Their antennae are prominent but usually much shorter than their body.

Shorthorned grasshoppers lay eggs in small masses (egg pods) in soil, and most produce one generation per year. In the great majority of grasshoppers, the egg is the overwintering stage, but some species survive winter as nymphs or adults. It is usually possible to find some kind of grasshopper active on any warm day of the year.

Shorthorned grasshoppers can produce sound but are not so noisy as crickets and katydids. Males of many species produce sounds by **stridulation**, rubbing a row of short pegs on the femur of the hindleg over the edge of the forewing, creating a buzzing noise. These sounds are received by other grasshoppers via auditory organs called **tympana** (sing. tympanum) located on the first segment of the abdomen. In addition, bandwinged grasshoppers (subfamily Oedipodinae) produce a loud rustling sound by snapping their hindwings as they fly. The bandwinged grasshoppers have brightly colored underwings underneath their dull-colored tegmina. When disturbed to flight, their unfolded wings produce a burst of color, which, combined with the

(a)

(b)

FIGURE 9-3
Examples of grasshoppers: (a) a slantfaced grasshopper;
(b) mating pair of grasshoppers, *Hesperotettix viridis.*

(a)

(b)

FIGURE 9-5
A grasshopper egg pods exposed from the soil. Photographs by
(a) Whitney Cranshaw/Colorado State University and (b) by
Alexandre Latchininsky/University of Wyoming.

FIGURE 9-4
A grasshopper ovipositing in soil. Grasshoppers lay eggs in the
form of pods, which they insert into the soil. Photograph
courtesy of William Hantsbarger.

FIGURE 9-6
Grasshopper nymphs are smaller than adults and lack functional
wings. Wing pads become increasingly evident as nymphs
mature. Photograph by Whitney Cranshaw/Colorado State
University.

FIGURE 9-7
Generalized life cycle of a grasshopper. Drawing from *Grasshoppers of the West*, Robert Pfadt/University of Wyoming.

crackling noise, can startle and distract a predator. They land abruptly with hindwings folded and hidden; these insects blend in nearly invisibly with the soil background.

Shorthorned grasshoppers are all plant feeders, but few limit their feeding to grasses. Some are highly specialized in what types of plants they will eat, while others feed on a mixture of plants. Only a handful of species, probably less than a dozen in North America, ever become numerous enough to seriously damage rangeland and crops. (Most of the pest species are generalist feeders in the genus *Melanoplus* along with the clearwinged grasshopper, *Camnula pellucida*.) The overwhelming majority of grasshoppers never become pests and, like all insects, are critical components of ecological food webs, serving as food for other animals and partially regulating plant populations.

FIGURE 9-8
The great crested grasshopper, *Tropidolophus formosus*, is one of a great many North American grasshoppers that are not considered pests. It is a rangeland species that limits its feeding almost entirely to wild mallows and winecup, non-economic plants in the family Malvaceae. Photograph by Whitney Cranshaw/Colorado State University.

(a)   (b)

FIGURE 9-9

Only a handful of the 620 North American grasshopper species damage cultivated plants with any regularity. Among these are (a) the differential grasshopper, *Melanoplus differentialis*, and (b) the clearwinged grasshopper, *Camnula pellucida*. Photographs by Whitney Cranshaw/Colorado State University.

A few species of grasshoppers may periodically become very abundant, band together, and migrate in massive swarms. Such "grasshoppers gone berserk" are called **locusts** and they can cause devastating crop losses during outbreaks. The best known of the locusts is the desert locust (*Schistocerca gregaria*) found in northern Africa, the creature featured as the Eighth Biblical Plague of Egypt and which, still to this day, erupts in occasional swarms that can extend into the Middle East, southern Europe, and occasionally farther. Southern Africa has periodic plagues of the red locust, *Nomadacris septemfasciata*. In Australia,

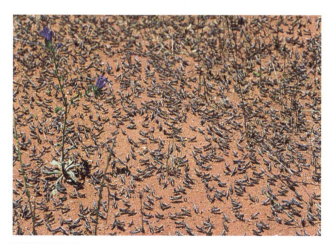

FIGURE 9-10

The Australian plague locust, *Chortoicetes terminifera*, during an outbreak phase. Photograph courtesy of Alexander Latchininsky/University of Wyoming.

the Australian plague locust, *Chortoicetes terminifera*, can sometimes produce large swarming populations.

Many swarming grasshopper species exhibit differences in development known as **phase polymorphism**, a sort of Dr. Jekyll and Mr. Hyde transformation. At low population densities the grasshoppers are "normal" grasshoppers. During this **solitary phase** they are usually light colored (greens and tans), have relatively short wings, and display little tendency to aggregate or migrate.

Certain environmental changes can trigger the grasshoppers to morph into a very different animal in the **gregarious phase**. A grasshopper in the gregarious phase generally has darker colors, develops long wings, and displays behaviors that cause banding together. Ultimately they may swarm. The transformation of a grasshopper population in appearance, behavior, and habits from primarily the solitary phase to the gregarious phase characterized by swarming masses usually occurs over several generations. Phase transformation occurs when several environmental factors come together. Gregarious phase grasshoppers are produced by the crowding of young stages, food availability, and weather patterns.

The desert locust has been the most consistent species to occur as a "plague locust." A relatively recent (2004) outbreak originated in a swath in western Africa from Senegal to Morocco. A combination of favorable weather events, notably timely rains, allowed several generations to be produced in quick succession, resulting in massive

TABLE 9-1 Grasshopper species that are generally recognized as having habits of locusts.

COMMON NAME	SCIENTIFIC NAME	GEOGRAPHIC DISTRIBUTION
Desert locust	*Schistocerca gregaria*	Africa, Southern Europe, SW and Central Asia (invasion area)
Migratory locust	*Locusta migratoria*	Africa, Eurasia, Australia, New Zealand
Moroccan locust	*Dociostaurus maroccanus*	Northern Africa, Europe, Central Asia
Italian locust	*Calliptamus italicus*	Europe, Asia
Red locust	*Nomadacris septemfasciata*	Africa
Brown locust	*Locustana pardalina*	Southern Africa
Australian plague locust	*Chortoicetes terminifera*	Australia
American bird locust	*Schistocerca americana*	South and Central America
Bombay locust	*Nomadacris succincta*	SE Asia
Rocky Mountain locust	*Melanoplus spretus*	North America (extinct)

Source: Acridids of Kazakstan, Central Asia and Adjacent Territories, Latchininsky et al. 2002.

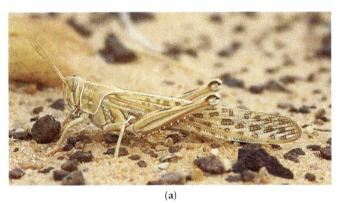
(a)

(b)

FIGURE 9-11

The desert locust, *Schistocerca gregaria*, is perhaps the most notorious of all grasshoppers that periodically aggregate and migrate in swarms. This occurs when a phase change occurs from (a) the solitary form to (b) the gregarious form. While in the solitary form, they behave like any other grasshopper and do not migrate in bands. Photographs courtesy of Jean-Francoise Duranton/CIRAD.

populations that then dispersed. One swarm observed in Morocco was estimated to be 230 km long and 150 m wide, containing 69 billion insects. The 2004 outbreak produced swarms penetrating to the Middle East, southwestern Europe, and Crete. Up to half of all agricultural production was destroyed in some sub-Saharan countries, and about $400 million was spent in an attempt to control these insects.

Flying with winds, the desert locust can often cover 100–200 km/day, and when they land, they generally consume an amount of plant matter equivalent to their body weight each day. Winds will affect their dispersal. Historically, swarms have extended to India. Winds moving west can cause them to land on Atlantic islands, such as those that occurred in Cape Verde and the Canary Islands in the 2004 outbreak. However, an even more remarkable occurrence happened in 1988 during a desert locust outbreak. Individuals from swarms descended on several Caribbean islands after a sustained flight of at least 7,200 km (4,500 miles), the longest sustained insect flight achievement on record.

"Locust" events have occurred in North America, most spectacularly involving the Rocky Mountain locust (*Melanoplus spretus*). The largest recorded swarms of any locust on the planet were produced by this species during the 1870s, yet changes in the ecology of the western United States taking place during this period soon drove the Rocky Mountain locust to extinction. During the Dust Bowl years in the 1930s, populations of another grasshopper species exploded, the High Plains grasshopper (*Dissosteira longipennis*). During the peak of the outbreak large migrating swarms of High Plains grasshoppers destroyed over four million hectares of croplands per year. Prior to the 1930s outbreaks and in the years since, the High Plains grasshopper has been in almost complete remission as a damaging pest grasshopper, remaining a relatively uncommon species among the 100-odd grasshoppers that occur in the western United States.

(a)

(b)

FIGURE 9-12
During the Dust Bowl years in the late 1930s, the High Plains grasshopper (a), *Dissosteira longipennis*, produced extraordinarily high populations (b) and migrated in locust fashion, destroying millions of acres of cropland. Since 1940 its numbers have receded to its relatively low normal levels, and it has not been a significant pest since. Photograph of High Plains grasshopper by Whitney Cranshaw/ Colorado State University. Historical photograph of grasshopper outbreak courtesy of the Colorado State University Archives.

● The Mystery of the Rocky Mountain Locust

Presently none of the short-horned grasshoppers of North America transform into the types of swarming locusts found in other parts of the world. However, this was not always the case. The Rocky Mountain locust (*Melanoplus spretus*) was one such insect that farmers and ranchers on the western plains of the United States feared greatly in the 1870s. Between 1873 and 1877 huge swarms of these insects were present, with eyewitness accounts describing clouds of grasshoppers so dense as to darken the sun. One fairly well-documented swarm occurred in 1874 and covered an area of over 500,000 km², an area a bit smaller than Texas but larger than California. If this report is accurate (and there is every reason to believe that it is), this is the largest concentration of non-microscopic animals ever recorded, with an estimated 12.5 trillion individuals.

(continued)

The Rocky Mountain locust swarms during this period were devastating to the people and farmers of the Great Plains. Wherever the locusts alighted, crops were consumed to the ground. The consequences were dire and resulted in the largest food relief effort in the United States at the time.

Then a very strange thing happened—the swarms ended. Over the next 20 years small outbreaks were seen, but none remotely approached those of the 1870s. A few years later, in 1902, the last living Rocky Mountain locust was recorded, and since then, not a single one has been seen. It abruptly disappeared, making it the only pest insect in history to have become extinct.

The cause of this extinction is unknown, but there are speculations. Jeffrey Lockwood at the University of Wyoming has proposed that during the late 1800s, the breeding grounds of these locusts, where eggs were laid, were incidentally destroyed during the course of rapid settlement expansion. These breeding grounds were areas of midgrass and tallgrass prairies, particularly in mountain valleys—lands that were subjected to intensive agricultural land development and massive cattle grazing (following the extirpation of the previous dominant grazing species, the bison). If this explanation is true, then the Rocky Mountain locust lost out to habitat destruction on a massive scale. Although very few mourn the passing of this insect, it causes one to wonder what else was lost during this period of western US history.

FIGURE 9-13

Life cycle of the Rocky Mountain locust, *Melanoplus spretus*. This species formerly produced immense swarms that devastated large swaths of Great Plains, with the last peak being between 1873 and 1877. When this figure was drawn for a USDA publication in the 1950s, the insect had been extinct for nearly 50 years. Illustration by Arthur Cushman/USDA Smithsonian Entomology Laboratory.

FAMILY STENOPELMATIDAE— JERUSALEM, SAND, OR STONE CRICKETS

Jerusalem crickets likely have the most bizarre appearance of all the North American orthopterans. They are large flightless insects (up to 5 cm) with spiny legs. They have bulbous, banded abdomens, but the most noticeable feature is their very large, round, and somewhat humanlike head. These odd insects commonly attract attention when encountered, and they have a host of common names, including "child of the earth," "potato bug," "skull head," and "old bald-headed man." They are found in most of the western states and in Mexico.

The front legs of Jerusalem crickets are thickened, allowing them to dig, and they have large and powerful jaws. They are omnivorous and will prey on other insects, scavenge, and chew roots and tubers. Despite some widespread legends, Jerusalem crickets are not poisonous, although they can pinch with their mandibles if handled. When disturbed they can also produce a hissing noise by rubbing their legs along the sides of their abdomen.

Mating calls are produced by drumming the soil with their abdomen, which produces vibrations through the soil detected by members of the opposite sex. Recently the number of described species of Jerusalem crickets has greatly increased with the recognition that most species cannot be distinguished by external features (i.e., they are **cryptic species**). However, the songs produced by drumming are unique to each species, and these can be used to separate many of the more than 20 Jerusalem crickets now known to occur in the southwestern United States.

FIGURE 9-14
A Jerusalem cricket. Photograph courtesy of Jim Kalisch/ University of Nebraska.

FIGURE 9-15
Front view of a Jerusalem cricket. Photograph courtesy of Tom Murray.

FAMILY TETTIGONIIDAE—KATYDIDS

Katydids possess hair-thin antennae often much longer than the insect's body, a feature that sometimes lends them the name "longhorned grasshopper." Their legs tend to be more delicate than shorthorned grasshoppers, and they also show more prominent differences between the sexes. The female has a prominent ovipositor (swordlike in some species) used to insert eggs into plants or the soil. However, some of the common species lay flat eggs resembling overlapping fish scales on the surface of twigs.

The majority of katydids live in trees or shrubs and have coloration that allows them to blend well with the vegetation. As a result, katydids are much less

frequently noticed than grasshoppers, although males can make loud, sometimes harsh-sounding, mating calls that announce their presence. Sound production by katydids involves stridulation using a "**file and scraper**" associated with the wings. The scraper is a sharp edge usually found on the front edge of the upper wing, which is drawn across a series of small ridges (file) on the front edge of the opposite wing. Katydids usually sing at night, and local groups may synchronize their songs. Sometimes two synchronized groups will alternate the mating call, with one producing the "katy did" call and the other responding "katy didn't" call. The net effect of such communication is a rhythmic strumming call that gives them their name. (This is characteristic of the true katydid, *Pterophylla camellifolia,* found in the northern and eastern United States.) Katydids hear these songs using the tympana located on the tibia of their front legs.

FIGURE 9-16
The true katydid, *Pterophylla camellifolia.* Photograph courtesy of Tom Murray.

FIGURE 9-17
In katydids, the auditory organ used to detect songs is located on the foreleg at the upper part of the tibia. Photograph by Whitney Cranshaw/Colorado State University.

One of the oddest katydids found in North America is the Mormon cricket (*Anabrus simplex*). It occurs over a broad area of the central and western portions of the United States but is most abundant in the intermountain regions of Nevada, Utah, southern Idaho, and western Colorado. Like locusts, the Mormon cricket also exhibits a form of phase polymorphism. The solitary phase

occurs predominantly within the Rocky Mountains and eastward. These insects occur at very low population densities and are bright green in color. West of the Rocky Mountains, the insect occurs predominantly in the dark-colored gregarious phase, periodically producing very high population densities (over 50 per m^2).

While in the gregarious phase, Mormon crickets will band together and move in massive swarms. However, they are flightless and travel by walking, running, and jumping. As with plague locusts, Mormon cricket swarms can exceed millions of individuals and can become quite serious pests in local areas. Swarms crossing highways have been known to temporarily slow automobile traffic when the road becomes slick with the crushed bodies of road-killed crickets.

The common name "Mormon cricket" is derived from their appearance in late May 1848 in the early Mormon settlements around Salt Lake City. Large swarms of this "black and baleful" insect threatened the critical first year crops of the settlements, and their numbers and damage reminded the pioneers of the plague locusts referred in the Bible. Fortunately, serious damage was averted, in part by flocks of California gulls from the lake that consumed large numbers of crickets. A commemorative statue to the gulls was later erected in Temple Square in Salt Lake City. On the other hand, the Native Americans of the Great Basin region once regarded Mormon cricket outbreaks as a boon—they represented large easily taken amounts of high-quality protein and fat.

FIGURE 9-18

A Mormon cricket, *Anabrus simplex*, in its solitary phase. The Mormon cricket is flightless but can undergo phase change and produce both solitary and gregarious forms. Photograph courtesy of Nathan W. Bailey.

FIGURE 9-19

Gregarious phase Mormon crickets on a rangeland shrub. Mormon crickets in the gregarious phase are dark colored and band in groups. Photograph courtesy of Nathan W. Bailey.

FIGURE 9-20

A Mormon cricket band crossing a highway. Photograph courtesy of Paula Krugerud.

● Temperature Forecasting with Crickets

Most insect activity is dependent on ambient temperature, including the sound production of crickets and katydids. For crickets to sing, the temperature usually needs to be above 12°C–13°C and below 38°C. Within that range, there is some association between temperature and activity, with increased temperature leading to increased frequency of cricket chirping.

This relationship was first quantified by Amos Dolbear, a physicist and inventor who taught in Ohio. (Among his many inventions were a telephone receiver

FIGURE 9-21

The snowy tree cricket, *Oecanthus fultoni*, was subject to an early study that showed that the rate of chirping was predictably correlated with air temperature. In his 1897 paper "The Cricket as a Thermometer," Amos Dolbear developed an equation for predicting air temperature by counting the number of chirps produced in a set time interval. Photograph courtesy of Tom Murray.

using a permanent magnet, preceding that of Alexander Graham Bell, and a wireless telegraph as a means of communication over considerable distance, preceding that of Guglielmo Marconi.) Rather late in his career, he turned his attention to the song produced by a local cricket, *Oecanthus fultoni* (probably), known as the snowy tree cricket. His article *The Cricket as a Thermometer*, published in the 1897 issue of *American Naturalist,* was the first to note the specific effect of temperature on cricket chirp frequency.

The essential equation he proposed, since known as **Dolbear's Law**, is as follows: $T_F = 50 + (N - 40)/4$, where T_F is the temperature in Fahrenheit and N is the number of chirps in 1 min. (This can be converted to the simpler equation $T_F = 50 + N$, where N is the number of chirps in 15 s. The metric conversion for temperature in Celsius (T_C) is $T_C = 10 + (N - 40)/7$.)

Not all crickets show such a direct relationship of chirping to temperature. Also, male field cricket singing can be affected by other factors, such as the length of time since mating.

(Unmated crickets will sing increasingly strenuously with time.) Formulae have been identified for other species as well, such as the common true katydid (*Pterophylla camellifolia*), which has a song chirp rate related to temperature that is approximated by $T_F = 60 + (N - 19)/3$.

These familiar examples help illustrate how insect activity is related to temperature and they easily capture the public imagination. The use of communicating sounds by insects is extremely widespread; although all except those made by certain particularly loud species

(continued)

are missed by the human ear, because of our limited auditory range. These can only begin to be fully appreciated by recording methods that can capture these sounds and produce observable "voiceprints." Studies of acoustical production of insects have shown that the miniature world that they inhabit can be a very noisy place. Studies of sound production have also identified many new species that cannot be distinguished by physical features (**cryptic species**) but produce unique songs. Furthermore, identification of other sounds generated by insect activity, such as those produced during feeding or tunneling, has allowed ways to better detect their presence in materials where they are hidden from view.

FAMILY GRYLLIDAE—CRICKETS

Crickets share many features with katydids, including the long, thin antennae and similar structures for sound production and perception. Some species are adapted to living in trees and shrubs, with an ovipositor designed for inserting eggs into plant stems. Female ground crickets, field crickets, and house crickets have a long, thin, straight ovipositor used to insert eggs into the soil.

Crickets produce the most musical songs of any insect, and each species of cricket has its own unique repertoire. Although to our ears they may sound lively and joyful, the songs' purpose is very serious; males are attempting to attract potential mates. In the common black-colored field crickets (*Gryllus* spp.), this may involve a series of songs, with the "calling song" the most familiar. When a female chooses to approach, the male shifts to a different "courtship song." If things continue in the right direction, the male will ultimately produce a "staying together song," and, from the glands on his back, he will evert a substance that the female feeds upon. Field crickets also produce rivalry songs that warn other males intruding on their territory.

Interestingly, some parasitic flies have evolved the ability to eavesdrop on cricket courtship songs. These flies detect and locate male crickets by their songs. Once located, the fly pounces on the male cricket and lays an egg on it. The fly egg hatches and the larva develops inside the cricket, eventually killing it.

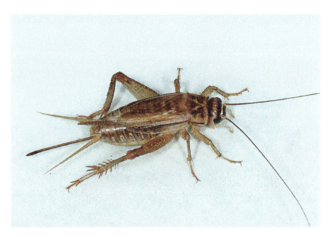

FIGURE 9-23
A house cricket, *Acheta domestica*. This European species has long been widely distributed in North America as a staple of the pet food trade. Photograph courtesy of Tom Murray.

Crickets are also important from a commercial standpoint, being widely used as fish bait and food for pets and zoo animals. The sole commercial species reared in the United States has been the European species known as the house cricket (*Acheta domestica*) and tens of millions are produced weekly across the country. In recent years, the producers have faced serious problems with the introduction and spread of a new disease caused by the cricket paralysis virus, which has devastated many cricket-rearing operations. As a result, alternate species of crickets are being considered for use in sustaining this large insect-based industry.

FIGURE 9-22
Field crickets, *Gryllus* spp., are some of the most common and familiar of the singing insects of summer, making melodious songs. Photograph courtesy of Tom Murray.

FAMILY ANOSTOSTOMATIDAE—WETA AND KING CRICKETS

Weta and king crickets include several of the largest and heaviest orthopterans. They are almost entirely restricted to the southern hemisphere, being most abundant in South Africa, parts of South America, Australia, and New Zealand. In the latter region, they are considered an iconic species along with the kiwi and silver tree fern, and the name "weta" is derived from a Maori word used to describe a particularly large New Zealand species (*Deinacrida heteracantha*)—"wetapunga," the embodiment of ugly and evil things. (The genus name for the insect, *Deinacrida*, means "terrible grasshopper.")

In New Zealand over 70 species are thought to occur (many undescribed) and they are popularly known as giant weta (*Deinacrida* spp.), tree weta (*Hemideina* spp.), ground weta (*Hemidranus* spp.), and Mercury Island tusked weta (*Motuweta isolata*). (Another commonly recognized group in the country is "cave weta." These belong in the family Rhaphidophoridae, known as the cave or camel crickets in the United States.) Developing in a region that was isolated 73 mya with the breakup of the supercontinent Gondwanaland and was mammal-free (except for bats) before humans arrived, the weta came to occupy many ecological niches that are filled elsewhere by small rodents. The giant weta and tree weta are primarily herbivorous, feeding on leaves, fruit, and seeds, occasionally scavenging on dead insects; ground weta are primarily predators of other insects.

FIGURE 9-25

Populations of tree weta, *Hemideina* spp., have been greatly reduced in New Zealand, primarily due to habitat loss. The availability of tree holes and other cavities where they can safely hide during the day is particularly limited in some areas. As a result, "weta motels" of various designs have been developed. These are designed to provide the types of cavities a hollow log or small tree cavity might provide. Photograph by Whitney Cranshaw/Colorado State University.

Extreme size is a common feature among the giant weta, with some exceeding 10 cm in length and 30 g in weight—considerably larger than many mice. (A pregnant female in captivity was once weighed at 71 g, making it arguably the heaviest insect of record.) Many of the tree weta are also impressively large and show a form of sexual dimorphism where adult males develop grossly enlarged heads with long mandibles used as weapons. Unfortunately, many of the larger weta have fared very poorly since humans began arriving about 750 years ago, bringing along a host of animal predators that have transformed the ecology of the region—various rats, mice, feral cats, and weasels being particularly destructive. Presently most giant weta can be found only in a tiny fraction of their original range, primarily on small islands that have not been colonized by rats and other predators. Tree weta populations have also been decimated and conservation efforts are in place including the employment of artificial tree cavities

FIGURE 9-24

A male tree weta, *Hemideina ricta*. Several genera of weta live in New Zealand with the giant weta, *Deinacrida* spp., reaching immense size. Although slightly smaller, the various tree weta (*Hemideina* spp.) are also impressively large and produce adult males with grossly enlarged heads and jaws. However, all the larger species of weta have been decimated since the introduction of rats, and the largest species of giant weta only survive on outlying islands of New Zealand that are rat-free. Photograph by Whitney Cranshaw/Colorado State University.

("weta motels") that some species use as a refuge. However, a few of the tree weta have adapted fairly well to human-altered habitats that provide wood piles and various types of suitably sized (and rodent-proof) cavities within which they rest during the day. Species that may be encountered in yards and gardens include the Wellington tree weta (*Hemideina crassidens*) and the Canterbury tree weta (*H. femorata*).

The term "king cricket" is typically used to describe anostostomatids that occur outside New Zealand. In Australia some 70 species are thought to occur, although only about a quarter of these have yet been formally described. They are scavengers that feed on the ground and occur in rural areas where they are rarely encountered. The largest species (*Anostostoma australasiae*) is 7–8 cm in length.

At least two dozen species of king crickets occur in South Africa, mostly in drier areas. As with the tree weta, adult males of some species develop enormous jaws, although they are only weakly muscled. South African species are predators of snails or slugs and scavengers. One king cricket known as the "Parktown prawn" (*Libanasidus vittatus*) has thrived in human-altered environments and is now a common garden resident in the Johannesburg area where its snail feeding is considered beneficial. They are often 4–5 cm long (sometimes considerably larger), possess very long antennae, and are strong jumpers—unlike most members of the family.

FAMILY GRYLLOTALPIDAE—MOLE CRICKETS

Mole crickets are highly adapted to subterranean life. Their **fossorial** front legs are greatly enlarged and have bladelike projections (dactyls) that aid in tunneling. The general body form is rather streamlined and hairy, and their antennae are short. Some are winged, while others are wingless. Seven species of mole crickets occur in North America, with the majority accidentally introduced from South America or Europe.

Three introduced *Scapteriscus* species now found in the southeastern and south central United States occasionally cause problems to lawns and other plantings. The most damaging is the tawny mole cricket (*Scapteriscus vicinus*), which chews plant roots, makes extensive tunnel systems just below the soil surface, and creates soil mounds during mating season. Less damaging, but present over a wider area, is the southern mole cricket (*Scapteriscus borellii*), a predator of other insects. Both are frequently attracted in large numbers to lights during mating season.

FIGURE 9-26

A northern mole cricket, *Neocurtilla hexadactyla*. Mole crickets spend most of their life tunneling through soil and have well-developed forelegs that enable digging. Photograph courtesy of Natasha Wright/Florida Department of Agriculture and Consumer Services/Bugwood.org.

Adults emerge shortly after dusk on warm, clear nights, usually following rain. To attract mates, males dig a trumpet-shaped tunnel in the soil surface and begin to stridulate, rubbing the forewings, with a file and scraper, together. The call of the tawny mole cricket is a buzzing sound, while that of the southern mole cricket is more musical, and the acoustically designed burrow greatly amplifies the calls. Females perceive the songs via the tympanum located on the foreleg tibia.

Order Phasmatodea— Walkingsticks and Leaf Insects

Entomology etymology. The order name is derived from the Greek word "phasm" meaning phantom, apparition, or spirit and refers to the cryptic habit and appearance of these animals. The common names "walkingsticks" and "leaf insects" refer to their ability to mimic sticks, twigs, and leaves. Collectively all may also be called phasmids.

In no other insect order is cryptic appearance so widespread as in Phasmatodea. Most phasmids look like either twigs (walkingsticks) or leaves (leaf insects) and are masters of camouflage. A few species are even able to change their color over the course of a day, enabling them to match their background

FIGURE 9-27

Walkingsticks may have fully functional wings (left) or nonfunctional wings (center) or lack wings altogether. The phasmids known as leaf insects (right) are winged and have a broad body that resembles a leaf. Photograph by Whitney Cranshaw/Colorado State University.

color or adjust temperature by absorbing more or less light. Currently there are more than 3,000 described species worldwide with 32 species occurring in four families within North America. Most species are found in the tropics.

Most phasmids are wingless or **brachypterous**, possessing wings reduced in size that are non-functional for flight. Others may have normal-sized membranous hindwings, but the front pair is reduced. The wings may be brightly colored and used to flash in a startle response to deter predators.

FIGURE 9-28

Diapheromera femorata is the most commonly encountered of the walkingsticks found in the United States. It feeds on the leaves of various trees and, on occasion, achieves populations that can produce noticeable defoliation of trees. Photograph courtesy of James Solomon/USDA Forest Service/Bugwood.org.

Their legs are designed for walking, but most members of the order move slowly and deliberately. If legs are lost, they can regenerate them during later molts, an ability that no other insect order is capable of doing. Most phasmids are large insects, and the order includes the world's longest insects. Measured by body length, *Phobaeticus kirbyi* of Borneo can be 32.8 cm in length from its head to the tip of the abdomen. When measuring the outstretched insect with front legs extended, the current record holder is the Malaysian species *Pharnacia serratipes*, which reaches 55.5 cm.

Walkingsticks and leaf insects have chewing mouthparts and all are phytophagous, feeding on leaves. Very few cause any significant plant injury, although the common walkingstick of the eastern United States, *Diapheromera femorata*, is an example of a species that occasionally causes significant defoliation of forested areas. Reproduction in many species is by parthenogenesis with males entirely unknown. Other phasmids have sexual reproduction but usually exhibit **sexual dimorphism**, with males being much smaller than the females.

The obvious main defense of leaf and stick insects is **crypsis**, appearing and behaving like twigs and leaves of their host plants. Moving slowly and swaying like a leaf in the breeze is a common behavior among walkingsticks which enhances their ability to avoid detection. Not only do these active insects appear as plant parts, but also their eggs deposited by the female mimic plant seeds. Indeed, this mimicry

FIGURE 9-29
Sexual dimorphism is quite extreme in many walkingsticks, with females being much larger than males. Differences in color and wing development may further distinguish some species. Photograph courtesy of Tom Murray.

FIGURE 9-30
Some walkingsticks are chemically defended. Walkingsticks in the genus *Anisomorpha* will eject a strongly scented irritating fluid when disturbed. Photograph courtesy of Hebert A. "Joe" Pase III/Texas Forest Service/Bugwood.org.

is so complete that certain phasmid eggs are collected and dispersed by ants that mistake the eggs for seed.

Other defenses include playing dead (**catalepsy**) and falling from their host plant. If grasped, some of the larger species can produce a painful kick with spiny legs. Finally, all members of the order possess two glands on the thorax and are capable of producing a caustic fluid when the insect is disturbed. This is often released as oozing droplets to cover parts of the body, but some species can forcefully spray the fluid, which can cause burning and temporary blindness if the eyes are hit.

Variety Is the Spice of Life—Some Minor, but Interesting, Insect Orders

Stoneflies, Webspinners, Angel Insects, Earwigs, and Rock Crawlers

There are 17 orders of insects with hemimetabolous development. Several of these fall into obvious groups of close relatives with many shared physical features. The orders containing grasshoppers and walkingsticks (Orthoptera and Phasmatodea) are one such grouping. Dragonflies and damselflies, and mayflies (Odonata and Ephemeroptera) have many similar features, including the way their wings fold and an aquatic immature stage. Another example would be the mantids, cockroaches, and termites, which are so similar that they are sometimes classified together in a single order, the Dictyoptera. Then there are those sometimes referred to as "minor orders," many of which are featured in this chapter.

Order Plecoptera—Stoneflies

Entomology etymology. The order name is a combination of the Greek words "pleco," meaning folded or plaited, and "ptera," winged, in reference to their ability to fold the large hindwings over their bodies and under their forewings. The common name "stoneflies" refers to the habitat of many nymphs being found under rocks or among small stones in flowing water. Collectively, these animals are sometimes referred to as plecopterans.

Stoneflies are often considered to be one of the "truly aquatic" orders (along with Ephemeroptera, Odonata, and Trichoptera) with immature stages that are restricted to development in water. (Rare exceptions sometimes occur. For example, the curious wingless stonefly genus *Holcoperla* found in New Zealand has terrestrial habits with immature stages that crawl about and hunt prey among damp vegetation and other moist sites in alpine regions.) As a group, stoneflies are also considered to be among the most sensitive species to poor water quality, and they are only ever abundant in running streams and rivers that are well oxygenated, clean, and cool. Worldwide, about 2,000 species are described, with 540 in North America.

Adults have an elongated body form with membranous wings that fold over the abdomen. They have long filamentous antennae and two large cerci ("tails") at the tip of the abdomen. Most adults hide during the day among streamside vegetation and are rarely seen. When disturbed, stoneflies can be difficult to observe as they are fast moving and rapidly flee to cover for protection.

The largest stoneflies are in the genus *Pteronarcys* (*P. dorsata* in the eastern United States and *P. californica* in the west). Large females from the head to the tip of their

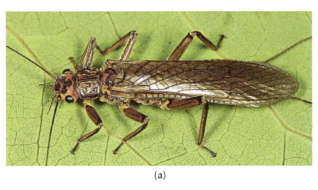

(a)   (b)

FIGURE 10-1

Examples of two stoneflies: (a) *Acroneuria abnormis* (Perlidae family); (b) *Isoperla dicala* (Perlodidae family). Photographs courtesy of Tom Murray.

abdomen may be 50–60 mm in length. Adults of these species, well known as "salmonflies" to anglers, emerge in late spring or early summer and are important sources of food for fish and birds. Salmonflies are

FIGURE 10-2

Salmonflies, *Pteronarcys* spp., are the largest of the North American stoneflies. Photograph by Whitney Cranshaw/ Colorado State University.

typically found in medium- to small-sized streams, and they are somewhat more tolerant of warmer temperatures than most other stoneflies.

Immature forms (nymphs) live in water and have a generally flattened body, and each leg bears two claws, features that allow them to cling to surfaces in fast-moving waters or sprawl on the bottom. At the tip of the abdomen are two tails, a feature that is particularly useful in separating them from most of the other aquatic insects. Unlike the mayfly nymphs, their gills are inconspicuous and clustered on the underside of the thorax, often around the base of the legs. Wing pads become apparent on older nymphs. A univoltine life cycle (completed in 1 year) is typical, but some species living in very cold waters may have considerably longer life cycles that require years to complete. Most stoneflies are well adapted to developing at low temperatures, some continuing to grow in waters that are barely above freezing. Often, it is the occurrence of warm water temperatures that cause many stoneflies to temporarily cease their development and go into diapause.

(a)   (b)   (c)

FIGURE 10-3

Examples of the nymphs of three stoneflies: (a) *Pteronarcys proteus* (Pteronarcyidae family); (b) *Isoperla montana* (Perlodidae family); (c) *Acroneuria carolinensis* (Perlidae family). Photographs courtesy of Tom Murray.

The majority of stonefly nymphs feed by chewing small pieces of living or dead plant material (shredders) or by grazing upon algae. A great many are predators of small insects, and some switch from a plant diet to an insect diet in later stages of development. Stonefly nymphs are well adapted to crawling about in running water but can swim by lateral undulations of the body. To avoid predation by fish, they may curl up and play dead or produce distasteful chemicals through their joints, a behavior called **reflex bleeding**.

When fully grown, nymphs crawl out of the water onto rocks or along the water's edge and molt to the adult stage. As they do so, they leave behind the shed exoskeleton—their distinctive "shuck." The adults usually rest during the day among leaves or on tree trunks close by the water. During their relatively brief life span (1–4 weeks), adults drink, and most feed a bit on soft plant materials (buds, tender leaves, blossoms, pollen).

The primary activities of the adults involve mating and egg production. Adult stoneflies communicate through distinctive "drumming" produced either by tapping the tip of their abdomens on the surface of their resting substrate (e.g., a tree branch) or by vibrating the abdomen itself. Each species, and each sex of each species, makes its own distinctive drumming pattern; males and females find one another by homing in on specific drumming patterns. After mating, the female deposits loose masses of eggs, usually directly onto the water.

FIGURE 10-4

A winter stonefly, *Taeniopteryx burksi*. Winter stonefly adults emerge in winter and nymphs may go dormant when water warms in summer. Photograph courtesy of Tom Murray.

In the two families of stoneflies known as "winter stoneflies" (Capniidae, Taeniopterygidae), the entire life history is seasonally reversed. Dark-colored adults emerge from icy streams in the middle of winter and are active during broad daylight. They are often seen crawling about the ice or resting on exposed rocks to warm themselves. Some may spend their entire adult life underneath the ice in cracks and crevices just above the flowing water. Eggs are laid during this period and the larvae develop until stream temperatures rise. As the water temperature increases, these immature insects enter diapause, becoming metabolically inactive. Only when water temperatures again drop in autumn do the nymphs resume development.

● Insects as Water Quality Indicators

The protection and improvement of surface waters (streams, lakes, and ponds) is a common environmental goal, and terrestrial waterways are routinely monitored to assess their health. Various chemical analyses of water samples are frequently used in water quality testing. In addition, monitoring the presence of various aquatic insects gives one a unique perspective on the water quality history of a river, stream, lake, or pond.

One reason that insects are useful in this regard is that many are very sensitive to water conditions. Some species only survive in highly oxygenated waters. Low water flows or organic matter pollution may deplete oxygen. Siltation can interfere with gill function, and most species have a temperature range outside of which

they cannot survive. Aquatic insects can also be sensitive to heavy metal pollutants or changes in alkalinity (pH).

Another factor making aquatic insects valuable as water quality indicators is that they cannot escape. A transitory pollution event that may be missed during water testing could devastate an insect population, providing a record of relatively long-term effects. By knowing the relative sensitivity of different insect groups to pollution, indices of water quality can be established for a waterway. Most of these are some derivation of the EPT Index, which measures the relative number of different kinds of mayflies (Ephemeroptera), stoneflies (Plecoptera), and caddisflies (Trichoptera).

(continued)

FIGURE 10-5

Examples of some aquatic insects that may be indicators of high water quality: (a) *Ephemerella* spp. mayflies (Ephemerellidae family); (b) winter stoneflies (Capniidae family); (c, d) larvae and adults of free-living caddisflies (Rhyacophilidae family). Photographs courtesy of Tom Murray.

Aquatic insects particularly sensitive to organic pollution include virtually all species of stoneflies, some mayflies (particularly some in the Ephemerellidae and Leptophlebiidae families), and many of the caddisflies (Glossosomatidae and Rhyacophilidae families). These insects are only found in water bodies with sustained high water quality.

Where water quality has begun to deteriorate, highly sensitive species disappear and are replaced by moderately pollution-sensitive species. These include many mayflies (Baetidae, Ephemeridae, and Tricorythidae families), caddisflies (Hydropsychidae, Leptoceridae, and Limnephilidae families), and aquatic species of crane flies (Diptera order, Tipulidae family).

(continued)

The common burrowing mayflies (Ephemeridae family), for example, are quite sensitive to the effects of organic pollution, which reduces oxygen levels, and to chemical pollutants concentrated in the sediments where they live. Due to the effects of pollution, most of these mayflies were largely eliminated from large tracts of the upper Mississippi River and western Lake Erie by the 1950s. Following controls on pollution and improved water quality, populations of burrowing mayflies have rebounded in these waters.

Where water quality has degraded further, only a few types of mayflies may be found (Caenidae and Siphlonuridae families) as well as various types of true flies such as midges (Ceratopogonidae family and most members of the family Chironomidae) and black flies (Simuliidae).

Streams or rivers in the worst condition may be able to support only the blood-red midges of the Chironomidae family—or no insect life at all.

FIGURE 10-6
The blood-red midges of the family Chironomidae are among the aquatic insects that are most adaptable to poor-quality water with low oxygen content. They are one of the few insects that produce hemoglobin in their blood, which allows them to store oxygen and produces their distinctive reddish color. Photograph by Whitney Cranshaw/Colorado State University.

Order Embiidina— Webspinners

Entomology etymology. The original order name "Embioptera" is derived from the Greek words "embio," meaning "lively," and "ptera," meaning wing, in reference to the way males fly in a fluttery manner. Most taxonomists now prefer the name Embiidina or Embiodea as the name of the order.

Webspinners are one of the smaller orders of insects with only about 360 described species, most of which are found in the tropics. Only 11 species occur in the United States, all in the more southerly areas. They are small insects, typically around 10 mm in length when fully grown, with a cylindrical body. Webspinners may be locally abundant, and the winged males are commonly attracted to lights.

Webspinners live in silk-lined burrows and most feed on dead plant matter. Their legs are well adapted for their burrowing habit and the hindlegs are enlarged so that they may be rapidly run backward in retreat through their burrows. Indeed, webspinners can run quickly both forward and backward within their tunnellike silken tubes. The most unique feature of the order Embiidina is the large silk glands housed in the enlarged basal segment of the tarsi on the front legs. (Silk glands in most other insects are located in

or near the mouth or anus.) Through specialized hollow setae (or hairs), the silk is drawn out and used to create the silk-lined interconnecting tunnels or galleries in which they live. Typically, these galleries are found in tree bark, or under rocks, logs, or large leaves.

FIGURE 10-7
Female webspinner, *Haploembia* sp., in its silk-lined tunnel. Photograph courtesy of Alex Wild.

There are major differences in the appearance and behavior of the adult males and females. Females are wingless, while the males possess membranous wings. Both sexes are uniquely adapted to living in narrow tunnels. The body is not heavily sclerotized, and its

FIGURE 10-8
Webspinner adult and nymph. Photograph courtesy of S. Dean Rider, Jr.

shape is largely maintained by the pressure of the hemolymph pressing outward on the fairly flexible exoskeleton. Because of this flexibility, the wings on the male may fold forward as it crawls backward in the burrow. Adult males also do not feed and instead use their mandibles solely to chew into the galleries of females and to grasp them prior to mating.

FIGURE 10-9
Winged male webspinner, *Oligotoma* sp. Photograph courtesy of S. Dean Rider, Jr.

Female webspinners show an unusual level of maternal care. Eggs are laid within the gallery of the mother's silken burrow. The adult female then covers her eggs with a protective paste of chewed plant matter and feces. When the eggs hatch, the female remains with the nymphs during the early stages of their development. During this time, she guards her young against ants and other predators.

Order Zoraptera— Zorapterans, Angel Insects

Entomology etymology. The order name is derived from the Greek words "zor" (pure) and "apteron" (without wings). When this order was described, only wingless individuals were known and it was thought to be a distinctive feature of this order. We know now that is not the case.

Zoraptera is one of the least diverse and most poorly studied of the insect orders. The order contains only a single family (Zorotypidae) and 32 known species. The majority of the described species are tropical and only three occur anywhere in the United States.

Zorapterans are small (ca. 3 mm), elongate, pale-colored insects. They have long bead-like antennae much like those of termites they somewhat resemble. They are almost always found in wood cavities or sawdust piles, where they typically occur as small colonies of 15–200 individuals. Blind, wingless forms predominate, and they feed on fungi, nematodes, mites, and perhaps dead arthropods. Zorapterans undergo simple metamorphosis with four instars suspected.

FIGURE 10-10
A zorapteran. Photograph courtesy of David Shetlar/The Ohio State University.

When overcrowded, or when food is depleted, dark pigmented winged forms, possessing light-sensing ocelli and compound eyes, are produced. These disperse to new areas, and, once settled, drop their wings in a manner similar to winged ants and termites. *Usazaros hubbardi* is the only species found broadly through North America, occurring in 33 states of the central and southern United States.

Order Dermaptera— Earwigs

Entomology etymology. Dermaptera means "skin wing," a reference to the insect's leathery forewing. This is not particularly descriptive as it was also originally used to describe other orders of insects with a similar forewing, such as grasshoppers and crickets (order Orthoptera) or mantids (order Mantodea). Collectively, members of this order are known as dermapterans.

The common name "earwig" is derived from the old English "ear wic," meaning "ear beetle," from the folktales that they would enter the ears of sleeping people (an extremely rare event). In other languages, common names reflect the prominent cerci present on their hind ends, such as *tijerata* in Spanish and *forficola* in Italian, words that roughly translate to "scissors."

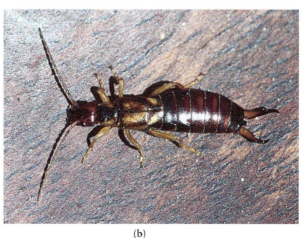

(a)   (b)

FIGURE 10-11

Male (a) and female (b) of the European earwig, *Forficula auricularia*. Male earwigs have cerci that are more bowed than those of females. Photographs by Whitney Cranshaw/Colorado State University.

Earwigs are among the more easily recognizable of insects. They are mostly small- to medium-sized insects with an elongated, dark-colored body. Most have winged adult forms, with the hindwing tightly folded underneath the short, semicircular forewing. The head and chewing mouthparts project forward.

The most conspicuous character, though, is the pair of prominent, unsegmented, pincerlike cerci on the tip of the abdomen. Worldwide, they are a moderately diverse order of about 1,900 species. Only 22 species are found in North America, and those most commonly encountered are European introductions.

● Bugs in Ears

In much of the English-speaking world, earwigs elicit undue fascination because of fears that they enter ears. That these concerns have very little basis in fact has become a casualty of long-standing urban legends. Indeed, the association of earwigs with ears is even elevated to a verb meaning "to fill the mind with prejudice by insinuations" or "to attempt to influence by persistent confidential argument or talk."

Of course, many insects can and do end up in ears by accident. One of the more remarkable episodes occurred during the 1957 Boy Scout Jamboree in Valley Forge, Pennsylvania, with a mass emergence from the soil of adult Japanese beetles (*Popillia japonica*), a type of scarab beetle. Several of the emerging beetles temporarily (and without lasting damage) found their way into the ears of sleeping scouts. But the insect

(continued)

culprit most likely to be found in an ear—by far—is a cockroach. In cities where cockroaches abound, it is not uncommon for these to be the cause of an emergency room visit, especially for young children and infants. They are easily extracted by medical staff (oil will drown them effectively, but lidocaine will cause them to back out and exit the ear canal), and while the experience is often traumatic—and can be painful—there is no serious injury. Insects can penetrate no further than the outer ear canal, as the hardest bone of the human body lies in the skull immediately beyond.

Regardless, fears of arthropods invading bodily orifices have a long history and are well exploited in fiction. A classic of short story horror was *Boomerang* (1931) by Oscar Cook. This tale featured an assassination attempt (which went terribly awry) using a caterpillar put into the ear of a sleeping victim. This story was resurrected in one of the most memorable of the *Night Gallery* television shows, *The Caterpillar* (1972). A decade later it was earwigs again (or was it antlions?), in the movie *Star Trek II: The Wrath of Kahn* (1982), when mind-controlling earwigs are introduced into Chekov's ear.

And if it is not the ear, it is some other body opening that people will obsessively worry about. Currently one of the more widespread urban legends has spiders entering the mouth as one sleeps—an absurdity that unfortunately gets wide traction in our current age of high-speed miscommunication.

Despite their rather fearsome appearance, earwig cerci are weakly muscled and do not produce a very effective pinch to a human. (The jaws are usually much stronger and bites may be felt.) Many earwigs use the cerci to help manipulate food, including insects, as the body is very flexible. Some earwigs use the cerci during mating, with those of the male modified to help grasp the female. The cerci may also be used to draw the wings out in preparation for flight.

Another common earwig behavior is their propensity to get into things. Tight, dark, somewhat humid spots are favored day-time resting areas. Such behavior is known as a **thigmotactic** response—often described more dryly as "searching close to surfaces" but a bit more elegantly as having a "propensity for pressure." Earwigs may also show some aggregation behaviors, so that often several can be found clumped together. The combination of numbers, unexpected encounters, and their pincerlike cerci makes earwigs among the least popular of insects.

Some are primarily phytophagous, many are predators, but most are omnivores. The latter feeding behavior is well demonstrated in the European earwig (*Forficula auricularia*), an insect accidentally introduced from Europe that is common in many gardens in the northern half of the United States. Although it will commonly feed on flower petals and tender leaves, it also eats a great many aphids and other garden pest insects. Several other characteristic features of the order are exhibited by this species, including the maternal care of the young—a high order of "family values" in the insect world. Some earwigs lay eggs, while other produce live young; regardless, the female stays with the young through at least the first molt.

The mammoth among the earwigs appears to have recently gone extinct. The St. Helena earwig (*Labidura herculeana*) was described in 1798 from the tiny mid-Atlantic island of St. Helena. This earwig reached a size of about 8.4 cm. Last found alive in 1967, expeditions after 1988 have failed to find evidence of surviving populations. Loss of habitat and predation by introduced mice, rats, and centipedes have been proposed to explain the

FIGURE 10-12
European earwigs seek out tight, dark crevices in which they hide during the day. This habit causes them to get into such things as flowers or cracks in fruit, or other similar sites where they may be encountered unexpectedly. Photograph by Whitney Cranshaw/Colorado State University.

FIGURE 10-13
A European earwig tending the eggs. Earwigs often show a high degree of maternal care, not only staying with the eggs but also protecting the newly hatched nymphs for the first couple of weeks of their life. Photograph courtesy of Ken Gray/Oregon State University.

apparent extinction. The St. Helena earwig has sometimes been described as the "dodo of the Dermaptera," a reference to its restricted island distribution, large size, and rapid (probable) extinction following human alterations. How such an unusual species could occur in such an isolated location has also attracted a great deal of speculation. It is quite similar in features to, although much larger than, the tawny/striped earwig (*Labidura riparia*), a species found along coasts throughout much of the world.

FIGURE 10-14
The St. Helena earwig, *Labidura herculeana*. This was the largest earwig, measuring up to 8.4 cm in length. It was restricted to the tiny island of St. Helena in the Atlantic Ocean and is now extinct probably because of predation by introduced animals. Photograph courtesy of the Manchester Museum.

Order Grylloblattodea— Rock Crawlers or Ice Crawlers

Entomology etymology. The order name is derived from the Greek terms for cricket (*gryll*) and cockroach (*blatt*), which reflects the confusion that surrounded the discovery of these insects and their unusual mixture of features. In many new taxonomic arrangements the rock crawlers are combined with the heelwalkers (Mantophasmatodea order), along with some extinct species known only from fossils, into the order Notoptera.

Rock crawlers are among the most rarely encoun. tered insects in North America. Only 25 species are known worldwide (with 10 of these being found in North America), making this group the second smallest order, only eclipsed in this regard by the recently described order Mantophasmatodea.

FIGURE 10-15
A rock crawler feeding on the surface of snow. Photograph courtesy of Alex Wild.

Although they lack jumping legs, rock crawlers probably most closely resemble pale-colored wingless crickets. They have an elongated, cylindrical body, a relatively large head, and long antennae. Eyes are absent or greatly reduced. Adults range in size from 14 to 30 mm.

The rarity of rock crawlers is due in large part to their habits, which are highly adapted to very cold environments. They are found in cold, temperate forested areas, typically at high elevations, where they live among leaf litter and under stones. The optimum temperature for their activity and development is around 1°C–4°C, well below the temperature when most insects can even move, much less successfully develop. Rock crawlers cannot tolerate warm temperatures and some have been observed to die soon after they are exposed to temperatures exceeding 10°C.

Little is known about the ecology and biology of rock crawlers. They are scavengers, primarily feeding on dead insects and other debris that accumulates in and around snow fields. Development is slow, taking up to 7 years to reach the adult stage.

● Insect Survivors—Meeting the Challenge of Freezing

Excepting a few regions with permanent ice cover, insects and springtails can still be found throughout the coldest sites on the planet. This is achieved by various adaptations to a cold temperature lifestyle that allow them to avoid the damaging effects of freezing.

Freezing poses many risks to all living cells. The most serious effects result from the expansion of water that occurs with ice crystal formation and is capable of producing lethal cell rupture. Severe tissue dehydration may occur with freezing, which draws free water from cells, causing destructive deformation and shrinkage of cells. To meet these challenges arthropods have developed several means to avoid freezing or to tolerate its effects.

Freezing may be avoided if insects have a means to depress the temperature of ice formation, a process known as **supercooling**. One method to achieve this is to produce compounds that increased fluid viscosity, such as polyols (glycerol, sorbitol, mannitol, ethylene glycol) that function as internal antifreeze. Other **cryoprotectants** may prevent freezing damage by inhibiting water binding to cell membranes. Also, since ice forms around particles or certain large molecules that serve as **ice-nucleating agents**, insects may actively eliminate these from their gut or hemolymph prior to freezing events.

Conversely, controlled freezing can allow insects to better tolerate exposure to low temperatures. This often is achieved by the production of proteins that actually function as ice-nucleating agents but allow a gradual freezing, avoiding the damage that occurs with abrupt ice crystallization. These compounds are present in the fluids between cells, and as freezing progresses, water may slowly migrate from within cells, increasing viscosity and depressing the freezing point to avoid internal cell freezing. The familiar banded woollybear (*Pyrrharctia isabella*) is one insect that utilizes this method of freezing tolerance.

The ability to survive freezing typically shows seasonal variation. Environmental cues that indicate the need for cold temperature adaptations—such as decreasingly day length or exposure to cooling temperatures—trigger the internal changes that are involved in making an insect "cold-hardy." As a result even insects that are potentially capable of surviving freezing may succumb when abruptly frozen following a period of warm temperature. This effect is frequently used when attempting to kill insects by freezing, which is usually most effective if the insects are moved to a deep freezer immediately from a warm environment.

Behavioral changes are also employed by insects to avoid being killed by freezing temperatures. The changes of diapause typically begin in late summer or early fall and extend through winter, which largely suspends development and activity. Prior to winter, insects often move to protected sites buried in soil, underneath leaves, behind loose bark, or in other cavities where they are better insulated against freezing temperatures. And a few, such as monarch and painted lady butterflies and some dragonflies, avoid winter altogether by their annual return migrations to warmer climates.

Order Mantophasmatodea— Heelwalkers, Gladiators, African Rock Crawlers

Entomology etymology. The name of this most recently described order of animals is intended to convey the notion that these insects appear to be constructed from bits and pieces of both mantids (order Mantodea) and walkingsticks (order Phasmatodea). Several common names have been applied to this order, including the African rock crawlers (as they occur in Africa but appear similar to the northern hemisphere order Grylloblattodea), the gladiators (as they are voracious predators), and the heelwalkers (as one group, due to the structure of the tarsi, appears to be walking on their "heels").

The discovery of the order Mantophasmatodea in 2002 created a sensation in the scientific world because it was the first completely new order of insects found since 1914 (Zoraptera). Originally detected while sorting through museum collections, live individuals were subsequently found in Namibia and South Africa. Presently 15 living species are known and at least one extinct species found in 45-million-year-old Baltic amber.

As their order name suggests, members of Mantophasmatodea look like somebody glued parts of mantids and walkingsticks together. They are small- to medium-sized wingless insects (less than 3 cm in length). The legs can best be described as "stocky" yet are designed for walking (cursorial) and equipped with spines. They lack ocelli, have chewing mouthparts, and are hemimetabolous. Mantophasmatodea are most closely related to the Grylloblattodea (rock crawlers) and the Phasmatodea (walkingsticks and leaf insects). Taxonomic arrangements that combine Grylloblattodea and Mantophasmatodea into a single order, Notoptera, have been proposed.

Little is currently know about the biology of these animals. They are found in southwestern Africa and occupy rocky, arid habitats. The heelwalkers are

FIGURE 10-16

Nymph of a heelwalker or gladiator insect. Heelwalkers are the most recently discovered (2002) order of insects (Mantophasmatodea) and are closely related to the rock crawlers, often combined into the single order Notoptera. Photograph by Whitney Cranshaw/Colorado State University.

fierce predators and hunt using both the sharp spined prothoracic and mesothoracic legs to grasp their prey. Prey items include any small arthropods they are capable of overpowering. Typically, they hunt at night and rest during the day under small stones or among plants near the ground.

Cockroaches, Termites, and Mantids

Cockroaches, termites, and mantids are all among the most familiar of insects. Despite their individual distinctions all are quite closely related insects. Most of the physical features of cockroaches and termites are quite similar, and some cockroaches even show social behaviors, the hallmark of termites. Mantids, which have modifications of the prothorax and forelegs related to their predatory habit, otherwise share most features with cockroaches, including the type of eggs that they produce. These similar features are cause for some taxonomic arrangements to combine cockroaches, termites, and mantids into the order Dictyoptera.

Order Blattodea—Cockroaches

Entomology etymology. The order name is derived from the Greek word (blatta) for cockroach. The common name in English is a corruption of the Spanish word "cucaracha."

Few insects have a poorer reputation in North America than cockroaches, largely due to the fact that a few species of the order are prominent household pests. Nonetheless, cockroaches are a large insect order (about 4,000 species) of diverse habits and should not be stigmatized by the tiny minority of pest species found in and around human dwellings. Most cockroaches feed on plant matter, somewhat like crickets, and they are particularly common in moist, forested areas. About 45 species are known to occur in North America, with all of the most commonly found pest species being accidentally introduced nonnatives. Even the inappropriately named American cockroach (*Periplaneta americana*), more innocuously known as the "palmetto bug" in southern states, is native to Africa.

● Run Roach Run!

Anybody who has turned on a light and startled a cockroach to scurry away is well aware that they are extremely fast runners. Indeed, the American cockroach has been clocked at speeds exceeding 1.5 m/s, running at speeds of 50 body lengths per second. If we were to scale up the same feat for humans, a person would need to run above 320 km/h. At the time of this writing, the documented fastest human on the planet is Usain Bolt, who ran the 100 m dash in 9.58 s. (35.58 km/h).

(*continued*)

In order to accomplish this remarkable speed, roaches do not run on all six legs. As it begins to move from a fast walk to a run, the roach appears to rear up at an angle such that the first two sets of legs no longer touch the surface. The roach actually becomes bipedal (just like Usain) using its longer and more powerful rear legs to drive it forward. Although while running, the cockroach is only angled slightly upright (about 30°), the air pressure that is derived from it running forward is sufficient to keep the animal suspended in this manner, preventing it from falling back down on its front sets of legs.

FIGURE 11-1

An American cockroach, *Periplaneta americana*. The common names of many cockroaches are misleading as to their origin. This common household pest (a.k.a., the "palmetto bug") is native to Africa. Photograph courtesy of Jim Kalisch/University of Nebraska.

As a group, cockroaches generally have an oval shape. The head is concealed, when viewed from above, by a shield-like plate of the prothorax (**pronotum**), and very long, filamentous antennae project outward from the head. From the hind end projects a pair of short cerci that are very sensitive to vibration and air movement. The body is dorsoventrally flattened, allowing them to move into cracks. This ability is enhanced by their possession of a greasy, rather than waxy, surface body covering.

Adult stages of most cockroaches are winged, with the front pair thickened into tegmina and the hindwings being more membranous.

A cockroach crawling across the floor may seem huge to a startled apartment dweller, particularly if it is a larger species like the Oriental cockroach (a.k.a., the "water bug," *Blatta orientalis*) or the American cockroach. These species range in size from 25 to 40 mm but considerably larger cockroaches can be found elsewhere. The Colombian species *Megaloblatta longipennis* may reach over 97 mm; *Megaloblatta blaberoides*, found in rainforest areas of the Amazon, may be even a bit larger. Both have wingspans that may reach 185 mm. Although somewhat shorter (about 80 mm), the giant burrowing cockroach of Australia (*Macropanesthia rhinoceros*) is the most massive, weighing in at up to 35 g. In addition to these monsters, this large order of insects also includes many that are quite small, such as the tiny (3 mm) *Attaphila fungicola* associated with the nests of leaf-cutter ants.

FIGURE 11-3

The Madagascar hissing cockroach, *Gromphadorhina portentosa*, is a common insect sold in pet stores—and a staple of many horror films involving cockroaches. When disturbed they may forcefully expel air through their spiracles, making a loud hissing sound. Photograph courtesy of Jim Kalisch/University of Nebraska.

FIGURE 11-2

The flattened body of cockroaches is modified to allow them to tuck into tight crevices. Photograph courtesy of Jim Kalisch/University of Nebraska.

Cockroaches have an ancient and distinguished lineage. Various roach-like insects began to appear in the fossil records 325 mya, and they were so abundant in the late Carboniferous period

FIGURE 11-4
A palebordered field cockroach, *Pseudomops septentrionalis*. Photograph courtesy of S Dean Rider, Jr.

FIGURE 11-5
An American cockroach with ootheca. Photograph courtesy of Joseph LaForest/University of Georgia/Bugwood.org.

FIGURE 11-6
A German cockroach, *Blatella germanica*, with ootheca. Photograph courtesy of Jim Kalisch/University of Nebraska.

that this geologic period is sometimes described as "The Age of Cockroaches." These extinct insects differed from modern cockroaches in some features, notably in possessing long ovipositors for egg deposition. The cockroaches that presently live on Earth are somewhat more "modern" insects, being first recorded from fossils with similar features from only the early Cretaceous period.

COCKROACH LIFE HISTORY AND HABITS

Reproduction in cockroaches typically involves chemical communication between the sexes. Females release a pheromone that is detected by the male, which he finds by moving upwind when the chemical is detected. Once the female is located, he will release his own pheromone, an **aphrodisiac pheromone**, which can trigger the female to become receptive to mating.

After mating the female produces eggs that she lays in a purse-shaped packet, known as an **ootheca**. Most species drop the ootheca as soon as it is produced, while others (e.g., German cockroach) carry them until the eggs in it are nearly ready to hatch. A few even carry them internally full term and may even nourish the developing eggs, allowing for live birth. As cockroaches undergo the simple pattern of metamorphosis, the immature forms generally resemble the adults but lack fully developed wings. (Wing pads may be prominent on later stage nymphs.) Many cockroaches can fly, but most of those found within buildings in the United States fly rarely, if at all.

FIGURE 11-7
Close-up of the ootheca of an American cockroach. Photograph courtesy of Gary Alpert/Bugwood.org.

Cockroaches employ a variety of defensive behaviors to avoid predation. Perhaps the best well known and observed is their speed; they are able to quickly run from danger and hide. This ability is

FIGURE 11-8
Life stages of the German cockroach. Photograph courtesy of
Jim Kalisch/University of Nebraska.

aided by acute senses that can detect ground and air vibration produced by an approaching predator. Some cockroaches may also chemically defend themselves by emitting foul-smelling liquids from specialized glands in the abdomen (e.g., *Eurycotis floridana*, Blattidae family). Still others actually eject caustic chemicals up to several inches away (e.g., *Diploptera* spp., Blaberidae family).

As with many insects, their potential to reproduce can be impressive. In one trial, 10 pairs of well-maintained German cockroaches were able to develop into a population of 51,000 in just 7 months. Under perfect conditions it has been speculated that in a year's time a single female could potentially have 10 million descendants. Of course, such numbers are never remotely approached under natural conditions because of predation, cannibalism, lack of food, and other natural controls that continuously take a toll. (The reproductive capacity of the German cockroach appears to be exceptional among cockroaches, most of which take substantially longer time to develop and reproduce.)

Cockroaches will feed on a wide variety of materials and they must be considered among the most polyphagous of all insects. Indeed it is hard to find anything that won't be eaten by some of the more annoying species that have adapted to human dwellings. However, most species that occur outdoors largely feed on dead leaves and fungi and should be considered beneficial. In some tropical areas where they are abundant, they have important roles as macrodecomposers of dead plant matter. For example, a very important group in the

Amazon region (*Epilampra* spp.) is estimated to chew up over 5% of the fallen leaves in tropical forests, making impressive contributions to the recycling of nutrients. Additionally, they serve as an important food resource for a multitude of other animals.

Their high rate of reproduction and ease of rearing—and lack of features that make them sympathetic—have made cockroaches an ideal model for all manner of research. Their nervous system has been of particular interest as they possess unusual giant nerve fibers that allow them to react extraordinarily quickly. A puff of air disturbing the sensory nerves on the cerci sets off a chain of events that can start a cockroach in movement in less than 1/20th (0.045) of a second. Once in motion they can run extremely fast, with an American cockroach covering 50 times their body length per second.

Cockroaches have been involved in some classic studies of learning. They have proved themselves at being quite adept at learning to run a maze to find food and shelter, a task few other insects manage. However, their memory is very short, and they must relearn this task daily. Experiments using electric shock treatments can cause them to avoid dark areas, to which they are normally highly attracted. Cockroaches have even been shown to be capable of learning things after decapitation, a clear demonstration of the decentralized nervous system of insects. This was achieved by suspending decapitated cockroaches over a surface that also provides mild electric shock when contacted. The headless cockroaches soon learned to hold their legs above this surface.

Cockroaches were also used as an early test model for sensitivity to radiation and, compared to humans, they show substantial tolerance. Whereas a human exposure to 400–1,000 rads for a couple of weeks is invariably lethal, German cockroaches (*Blatella germanica*) can typically survive 6,400 rads, sometimes even higher exposures. High tolerance to radiation is apparently widespread among insects and is likely due to their ability to better tolerate and repair cell damage; their smaller chromosomes have relatively low amounts of DNA that may be damaged by radiation. Several other kinds of insects, and even some insect cell lines in culture, have been demonstrated to be even more insensitive to radiation effects than are cockroaches.

FIGURE 11-10
Cockroaches in the genus *Cryptocercus* are semisocial with the parents and young living together. They are also capable of digesting cellulose, and their social behavior allows them to retain the symbiotic microbes that help them digest cellulose, much in the manner of the closely related termites. Photograph courtesy of S. Dean Rider, Jr.

Because these microorganisms are lost each time the insect molts, they must be reacquired, which is done by eating the microbe-contaminated feces of their nestmates. A similar requirement occurs with many primitive termites and the use of symbiotic protozoa to digest cellulose is a shared feature that illustrates how closely related termites and cockroaches are. Because of this close relationship, many taxonomists consider them to be a single insect order, with the termites being the more social branch of the order.

COCKROACHES AS PESTS OF HUMANS

Only a tiny percentage of all cockroaches (about 30 of the 4,000 species, or 0.75%) are "domestic" (**synanthropic**) species well adapted to living in and around human dwellings and occurring as household pests. A handful of others are "peridomestic," normally found outdoors but occasionally wandering inside, a habit they share with many other insects known as "nuisance invaders."

Several species of cockroaches can thrive in and around homes where conditions are suitable: adequate warmth, some food, and, the most critical, access to water. The most important pest species in the United States is the German cockroach (*Blatella germanica*), but the American cockroach (*Periplaneta americana*), Oriental cockroach (*Blatta orientalis*), brownbanded cockroach (*Supella longipalpa*), Asian cockroach (*Blattella asahinai*), and smokybrown cockroach (*Periplaneta fuliginosa*) are also common pests.

FIGURE 11-9
Brownbanded cockroaches, *Supella longipalpa*, in a harborage. Most cockroaches tend to aggregate when resting. Photograph courtesy of Gary Alpert/Bugwood.org.

Cockroaches like to wedge themselves into dark crevices when at rest, an ability that is eased by their flattened body and greasy cuticle. They are **thigmotactic** creatures that like to rest in sites that allow many points of body contact. This behavior is enhanced by their production of aggregation pheromones (occurring within their feces) that cause cockroaches to come together and aggregate. As a result, cockroaches often occur in groups, known as **harborages**. For reasons incompletely understood, cockroaches may even grow faster when crowded together.

Among these insects, the "ideal family"—a stable unit of two parents with their young—is only approached with the wood-dwelling cockroaches of the genus *Cryptocercus*. A "typical household" of these animals will have a pair of long-lived parents and perhaps 15–20 young. It may take 6 years for the young to become full grown, during which time the family remains together. When full grown, the young disperse to establish new colonies.

The reason for this unusual behavior is based on their diet. These semisocial cockroaches live in decaying wood and digest cellulose, a feat made possible by symbiotic cellulose-digesting protozoa in their hindgut.

Although their mere presence in home is enough to upset many people, cockroaches can produce some serious problems. Because of their extremely wide tastes, they may feed upon almost any food left on counters, in pantries, and in other areas they can reach. Seedling plants may be damaged and nonfood items such as paste, some kinds of artwork, paper, and draperies can be chewed on. They have also been known to chew eyelashes and even fingernails of sleeping humans and sometimes will enter ears. Where large numbers are present, they also can produce a noticeable and unpleasant odor from a combination of their excrement, fluid from abdominal glands, and their regurgitant.

FIGURE 11-11

The most important "domestic" species of cockroaches that infest buildings in North America: (a) American cockroach, *Periplaneta americana*; (b) German cockroach, *Blatella germanica*; (c) Oriental cockroach, *Blatta orientalis*; (d) brownbanded cockroach, *Supella longipalpa*; and (e) smokybrown cockroach, *Periplaneta fuliginosa*. Photographs of the American and German cockroaches courtesy of Clemson University/Bugwood.org. Oriental cockroach photograph courtesy of Jim Kalisch/University of Nebraska. Brownbanded cockroach photograph courtesy of Gary Alpert/Bugwood.org. Smokybrown cockroach photograph courtesy of S. Dean Rider, Jr.

A common medical concern involving cockroaches is the possibility of cross contamination as cockroaches visit filthy material and then transfer bacteria, viruses, or other microorganisms as they subsequently walk across other surfaces. Fortunately, despite the filthy food habits of cockroaches, their importance in transmitting human pathogens appears to be fairly minor. This is because they are fastidious insects, constantly grooming themselves, and usually do not carry very high numbers of bacteria on their bodies. The most common diseases associated with cockroaches are some of the food-borne diseases caused by *Salmonella*, although there are numerous reports where circumstantial evidence implicates them in the transfer of other pathogens.

A much more important medical issue can be the involvement of cockroach debris (body parts, feces, secretions) as a potential allergen. Several studies have indicated that sensitivity to cockroaches is extremely widespread and among the most common allergens, including house dust mites, grass pollen, and allergens associated with cats. In inner city areas where cockroaches are common, cockroach allergens are considered to be the most important trigger for asthma.

Order Isoptera—Termites

Entomology etymology. The order name is derived from the Greek words "iso" (equal) and "ptera" (winged), a reference to both the forewings and hindwings being of approximately equal size.

It has been argued that, next to humans, termites have the greatest impact on the ecology of the planet of any animal group. There are about 2,900 identified species of termites found worldwide with the great majority in the tropics, and tremendous local effects are produced by many species. About 50 species of termites occur in the United States, some of which are well known to damage wooden structures or other cellulose-based materials.

Termites are small- to medium-sized insects with chewing mouthparts and filament-type antennae that appear as a chain of beads. Many lack eyes altogether, a result that has evolved from a life entirely in darkness. However, compound eyes are present on the reproductive adults that briefly leave the colony, and these adults also possess two pairs of wings, both of which are similar in size, shape, and venation. The wings are dropped after mating and even reproductive forms of termites spend most of their life wingless, when they serve as the queen or king of a colony.

FIGURE 11-12
Worker caste subterranean termites, *Reticulitermes* sp. Photograph courtesy of Tom Murray.

Because they produce colonies and are often approximately the size of some ants, termites sometimes are referred to as "white ants." The comparison between these two very different insect groups breaks down quickly once the details are examined. Perhaps the simplest way to tell the two orders apart is to examine the junction between the abdomen and thorax. In termites, the abdomen is broadly joined at the thorax; in ants, the abdomen is narrowly attached to the thorax. Regardless, the confusion with ants and termites has been perpetuated in many popular images, including those involving anteaters that, despite their name, almost exclusively feed on termites.

Termites are one of the few insects outside the order Hymenoptera (ants, bees, and wasps) that are truly social (**eusocial**), characterized by having reproductive division of labor (fertile and sterile forms), collective care for the young, and overlapping generations such that progeny can assist in rearing young. Indeed, termites were likely the first animals to develop truly social behavior.

THE TERMITE CASTES

A broad range of forms occur among individual termites found within any nest. This is because termites produce various **castes**, each caste taking care of different colony functions. The primary castes are reproductives (queen, king, secondary reproductives, winged reproductives), workers, and soldiers. The proportions of each caste are largely controlled by pheromonal communication between the queen (see below) and the workers. Together the various termite castes interact within a nest to produce the most complex colony interactions of any insect excepting, perhaps, some ants.

Egg laying is performed by the **queen**, a fertilized female and the largest individual within the colony. She, along with the king, is the founder of the colony but her role shifts entirely to egg production after the colony becomes established with workers to assist in colony maintenance. The size of the queen, distended with eggs, is often about 1.5 cm among the subterranean termites that commonly damage wooden structures in North America. Queens that support the huge nest of mound-building termites in Africa are enormously distended, may exceed 8 cm, and are capable of producing up to 30,000 eggs/day. Termite queens are among the most long-lived of all insects and may survive a decade or more.

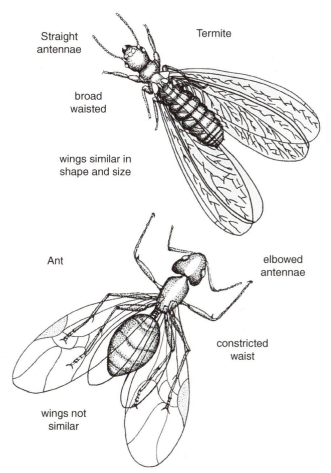

FIGURE 11-14
Illustration of the features useful in distinguishing a winged ant from a winged termite. Figure courtesy of USDA Forest Service Archives/Bugwood.org.

FIGURE 11-13
A winged reproductive caste of a subterranean termite, *Reticulitermes* sp. One feature of the order Isoptera is that the wings are all of approximately equal size and shape. Photograph courtesy of David Cappaert/Michigan State University/Bugwood.org.

Attending the queen throughout her life is a **king**, the primary male reproductive. He pairs with the queen during the swarming flight when both were winged following dispersal from their respective colonies of origin. Together they establish the new nest, and the king will then remain and periodically mate with the queen to continuously fertilize her eggs.

FIGURE 11-15
Queen of an African mound-building termite tended by workers. Photograph courtesy of Scott Turner/State University of New York.

The size of the king is often not much larger than those of the other castes and may be much smaller than the queen after she has begun egg production.

Secondary reproductives are fertile males and females that develop and remain within the colony. Their role in reproduction may increase if the original queen/and or king dies or if a portion of the colony becomes separated and requires a new reproducing caste. Usually the females of these secondary reproductive forms will produce fewer eggs per individual than the original queen, but where multiple secondary reproductives occur, their combined contribution to colony reproduction may exceed the queen.

The overwhelming majority of the colony is comprised of **workers**. Most of the basic colony functions are tended to by the workers including foraging, nest construction, care of the young, and feeding of the reproductive and soldier castes. Workers are wingless, blind, and soft bodied but possess powerful jaws. They are comprised of sterile males and females in about equal proportion. They typically may take about a year to develop and 2 years to live, and the immature workers are also involved in colony functions. Indeed, among the dampwood and drywood termites (Hodotermitidae and Kalotermitidae families, respectively), all workers are nymphs that can molt almost indefinitely and may ultimately change to soldiers, or even reproductives, depending on colony needs.

Soldiers serve the role of colony defense, and they are highly specialized for this purpose. Their head is usually grossly enlarged and, in some species, is used to plug nest openings. Most soldiers have very large mandibles designed to bite, pierce, or slash intruders. Chemical defenses are used by other species. For example, the Formosan subterranean termite (*Coptotermes formosanus*) has a large, forward-oriented pore on the head used to emit defensive secretions. Termite soldiers of the genus *Nasutitermes* are sometimes described as "spray machines" because of their ability to shoot irritating, entangling fluid from a tubular snout. Ants are the most common enemy of the termite colony that the soldiers engage.

FIGURE 11-17
Soldiers of *Nasutitermes* sp. have pores on the front of the head that can shoot entangling fluid for defense. Photograph courtesy of David Cappaert/Michigan State University/Bugwood.org.

The percentage of soldiers in a colony varies from about 1% up to 15%, and both males and females can and do develop into soldiers. Despite their formidable defense abilities, soldiers are helpless for most tasks and must be fed by the workers.

LIFE HISTORY OF TERMITES

New termite colonies are initiated when winged reproductive stages are pushed out of their parental colony for a mating flight. Sometimes known as "swarmers" or **alates**, these consist of fertile males and females, potentially new kings

FIGURE 11-16
Worker and soldier subterranean termites, *Reticulitermes* sp. The soldiers of subterranean termites have enlarged heads and strong jaws that are used in defense. Photograph courtesy of Gary Alpert/Bugwood.org.

and queens. Thousands, sometimes tens of thousands, of these winged reproductives may suddenly emerge from a nest and take to flight. Swarming events are usually triggered by weather patterns (e.g., heavy rain followed by warm evening), and within an area swarming events will be synchronized among most colonies of the same species. Depending on the species, flights may occur during daytime or night.

These winged termites are poor fliers and usually remain airborne for only a few minutes, although winds may carry them to considerable distances. When a winged female has landed, she breaks off her wings and releases a sex pheromone to attract a male. If a male finds her, they pair and attempt to cooperatively excavate a nest nearby under a piece of wood or other suitable site. When complete, the nest is sealed and the pair never again emerges. The first mating can then occur followed by the egg laying that initiates the new colony. The first young produced will be nourished by the founding pair, but as these develop into workers and the colony grows, the workers gradually take over all the chores of foraging, nest maintenance, and other colony functions, aside from reproduction.

FIGURE 11-19
A king and queen termite shortly after the mating flight. They have since dropped their wings and will attempt to initiate a new colony. Photograph courtesy of Tom Murray.

FIGURE 11-18
Winged reproductives of the eastern subterranean termite, *Reticulitermes flavipes*. Photograph courtesy of Gary Alpert/ Bugwood.org.

The colonies grow slowly during the first few years. For example, a colony of the eastern subterranean termite (*Reticulitermes flavipes*) may only include about 50 individuals after 1 year and less than 400 after the second. However, in time the colony

population becomes large enough to allow production of winged reproductive stages that will later emerge as swarmers to attempt establishment of new nests.

The ultimate size of a colony varies widely. Drywood termites (*Incisitermes* spp.) produce colonies that rarely exceed a couple of thousand individuals. Typical subterranean termites (*Reticulitermes* spp.) found around yards in the United States average perhaps 50,000 individuals/colony. However, the Formosan subterranean termite, an introduced species that is now well established in many states around the Gulf of Mexico, Southern California, and Hawaii, not uncommonly will have colonies that exceed a million or more workers and soldiers. Even these can be dwarfed by the enormous colonies that mound-building species of Africa may produce.

FEEDING HABITS OF TERMITES

Termites have a tremendously important ecological role as macrodecomposers. In environments as diverse as Arizona deserts, African savanna lands, and South American tropical forests, termites may be

the single most important organisms in recycling the nutrients locked within dead plant matter. Because of the widespread recognition of termites damaging buildings, wood is the best recognized of all foods that termites will consume. Although wood is used by a great many species, termites will feed on other materials such as dry grass, fallen leaves, humus-rich soils, or dried animal dung.

FIGURE 11-20
Arid land subterranean termites, *Reticulitermes tibialis*, feeding on dried cattle manure. Photograph by Whitney Cranshaw/Colorado State University.

FIGURE 11-22
Carton nests of *Nasutitermes* sp. termites are a common sight in tropical forests of Mesoamerica. Photograph by Whitney Cranshaw/Colorado State University.

FIGURE 11-21
Termite damage to books that had been stored on a basement floor that was accessible to foraging subterranean termites. Note the "dirt" associated with the feeding site, which is a mixture of feces and soil used to seal the feeding area. Photograph by Whitney Cranshaw/Colorado State University.

A major constituent of these foods is lignocellulose, the primary material of plant cell walls. Lignocellulose is indigestible to essentially all other insects. That termites can use this very abundant food source is largely achieved with the assistance of symbiotic microorganisms—protozoa, bacteria, and/or fungi. Enormous numbers of the symbionts may be present and it is estimated that the symbiotic protozoa of some wood-feeding termites may compose up to one fifth of the insect's body mass. These symbionts not only help to degrade the lignocellulose but can also provide many essential nutrients and may help recover usable nitrogen products from their waste. Although little can be absorbed back into the body through the hindgut, these nutrients become available through excreted liquids and feces that are then consumed. Termites continuously share in the production and consumption of anal excretions, a behavior known as **proctodeal trophallaxis.**

Not all termites require symbionts to degrade lignocellulose. In the largest termite family, Termitidae, cellulase is produced in the midgut to help degrade cellulose; bacteria that colonize the hindgut further digest the foods and convert nitrogen to usable forms. In addition, the fungus-growing termites (Macrotermitinae subfamily) use external symbionts, culturing specialized fungi on plant matter. The fungi help break down the plant material and these termites feed on the fungi and the fungi-degraded products.

In all of these cases, the permanence of a colony is essential to the use of microorganisms. For example, the protozoa and bacteria housed in the hindgut are shed, along with the hindgut lining, every time the termite molts. Therefore, they must be reacquired after molting, which they do by feeding on the microbe-rich feces of nestmates.

Termites are an extremely important group of insects to ecological processes. Their critical role in global carbon cycles cannot be overemphasized. A tremendous amount of energy and nutrients bound in the lignocellulose of dead plant matter remains unavailable until it is degraded. Without termites, such material would build up in massive amounts in many ecosystems, which would then rely on the substantially slower activities of microorganisms for its ultimate degradation and recycling. Incidentally, termites also release substantial amounts of greenhouse gases (methane, CO_2) as the lignocellulose is degraded by digestive enzymes or the activities of symbionts.

As they forage and feed, termites have also substantial roles in soil formation through their tunneling and recycling activities. Termites also play very important ecological roles as food resource for hundreds of other types of animals, including humans. Additionally, their mounds serve as homes and refugia for myriad other animals.

FAMILY RHINOTERMITIDAE— SUBTERRANEAN TERMITES

The primary wood-destroying termites in most of North America are various species of subterranean termites. Most widespread and damaging is the eastern subterranean termite (*Reticulitermes flavipes*) that occurs over a broad area east of the Rockies. The western subterranean termite (*Reticulitermes hesperus*) is the dominant pest species in the far western states. In much of the Gulf States and Hawaii, the most important termite is now the Formosan subterranean termite (*Coptotermes formosanus*), a species thought to have reached mainland North America in the 1940s.

Subterranean termites establish their nest in the soil, usually choosing sites of relatively high moisture. To forage for food they tunnel through the soil looking for buried wood, wood in contact with soil, or other cellulose-rich materials. The extent of area where the foraging workers will travel varies in

FIGURE 11-23
Fungus comb being cultivated by fungus-growing termites. Some very young termites are present in the upper right. Older soldiers and workers are also present. Photograph courtesy of Scott Turner/State University of New York.

FIGURE 11-24
Figure of life stages and associated injury produced by subterranean termites. Artwork by Arthur Cushman and figure courtesy of the USDA/Smithsonian Systematic Entomology Laboratory.

FIGURE 11-25
Formosan subterranean termites, *Coptotermes formosanus*, repairing a breach in nest. Photograph courtesy of Scott Bauer/ USDA ARS.

FIGURE 11-26
Shelter tubes constructed by subterranean termites to bridge across a cinder block wall. Photograph courtesy of USDA Forest Service Archives/Bugwood.org.

part with the colony size. For example, a very large colony of Formosan subterranean termites was found to have produced tunnels throughout an area of over 5,500 m² and individual tunnels have been observed that traveled almost 100 m from the nest. A colony of typical size of the eastern subterranean termite will forage through a much smaller area of about 20 m².

Subterranean termites are very sensitive to drying and never will openly expose themselves aboveground as they forage. Therefore, to reach wood that is beyond a barrier, such as a concrete building foundation, they may build **shelter tubes**. These are constructed of soil and bits of debris that are cemented together with feces and secretions of the termites. Given time very long shelter tubes can be produced, with construction making daily progress of several centimeters. Subterranean termites can also penetrate through noncellulose materials including drywall, plaster, stucco, and rigid board insulation.

When feeding within wood, subterranean termites prefer the sapwood to the springwood, causing a hollowing between the growth rings. The foraging workers do not carry back pieces of food but instead ingest it at the site and later regurgitate it to nestmates. Incidental to this process is that these termites *do not* produce sawdust as the wood is consumed.

(a)

(b)

FIGURE 11-27
Injury to wood produced by subterranean termites: (a) typical damage to lumber, with feeding largely confined to the sapwood and the presence of feces/soil; (b) extensive injury to a structure. Photographs courtesy of Jim Kalisch/University of Nebraska and USDA Forest Service Archives/Bugwood.org, respectively.

The amount of wood eaten depends on the number of foragers. It has been estimated that a colony of eastern subterranean termites containing 60,000 individuals consumes about 11.2 g of wood per day, if that is their entire diet. At this rate it would take 157 days to consume a 25 cm length of a 2 × 4 beam. Larger colonies can be expected to consume proportionately more. The Formosan subterranean termite, which frequently produces very large colonies, is considered to be the most aggressively damaging termite species in the United States.

Historically, the management of subterranean termites has largely involved the use of long-lasting insecticides injected into the soil such that a lethal barrier between the colony and the building is created. This technique is still widely used but more recently has been supplemented with the use of baits. Baiting techniques involve insertion of "bait stations" around an area of termite activity. These are checked periodically, and if termites are observed to be visiting the station, then the bait (usually some wood-based product) is switched over to one that contains a slow-acting insecticide. The latter is ingested by the foragers, returned to the colony, and fed to nestmates. In this manner colony populations can be greatly reduced and sometimes eliminated.

FIGURE 11-28
A bait station used to control subterranean termites. These are inserted in the soil and have material attractive to termites as bait. Once termites begin to forage at the bait station, an insecticide-treated bait is substituted. The insecticide is carried back to the colony and communally shared, killing many of the termites. Photograph courtesy of Scott Bauer/USDA ARS.

An analysis of the US termite market determined that the annual cost of providing termite protection services was $1.2 billion in 2001. The damage that results despite these expenditures is much more difficult to quantify but likely greatly exceeds this value. In pre–Hurricane Katrina Louisiana, the Formosan subterranean termite alone was estimated to cause $500 million in property damage statewide, with over half of that just in New Orleans.

FAMILY KALOTERMITIDAE—DRYWOOD TERMITES

In warmer areas of the United States and subtropical/tropical areas throughout the world, serious damage to wood can also be inflicted by drywood termites. Drywood termites nest directly within wood and do not need to have soil contact as do subterranean termites. They are much less sensitive to drying and can acquire sufficient moisture directly from metabolizing their food. Drywood termites can feed on wood with as little as 2.5%–3% moisture, but wood with about 10% moisture is preferred.

Drywood termite colonies are much smaller than those produced by most other termites and are slow to develop. At full size they may only include a maximum of a couple thousand individuals, allowing them to sustain a colony within a single piece of furniture or large chunk of work. There is no special worker caste, and colony duties are performed by juveniles that cease growing after two molts; later

these may resume growth to produce other forms depending on colony needs. Specialized soldiers and reproductive castes also are produced.

Wood damage by drywood termites differs from that produced by subterranean termites. Tunneling tends to cut across the wood grain rather than being restricted to the spring growth. Drywood termites

FIGURE 11-30
Winged reproductives of the western drywood termite, *Incisitermes minor.* Photograph by Whitney Cranshaw/Colorado State University.

FIGURE 11-31
Characteristic pellets excreted by drywood termites. Photograph by Whitney Cranshaw/Colorado State University.

FIGURE 11-29
West Indian drywood termites, *Cryptotermes brevis.* Photograph courtesy of Rudolf Scheffran/University of Florida.

also excrete characteristic dry pellets that they push out of the galleries through "kick holes" and the presence of these is often the first evidence that dry-wood termites are active at a site.

The most important drywood termite in the United States is the western drywood termite, *Incisitermes minor*. More recently, the West Indian drywood termite (*Cryptotermes brevis*), the most important species in the tropical areas of the world, has become established in peninsular Florida and areas along the Gulf of Mexico. Together, it is estimated that $300 million is spent annually on drywood termite control in the United States. Since nests are located within buildings, control approaches are very different for drywood termites than for subterranean termites, involving location and destruction of individual nests through insecticide injections, microwave treatment, and, in severe cases, tenting and fumigation.

FIGURE 11-32
Drywood termites, unlike subterranean termites, can establish nests within buildings. Fumigation of buildings is sometimes done to control infestations of drywood termites. Photograph courtesy of USDA Forest Service/Bugwood.org.

FAMILY TERMITIDAE—MOUND-BUILDING TERMITES

The largest family of termites is the Termitidae, with some 1,958 species. Known as the higher termites, these include termite species that have evolved several unique features of colony construction and specializations among the castes. A few build nests that involve large aboveground mounds with unusual features. Those that make the most elaborate structures are species found in parts of tropical Africa and northern Australia, but the forms of these mounds and their functions differ completely.

Mounds found in northern Australia are produced by either *Amitermes laurensis* or *A. meridionalis*, the "magnetic termites" (or "compass termites") that build thin, wedge-shaped mounds with a distinct north-to-south orientation. These termites live within the mound above the surface of the ground, a habit that avoids colonies being destroyed by the seasonal flooding that occurs where they are present. In so doing, they must then carefully regulate the temperature of the mound, which is exposed to the hot tropical sun, and the orientation of the mound greatly moderates the effect of direct sun during the hot summers. The thin walls of the mounds may also help with air exchange.

FIGURE 11-33
Mounds of magnetic termites, *Amitermes* sp., are constructed aboveground with a north-to-south orientation to moderate the effects of sunlight. Photograph courtesy of Howard Ensign Evans.

In contrast, the mound builders of Africa are fungus-growing species belonging to the large (approximately 350 species) subfamily Macrotermitinae. These termites do not harbor the internal symbiotic bacteria and protozoa that allow them to utilize cellulose. Instead, they have an ectosymbiotic relationship (outside the body) with various fungi that they culture on the dried grass, leaves, and woody material brought into the nest by foraging workers. These fungi (various *Termitomyces* spp.) grow in special belowground fungus-comb gardens that are maintained by worker caste members. Prior to dispersal from the colony in their mating flights, the future potential queens and/or the kings will collect stages of these fungi that will allow them to reestablish a fungal garden in the new nest.

Large aboveground structures, sometimes in the form of towers that may extend 9 m aboveground, are primarily used to help ventilate the fungus gardens and the nest areas. The structure of the belowground colony can be extensive, involving hundreds of chambers for fungal gardening and enough space to house many millions of individual termites. Sunlight heating the mounds helps to warm air, drawing it upward, providing air circulation. Rising heat from fermentation activities in the fungus gardens also provide sources of air movement that help ventilate the mound.

That this system is effective is illustrated by the incredible densities these termites may produce. Up to 800 mounds/ha have been reported and as much as 30% of the land surface may be taken up by termite mounds (active and dead). They have tremendous effects on local ecology and involve what likely are the most complex colonies and intricate constructions produced by any organism aside from humans.

Order Mantodea— Mantids

Entomology etymology. The name of the order is derived from the Greek word "mantis," meaning prophet, seer, or diviner. This is in reference to the common stance taken by these insects in which the grasping front legs are held in front of the body

FIGURE 11-34
Large mounds produced by termites in the savannah of eastern Africa. Photograph courtesy of G. Keith Douce/University of Georgia/Bugwood.org.

giving them a superficial resemblance of being in prayer. Because of this feature the insects are also known as "praying mantids"—and sometimes incorrectly as "preying mantids." The term "mantis" is interchangeable with "mantid."

There are over 2,300 species of mantids worldwide and most are found in the tropics. Within the United States there are 17 species that range in length from less than a centimeter to over 10 cm. A few of the most common species now encountered in the United States are nonnative, including the European mantid (*Mantis religiosa*) and the Chinese mantid (*Tenodera aridifolia sinensis*). The latter is commonly sold at nurseries and through garden catalogs.

All mantids are predators of other insects, and they have many physical features that make them among the most distinctive of all insect orders. The most immediately obvious is the prominent pair of spiny front legs that are designed to grasp prey. These unique legs are attached to an extremely elongated prothorax. The remaining legs are long and thin, allowing the insect to lunge at prey. Mouthparts are used to chew. The mantid head is triangular with large, widely spaced eyes that provide both binocular vision and a wide field of vision. Also the mobile head of mantids can twist, allowing them to see completely around them without moving their body.

FIGURE 11-36
Chinese mantid feeding on a grasshopper. Photograph by Whitney Cranshaw/Colorado State University.

flowers, leaves, and twigs, and some flower mimicking species can actually change color over a period of days as the flowers around them change. Prey often includes bees, wasps, flies, and other relatively large insects such as grasshoppers. Few terrestrial arthropods can withstand an attack from a mantis, and larger tropical species have been reported to successfully attack small reptiles, frogs, rodents, and the occasional hummingbird.

Most mantids can fly, and this is done primarily with the membranous hindwings. The front wings are thickened and protective tegmina that cover the hindwings when at rest. Males are smaller than the females and often more likely to fly, particularly when females swell with developing eggs. The hindwings may also be patterned or brightly colored, a feature that they can use to make a startle display or threatening bluff to deter predators.

FIGURE 11-35
The Chinese mantid, *Tenodera aridifolia sinensis*, is the largest mantid found in North America. It is native to Asia but has been widely distributed, in part through the sale of egg cases through gardening outlets. Photograph courtesy of Jim Kalisch/University of Nebraska.

Mantids capture their prey as ambush predators. They wait motionless on plants and may camouflage well with the background. Many are excellent mimics of

FIGURE 11-37
The ground mantid, *Litaneutria minor*, is a small species native to the shortgrass prairies of North America. Photograph by Whitney Cranshaw/Colorado State University.

FIGURE 11-38
Ghost mantids, *Phyllocrania* spp., have body features that allow them to blend in with the plants on which they hunt. Photograph by Whitney Cranshaw/Colorado State University.

Reproduction in the mantids first involves females calling to the males through the use of sex pheromones. Upon detection, the male will eventually locate and carefully approach the female, often from behind. Since mantid females are potentially cannibalistic, males may be mistaken for prey and eaten, particularly if the female is very hungry. Males of at least some species can gauge the relative hunger of females and will use greater caution if she is starved.

The male usually jumps upon the back of the female and mating progresses. On some occasions, during the mating process the head of the male will be grasped, removed, and eaten by the female. This does not interrupt mating and the headless male may actually copulate more vigorously and will transfer the spermatophore more successfully. Ultimately, the male may be almost entirely consumed, serving as an important food resource

for the female that allows better nourishment of the fertilized eggs. However, such dramatic cannibalism during mating is not the norm and most mantid males will successfully be able to mate multiple times with different female mantids.

FIGURE 11-39
A mating pair of European mantids, *Mantis religiosa*. In this species the male is considerably smaller than the female. Coloration of this species is variable and both sexes can be either green or brown, a feature that is determined largely by temperature and humidity. Photograph by Whitney Cranshaw/Colorado State University.

A few days after mating, a mass of eggs is laid by the female that is covered with a creamy material from the accessory glands. This quickly hardens, producing an egg-containing ootheca that in many common species has an approximate appearance of a "packing peanut." The ootheca helps insulate and protect the eggs.

Dozens to a hundred or more eggs occur in each ootheca, and most mantid species spend the winter in the egg stage. Upon emergence in the next spring, nymphs quickly disperse and begin hunting small insect prey, which may include some brothers and sisters.

(a)   (b)   (c)

FIGURE 11-40
Oothecae of different mantids; (a) European mantid, *Mantis religiosa*; (b) Chinese mantid, *Tenodera aridifolia sinensis*; and (c) ghost mantid, *Phyllocrania* sp. Photograph of the Chinese mantid ootheca courtesy of Sturgis McKeever/Georgia Southern University/Bugwood.org. Other photographs by Whitney Cranshaw/Colorado State University.

As they are large insects, mantids can be susceptible to predation by birds, mammals, and other vertebrates. When attacked, a mantid may display a startle response, simultaneously rearing up on its hindlegs, striking out with raptorial front legs, unfolding the wings, and "flashing" the colored underside of the rear wing. Additionally, while it is going through the display, it may produce a hissing noise by rapidly pushing air through the abdominal spiracles. Should the display and noise fail to deter the predator, the mantid may simply attempt attack with its spiny front legs.

Mantids may also be attacked in flight. Mantids often fly at dusk or night and at this time their predominant predators may be bats. To avoid bat predation mantids possess an "ear" that appears as a grove located on the metathoracic segment just in front of the leg coxae. The mantid ear cannot discriminate either sound direction or frequency but is highly tuned to the sound frequencies that bats use to echolocate their prey. Should a bat call be detected while the mantis is flying, the mantis goes into an avoiding maneuver and dives to the ground.

FIGURE 11-41

A European mantid in startle display. To deter predators, this mantid spreads its wings and displays the eyespot markings on the underside of the forelegs. Photograph by Whitney Cranshaw/ Colorado State University.

● **Lousy Nitpickers**

Lice and Psocids

Some early attempts to classify the different insect orders used mouthpart designs and feeding habits as a key feature. In the case of the peculiar insects known as "lice," such efforts stumbled as a wide variety of feeding habits are present, ranging from scraping and chewing to blood sucking. Several other features clearly indicate that at least two currently recognized insect orders—Psocoptera (booklice, barklice) and Phthiraptera (true lice)—are closely related, often even discussed collectively as the superorder Psocodea. Derived from the long-extinct insects that thrived during the Permian period, the present-day lice are much more recently evolved, particularly the parasitic true lice (Phthiraptera) that developed only after their bird and (later) mammal hosts had emerged.

Order Phthiraptera—Lice

Entomology etymology. The word "phthir" is Greek for louse. The suffix "aptera" means without wings. (In pronunciation the "ph" is largely silent—"fthir-ap-ter-a.") As the term "lice" can sometimes have more broad usage to describe various other small insects (e.g., booklice, barklice, or, the British term for aphids, "plantlice"), the animal parasites of the order Phthiraptera are often distinguished as being the "true lice." The singular form of lice is louse.

The order Phthiraptera includes a diverse group of insects, all of which are parasites on the bodies of mammals or birds. All are wingless, are flattened dorsoventrally, and have a heavily sclerotized body—adaptations to survive on a vertebrate host. Their legs are highly modified for grasping the hairs or feathers of their host, and these insects show extreme specialization in what kind of animal

FIGURE 12-1
A body louse engorging on a blood meal.
Photograph courtesy of the Centers for Disease
Control and Prevention Image Library.

(host) they can inhabit. Worldwide, there are about 4,900 named species of lice, but this is considered low because collecting lice for identification requires examination of often unwilling animal hosts.

Beyond the basic body outline and parasitic habit shared by all lice, things get much more difficult when one attempts to categorize members of the order. This is largely because two very different types of mouthparts occur. Most familiar and important to human interests are the "sucking lice" (suborder Anoplura) with mouthparts designed to pierce the skin and suck blood from mammals. A great many other lice (suborders Rhychophthirina, Amblycera, and Ischnocera) have mouthparts of different design that are capable of chewing the skin or feathers of birds and mammals.

All lice spend their entire life cycle on the host, and movement to a new host animal requires sustained close contact. In most species, eggs are strongly glued to the base of a hair or feather; typically, there are three nymphal instars preceding the adult stage. Metamorphosis is of the simple type (i.e., hemimetabolous).

CHEWING LICE

The three suborders of lice with chewing mouthparts include about 4,350 species, over 90% of which feed on birds; the remainder feed on mammals. The head of a chewing louse is broad, wider than the prothorax, and the eyes are usually entirely absent. A minor exception is the three species in the suborder Rhychophthirina; these have a narrow head with chewing mouthparts located at the end of the snout. Species in the latter suborder feed on warthogs, bushpigs, and elephants.

Every family of birds has at least one species of associated chewing louse. Most of these associations are highly specific such that it is often possible to

(a)   (b)

FIGURE 12-2
Adult (a) and nymph (b) of the large turkey louse, *Chelopistes meleagridis*, a type of chewing louse. Photographs courtesy of Ken Gray/ Oregon State University.

FIGURE 12-3
Chewing lice infesting a pigeon. Photograph courtesy of David Shetlar/The Ohio State University.

identify a bird by looking solely at the lice collected from it. Some may also be highly specific in where they feed; *Piagetiella* species occur within the mouth pouches of pelicans and cormorants. However, high specificity is not always the case, and some chewing lice have a broad host range. For example, the two lice species *Anatoecus dentatus* and *A. icterodes* can be found on 60 different species of ducks, geese, and swans.

A few chewing lice are important pests of poultry. At least six species are associated with chickens and these occur worldwide. Most important is the chicken body louse, *Menacanthus stramineus*, which primarily feeds on the barbs of feathers. Feeding by this species often ruptures the quills of new pinfeathers and these gnawing injuries may draw blood. When high populations develop on a bird, the irritation causes the bird to become restless, feed less, and digest more poorly—all reducing weight gain and egg production. Turkeys, too, have their own species of chewing lice that they have carried with them worldwide from their origin in Mexico and Central America.

SUCKING LICE

Sucking lice have unusual piercing mouthparts designed to suck blood from mammals. Very few families of mammals escape having at least one kind of sucking louse—bats, whales, sea cows, and platypus are among the few exceptions. Lice are even found among seals and sea lions, surviving under the animal's dense fur even during the extended periods these animals may be at sea. Humans have three unique louse associates—the head louse, body louse, and crab louse.

The head louse (*Pediculus humanus capitis*) has an ancient association with humans. They have been found entombed with Egyptian mummies and are found in even older gravesites in the Americas. Lice combs, still used today to extract louse eggs from the hair, have been in use for over 2,000 years.

Head louse eggs are known as **nits** and they are usually cemented onto hair shafts immediately next to the scalp. Over the next 7–10 days, the eggs develop and swell before hatching. (Nits found on the hair shaft more than 2.5 mm above the scalp have already hatched. Human scalp hair grows at a rate of about 1 cm/month.) The newly emerged nymph immediately begins to feed and will subsequently feed intermittently over the course of the next couple of weeks. Following egg hatch, developing lice go through three molts before reaching the adult stage. Adult males are about 2 mm in length, females being 3 mm. Recently fed lice have sometimes been described as appearing like tiny rubies because the blood meal is easily visible through the exoskeleton. Ultimately the overall body color tends to take on that of the host's hair.

FIGURE 12-4
A head louse, *Pediculus humanus capitis*. Photograph courtesy of the Centers for Disease Control and Prevention Image Library.

FIGURE 12-5
Louse eggs (nits) attached to hair. Photograph courtesy of Ken Gray/Oregon State University.

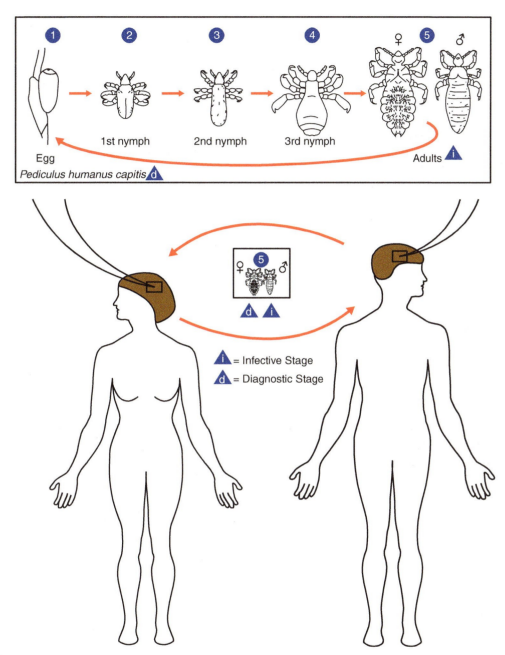

FIGURE 12-6
Life cycle of the head louse. Figure courtesy of the Centers for Disease Control and Prevention.

The overall effect of infestation with head lice is usually mild; an itchy scalp is the most prominent symptom. Fortunately, the head louse does not transmit any diseases, an issue that does arise with the closely related body louse (following). The overall incidence of head louse infestation is difficult to determine, largely because of the social stigma associated with infestations.

Regardless, infestations are widespread, particularly among early grade school children. This is because young children are often in particularly close contact with each other and also because they tend to share items that may transfer head lice, such as clothing, stuffed animals, and bedding materials. There are no differences in the susceptibility of infestation between boys and girls, but differences do occur among races.

The latter is due to various strains of head lice developing around the world with minor modifications of the claws that allow precise fit to the hair shape of regional human populations. The strain of head lice that predominates in the United States originated from Europe and is best adapted to the shape of a Caucasian hair shaft. The slightly different hair shape found in people of African or Asian descent make it more difficult for the US strain of head louse to establish on these individuals.

The body louse (*Pediculus humanus humanus*) is essentially indistinguishable from the head louse and arguments continue over whether or not it is a different species altogether from the head louse; it is usually recognized as a distinct subspecies, largely based on habit. Although the body louse tends to be slightly larger, the biggest difference between head and body lice is that body lice prefer to develop off the human body. The eggs are laid on clothing, and, when not feeding, the adults and nymphs reside in the clothing; a slang term for body lice is "seam squirrels." More importantly, body lice have a greater reproductive capacity than head lice, and they are involved in the transmission of some very important human pathogens.

FIGURE 12-7

An adult body louse, *Pediculus humanus humanus*. Photograph courtesy of Ken Gray/Oregon State University.

Infestations of body lice are presently rare but, historically, they were just a part of everyday life for people who did not have access to regular changes of clean clothing. Their occurrence is best documented in Europe, where body louse infestations were common in the last century. With the advent of regular laundering and clothing changes, the body louse has largely been eliminated in modern societies. Where good hygiene

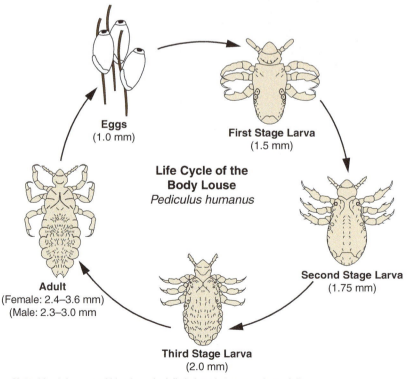

Life Cycle of the Body Louse
Pediculus humanus

Eggs
(1.0 mm)

First Stage Larva
(1.5 mm)

Second Stage Larva
(1.75 mm)

Third Stage Larva
(2.0 mm)

Adult
(Female: 2.4–3.6 mm)
(Male: 2.3–3.0 mm

Note: Lice take several blood meals daily in larval stages and as adults.

FIGURE 12-8

Generalized life cycle of the head louse. Figure by Scott Charlesworth and provided courtesy of Purdue University.

practices break down (e.g., refugee camps), body louse populations can rapidly build.

A sustained infestation with body lice can produce a variety of symptoms known as **pediculosis** that result from itching and swelling at the bite site. In extreme cases where heavy infestations are prolonged, skin discoloration ("vagabond's disease") may develop. The most important medical issue with body lice is their role as vectors of human pathogens. Body lice transmit the organisms responsible for producing trench fever, epidemic relapsing fever, and, most importantly, epidemic typhus.

The crab louse (*Phthirus pubis*), or "crab," gets its name because of the very large tarsal claws on the second and third pairs of legs and its "crab-like" body shape. The claws are designed to hold onto coarse, thick body hairs. As a result, it is most always found in the pubic region, although it is occasionally found among hairs of the eyebrows, beard, or mustache. There are only two species in the genus *Phthirus*; the other, *Phthirus gorillae,* is found on gorillas.

FIGURE 12-10
Crab lice clinging to a hair. Photograph courtesy of the Centers for Disease Control and Prevention Image Library.

Crab louse populations on humans are always small, typically averaging 10 or less, but even these low levels can cause very annoying irritation (and no small portion of embarrassment). Crab lice move very little during the 1 month or so that it takes them to become fully grown, and a few centimeters may be the extent of its lifetime travels, giving lie to the bathroom graffiti:

> Don't bother to stretch
> Or stand on the seat,
> The crabs in this place
> Can jump thirty feet.

Instead, movement from one human to another almost invariably involves intimate sexual contact, leading to the French term for this insect "papillon d'amour"—butterfly of love. Crab lice, like head lice, have not been implicated in disease transmission.

FIGURE 12-9
A crab louse, *Phthirus pubis*. Photograph courtesy of Ken Gray/ Oregon State University.

● Epidemic Typhus

Epidemic typhus is a disease resulting from infection with *Rickettsia prowazekii*, an odd type of bacterium known as a rickettsia. (The typhus-producing organism was named after two individuals who worked with and ultimately died from the disease: Howard Taylor Ricketts and Stanislaus von Prowazek.) The onset of symptoms in humans resembles those typical of influenza but then progresses to high fever, excruciating headache, mental disorientation, and a "besotted expression" produced by dark red spots. Historically, it has resulted in the death of 10%–100% of those individuals unfortunate enough to become infected.

The louse acquires the bacteria when it feeds on the blood of an infected human. The bacteria then reproduce within the midgut cells of the louse, producing a disease of the insect that ultimately is lethal as the insect's digestive track disintegrates. During the course of the infection within the louse, viable *Rickettsia prowazekii* bacteria are continuously excreted in the feces and remain viable after expulsion.

Human infections do not occur directly from the louse bite; rather, transmission occurs when the excreted bacteria (or crushed lice) are rubbed into wounds on the skin (such as an old louse bite) or breaks in the membranes around the eyes, nose, or mouth. The involvement of the body louse in typhus was first identified by Charles Nicolle in 1907, who later (1928) received the Nobel Prize in Medicine for the discovery.

Epidemic typhus has been a very important disease during periods when people are crowded together, conditions that allow the body louse to move from person to person. Before the identification of the pathogen and its disease cycle, typhus was sometimes referred to as "jail (gaol) fever" or "war fever" since outbreaks were common in prisons and among soldiers in wartime.

Typhus has had an extraordinary impact on the outcome of European wars. The first well-documented report of its occurrence was in 1489 when Spanish soldiers were sent by King Ferdinand and Queen Isabella to secure Granada from the Moors. During this campaign 3,000 troops died from wounds—and 17,000 died from typhus. This established a pattern that persisted in European wars for over 400 years. Over this span of time, typhus consistently killed more soldiers than did battlefield injuries, and the successful outcome of wars was often largely determined by which side

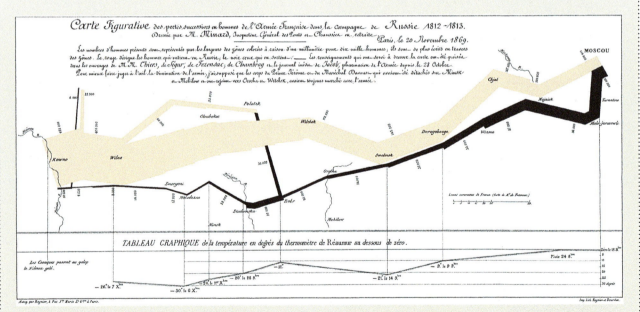

FIGURE 12-11

A graphic illustrating Napoleon's disastrous 1812 Invasion of Russia, which has been called "the greatest military victory of typhus." This graphic, produced by Joseph Charles Minard (1781–1889), illustrates the relative size of the French army as it moved across Poland and into Russia. Shortly after they crossed the Neiman River, the troops began to be devastated by typhus, which persisted throughout the military campaign. Following the retreat from Moscow, extremely cold temperatures (indicated at the bottom of the graphic) and starvation further decimated the army. Only 3,000 troops, out of the original force of 600,000, returned to France.

(continued)

FIGURE 12-12
US Army troops being dusted with DDT to prevent typhus during World War II. Photograph courtesy of the US Armed Forces Archives.

caught typhus last. Through World War I, when typhus resulted in about three million deaths, it continued to take more lives than did swords, lances or bullets. It has been said that "the greatest military victory of typhus" involved the 1812 invasion of Russia by Napoleon Bonaparte. Setting out with 600,000 troops—to only about 250,000 Russian defenders—the invasion progressed well until the French army crossed the Niemen River between Prussia and Poland. By the time of the first major battle in Ostrovno, 80,000 of the French troops had contracted typhus. Typhus continued to take its toll as the French army moved toward Moscow, while the Russians retreated and destroyed supplies in their wake. Although Napoleon's army ultimately took Moscow, they then were forced to make a wintertime retreat due to the lack of food. Extremely cold weather further burdened the retreat, and many soldiers died from cold and starvation. Throughout the campaign, typhus was present, transferred continuously as the clothing of the dead was retrieved by survivors— often with infected body lice. Although the number is debatable, it has been suggested that ultimately only 3,000 troops may have survived to return to France; louse-transmitted typhus and the Russian winter were responsible for most of the fatalities.

Our Lousy English

Lice have made several incidental contributions to the English vocabulary, although rarely are their origins recognized. Among these are:

Lousy—originally meaning to be infested with body lice. The term has developed into an adjective that describes one feeling bad or miserable or something done that is mean or contemptible.

Nit-picking—originally referencing the practice of grooming louse eggs (nits) from hair. The term is used now to describe activity that is excessively concerned with or critical of inconsequential details.

Cootie—originally referencing a head louse or body louse. In popular culture, it has morphed into some generic contagion that girls (if you are a boy) or boys (if you are a girl) have.

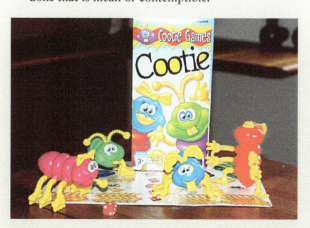

FIGURE 12-13
The child's game Cootie has been a longtime staple on toy shelves. This game involves putting together a simulated insect, the cootie. The origin of the slang term "cootie" refers to body lice. Photograph by Whitney Cranshaw/Colorado State University.

Order Psocoptera— Barklice and Booklice

Entomology etymology. The order name is derived from the Greek word *psokos*, meaning "rubbed small" or "gnawed." The name refers to the presence of some species found feeding on papers and cereal products. The suffix "ptera" (winged) is not well connected in this order name, but together the order name can be loosely translated to mean "gnawing insect with wings." Insects of this order are collectively known as psocids.

Although booklice are sometimes common indoors and barklice are often abundant in yards, Psocoptera is one of the most overlooked and poorly studied of all the insect orders. This is in part due to their small size; booklice typically are only 1–2 mm in length, and a monster-sized barklouse only reaches 7 mm. Wingless forms vaguely resemble miniature termites, but unlike termites, psocopterans have long filamentous antennae (rather than beaded antennae). Other distinguishing features include a prominent head, well-developed eyes, and a constriction behind the head. Winged stages hold their wings roof-like over their body. Worldwide, there are about 4,400 described species, mostly known from the tropics and at least 340 occurring in the United States.

Members of this order are usually called psocids (so' sids) and feed by using their chewing mouthparts in a scraper-like fashion to feed on films of yeast, fungi, algae, and lichen. Typically they are found in dark moist sites, such as under rocks or among rotting leaves. They are also able to extract atmospheric moisture through their pharynx of the foregut, an adaptation (also shared with the true lice) that does allow them a wider range of habitats. Psocids lay eggs in small clusters, sometimes covered with silk or digested material. The nymphs go through a simple form of metamorphosis, and six instars are typical before reaching the adult stage.

FIGURE 12-15
A wingless female barklouse, *Mesopsocus unipunctatus* (Mesopsocidae family). Photograph courtesy of Tom Murray.

FIGURE 12-14
Booklice, *Liposcelis* sp. Photograph courtesy of David Shetlar/ The Ohio State University.

FIGURE 12-16
A winged barklouse, *Psocus leidyi* (Psocidae family). Photograph courtesy of Tom Murray.

Some booklice can be found in insect tunnels and/or in the nests of birds and mammals, where they may occur in close proximity to their parasitic cousins (Phthiraptera) that feed on these vertebrates. Booklice also include the species most commonly found in homes, mostly from the genera *Liposcelis* and *Lepinotus*; in grain and other food storage facilities, sometimes in enormous numbers; or usually only in areas where high-humidity conditions exist that allow the growth of the molds on which they graze. Booklice are known to incidentally damage books, papers, and fabrics by feeding on the glue and starches used as binding agents in these materials.

Barklice are found outdoors and most all develop wings as adults. The winged adults easily disperse, and barklice are often among the very first insects found colonizing newly formed islands. Barklice are usually found associated with plants or on shaded rock outcroppings, where they consume algae, lichens, or molds associated with bark, dead leaves, and other surfaces. They can produce silk from their labial glands, and some aggregate to produce large silk webs extensively covering tree trunks and branches. These webs protect the eggs and immature stages, and large numbers of barklice, in all life stages, may live together in these aggregations.

FIGURE 12-17
A mass of barklice, *Cerastipsocus venosus* (Psocidae family). Photograph courtesy of Tom Murray.

FIGURE 12-18
Sheets of silk produced by barklice covering a tree trunk. Photograph courtesy of David Held/University of Mississippi.

• Life on an All-Fluid Diet

True Bugs, Aphids, Cicadas, Whiteflies, Thrips, and Other Sap Suckers

The overwhelming majority of insects feed on solid matter that they handle with some sort of mouthpart designed to chew or scrape. Liquid feeding arose within several insect groups—including flies (e.g., mosquitoes, deer flies), lice, and fleas—most of which used this new feeding approach to feed on the blood of vertebrates. Two orders of insects—Hemiptera and Thysanoptera—developed their own unique type of mouthparts to feed on fluids, and most of these insects feed on plants.

Order Hemiptera—True Bugs, Cicadas, Hoppers, Psyllids, Whiteflies, Aphids, and Scale Insects

Entomology etymology. The order name is derived from the Greek words "hemi" (half) and "ptera" (wings) in reference to the unusual forewings of true bugs (suborder Heteroptera). In these animals, the base of the wing is thickened and the tip or distal area is membranous. Collectively, members of this order may be referred to as hemipterans.

Insects of the order Hemiptera all feed by sucking fluids, and their mouthparts share a common design. Both the mandibles and the maxillae are very elongate and form a **stylet bundle** with the maxillae nested within the mandibles. Primary penetration of their food item is usually done with alternating movement of the mandibles on the outside of the stylet bundle. The maxillae interlock and have both a food channel for sucking fluids and a smaller, parallel salivary canal for introducing saliva into their food. The labium is greatly enlarged to support and cover the stylets, together with the labrum, giving the combined mouthparts a beaklike appearance, sometimes referred to as a **proboscis**.

The order Hemiptera is a large order with about 82,000 described species exhibiting very diverse habits. Historically, the order has often been subdivided; a common arrangement recognized two separate orders, the Hemiptera (or Heteroptera) and Homoptera. It is now well agreed that all of the insects share important physical features and evolutionary history so closely that they should be combined into the single order Hemiptera, and three suborders are generally recognized—Heteroptera, Auchenorrhyncha, and Sternorrhyncha.

FIGURE 13-1
Front view of a shield bug (Scutelleridae family) showing the proboscis within which the stylets used in feed rest. Photograph courtesy of Brian Valentine.

FIGURE 13-2
Side view of a leafhopper. The proboscis with the stylet bundle runs along the underside of the insect. Photograph courtesy of Brian Valentine.

Suborder Heteroptera— The True Bugs

The most distinguishing feature that separates members of the suborder Heteroptera from other Hemiptera suborders is their distinctive wing type. The front wing is of two textures, with the base thickened with a more membranous tip, producing a hemelytron (plural, hemelytra). The hindwings are uniformly membranous and fold under the protective hemelytra. A large triangular plate (**scutellum**) is present between the bases of the wings, just behind the prothorax.

The basic structure of the mouthparts is similar to that of other members of the order, although the mouthparts form a prominent beak that can be clearly seen to originate from the front of the head.

(a)

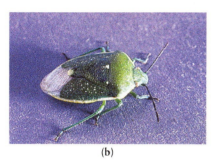

(b)

(c)

FIGURE 13-3
Three bugs illustrating variations of the hemelytra forewings. (a-Top) Minute pirate bug (Anthocoridae family). Photograph courtesy of Tom Murray. (b-Center) Stink bug (Pentatomidae family). Photograph courtesy of Ken Gray/Oregon State University. (c-Bottom) Lace bug (Tingidae family). Photograph courtesy of Tom Murray.

The great majority of true bugs are plant feeders, but a few have adapted to feed on blood of vertebrates (e.g., bed bugs, kissing bugs) or have evolved as predators of other insects and small animals (e.g., giant water bugs, assassin bugs).

Technically speaking, "all bugs are insects but not all insects are bugs." Only insects in the suborder Heteroptera, possessing the above features, are called "true bugs." Furthermore, when the common name of a heteropteran includes the word "bug," the word "bug" properly stands alone when written (e.g., "bed bug," "boxelder bug"). Common names for other types of insects (non-heteropterans) that include the word "bug" properly combine it into one word (e.g., "ladybug" to describe a type of beetle, "spittlebug" to describe a different type of hemipteran).

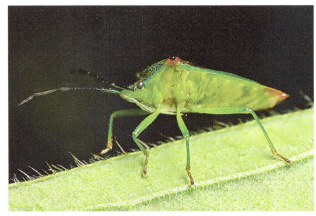

FIGURE 13-4
Mouthparts of true bugs clearly originate from the front of the head and are conspicuous when viewed from the side or underside of the insect. Photograph courtesy of Brian Valentine.

FAMILY GERRIDAE—WATER STRIDERS

Water striders (or "pond skaters") are among the most enchanting insects to watch as they effortlessly glide across the surface of a water pool. This trick is largely achieved by a special modification of the tarsi at the tip of their two pairs of hindlegs that are clothed with fine, precisely oriented hairs. The hairs trap air and are difficult to wet, producing a meniscus on the water surface that they use to move across the surface in a rowing motion. The long legs of the water strider are also important; they spread out the mass of the insect across the water surface such that it does not break the surface tension. To maintain this balance, larger and heavier species of water striders have proportionately longer legs.

The front pair of legs is much shorter and is adapted for grasping; water striders are predators and scavengers of insects trapped on the water surface. The middle pair of legs is used for locomotion, driving backward with much the same effect as a pair of oars. Water striders can be quite fast and have been clocked at 1.5 m/s over short distances. Adults of most species develop wings, at least during some times of the year, and these able

FIGURE 13-5
A water strider feeding on an insect trapped on the water surface. Photograph by Whitney Cranshaw/Colorado State University.

fliers can disperse to new water bodies or to edges of water sources where they survive winter.

Within the family are some of the very few insects that appear completely adapted to life in oceans—the sea skaters. Five (out of an estimated 40) *Halobates* species are found off coastal areas in warmer water areas worldwide, where they live a life similar to those found inland, laying eggs on floating debris and feeding on small, dead, or dying animals on calm ocean surfaces.

FAMILIES CORIXIDAE AND NOTONECTIDAE—WATER BOATMEN AND BACKSWIMMERS

Water boatmen and backswimmers are aquatic bugs that have specially modified legs that allow them to swim through freshwater ponds. The water boatmen and backswimmers are about the same size and thus can sometimes be confusing at first glance; however, other features allow the two families to be easily distinguished.

Water boatmen are generally brown on their dorsal surface, light on the ventral surface, and swim in a dorsum-up position using their broadened hindlegs. Most of their time is spent along the pond bottom, where they feed on living material stirred up with their front legs. Food can include a wide variety of materials such as diatoms, protozoa, algae, and small insects.

FIGURE 13-6

A water boatman (Corixidae family). Photograph courtesy of Tom Murray.

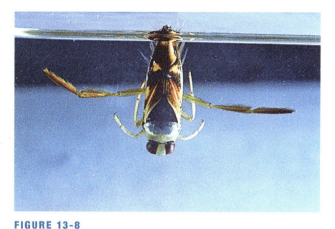

FIGURE 13-8

A backswimmer at surface reacquiring an oxygen bubble for its plastron used in underwater respiration. Photograph courtesy of Ken Gray/Oregon State University.

FIGURE 13-7

Underside of a water boatman. Photograph courtesy of Tom Murray.

Backswimmers similarly possess hindlegs adapted for swimming and these are typically angled forward when at rest. These insects swim on their back, and so their colors are reversed from that of water boatmen; the back is light colored and the belly dark. Backswimmers are predators of insects, other arthropods, and occasionally tiny fish and tadpoles. They capture prey by grabbing it with their spiny front legs and injecting it with a paralyzing saliva using their piercing mouthparts. The saliva also begins digesting the prey (**exodigestion**); they then feed by sucking back the fluids.

Insects from both families can remain underwater for long periods of time. They do this by carrying an air bubble on the undersides of their body or, in the adults, under the wings. The bubble (or plastron) covers their spiracles, and dissolved oxygen from the water moves into the bubble through diffusion as it is diminished from respiration. The insects must periodically briefly return to the surface to replenish the bubble, as a new bubble can sustain them for hours. In addition, backswimmers use the bubble as a sort of "buoyancy vest"; by regulating the bubble's size, they can remain suspended at a preferred depth with minimal swimming effort.

FAMILY BELOSTOMATIDAE—GIANT WATER BUGS

Giant water bugs (belostomatids) are the largest of the order Hemiptera, with some reaching lengths of up to 6.5 cm in the USA and up to 12 cm in Southeast Asia. They have a flattened, elongate-oval body and hindlegs modified for swimming. These insects are also found in freshwater ponds, lakes, rice paddies, and other still-water habitats. Belostomatids acquire oxygen through spiracles at the tip of two tail-like structures at the end of the abdomen.

Giant water bugs possess very large raptorial front legs for catching prey. They prefer to hang upside down from the water surface with their spiracles protruding into the air. Here, they wait for prey (insects, snails, small fish) to swim below and grab them from above. Like backswimmers, these insects bite and inject saliva into the prey. The saliva paralyzes the prey and then produces some exodigestion of the host, enabling giant water bugs to suck out liquefied internal tissues and body fluids of its captured prey.

FIGURE 13-9
A giant water bug, *Lethocerus* sp. Photograph courtesy of
Tom Murray.

FIGURE 13-10
A giant water bug feeding on a small fish. Photograph courtesy
of S. Dean Rider, Jr.

If threatened, water bugs will normally drop to
the bottom of the pond and may feign death.
Others may secrete a noxious compound from the
anus as a defensive mechanism. When defending
themselves, giant water bugs can deliver a very painful
bite. People sometimes accidentally come into contact
with these insects when wading in ponds or pools and
a common name for giant water bugs is "toe biter."

Mating in these insects follows an elaborate
courtship behavior involving visual, touch, and sound
cues. In some species of giant water bugs, the female
mounts the male from behind, copulates, and then lays
fertilized eggs on his back. The male only allows egg
laying after he has mated, and this behavior ensures
that the eggs he is carrying are indeed his. The male
keeps water flowing over the eggs and incubates them
until they hatch. If the eggs are removed from the male
before they hatch, the developing embryos die.

FIGURE 13-11
A male giant water bug, *Abedus* sp., carrying eggs. Photograph
courtesy of Tom Murray.

● Water Bugs as Food

Certain Asian countries use giant water bugs as a food
and a condiment. In Vietnam, male water bugs are
collected and their pheromone gland secretions are used
as a condiment for spring rolls. In Thailand, giant water
bugs are simply served up whole and fried.

Try this recipe for Maeng Da at home.

Ingredients:

6–8 large giant water bugs (found in the frozen food
   section of your local Thai or Vietnamese grocery)
2–3 Tbl soy or peanut oil

¼ cup chopped onions
2–3 chopped garlic cloves
½ Tbl crushed Thai red chili (more or less depending
   on your heat tolerance)
½ Tsp of white pepper

Heat oil in a wok until just barely smoking. Add in
chilis and water bugs and stir fry for a few minutes.
Add remaining ingredients and quick fry until onions
are tender. Dust with white pepper. Eat only the
abdomen; the rest is too crunchy. Serve with
cold beer.

FAMILY PENTATOMIDAE—STINK BUGS

Stink bugs include some of the largest and the most frequently seen true bugs. Although they have some odoriferous cyanide-based defensive fluid, they are by no means the only arthropods that do this. They aren't even among the "stinkiest" bugs. Their family name, Pentatomidae, is more descriptive as it means "five sections" and can refer either to their five-segmented antennae or the five areas of the body visible when viewed from above (head, thickened base of both pair of wings, v-shaped scutellum, and the overlapping tips of the wings). There are over 200 known species of stink bugs in North America and 4,100 worldwide.

FIGURE 13-13
Newly hatched nymphs of the brown marmorated stink bug clustered around the egg shells. Photograph courtesy of Gary Bernon/USDA APHIS/Bugwood.org.

FIGURE 13-12
A green stink bug, *Chinavia hilaris*. Photograph courtesy of Tom Murray.

Stink bugs produce unusual eggs laid in masses. Typically, they are barrel-shaped and often have ornamentations of patterning and/or associated spines. After hatching, the nymphs remain clustered about the egg mass for several days, a behavior that is thought to allow them to pool defenses during this vulnerable life stage. (Among some tropical species, the mother guards the eggs as well as the newly hatched young.) Immature stink bugs somewhat resemble the adult and feed in the same manner but usually have a more rounded body form, and their coloration is often brighter than that of the winged adults.

The great majority of stink bugs feed on plants, and a few are serious crop pests damaging plant tissues both by mechanical injuries produced by the piercing mouthparts and by cell-damaging saliva. The most important are species that feed on developing seeds and fruits, producing distorted

growth, seed destruction, and fruit abortion. The southern green stink bug (*Nezara viridula*), a common North American species, is extremely important in this regard; such injuries are also commonly produced by the green stink bug (*Chinavia hilaris*) and the brown stink bug (*Euschistus servus*). A gaudily colored species known as the harlequin bug (*Murgantia histrionica*) is sometimes an important pest on cabbage family plants: its feeding causes new leaves to twist. Brown marmorated stink bug (*Halyomorpha halys*), a species recently introduced into the mid-Atlantic states, is currently fast expanding its range in North America. In addition to crop injuries, this stink bug has the unfortunate habit of often moving into the upper stories of buildings at the end of the growing season.

FIGURE 13-14
Harlequin bugs, *Murgantia histrionica*, a brightly colored species that feeds on mustards and other cabbage family plants. Photograph courtesy of Tom Murray.

FIGURE 13-15
The brown marmorated stink bug, *Halyomorpha halys*. This insect recently was accidentally introduced into North America and has become a major household pest in some areas when the insect moves indoors prior to winter. Photograph courtesy of Susan Ellis.

A few other species of stink bugs are important predators of other insects. In these species, the stylets are hinged at the tip of the head and can be directed forward as they stalk their prey. Among the better-known predatory stinkbugs are the spined soldier bug (*Podisus maculiventris*), a caterpillar specialist, and the twospotted stink bug (*Perillus bioculatus*), a beetle larvae specialist.

FIGURE 13-16
A spined soldier bug, *Podisus maculiventris*, feeding on a sawfly larva. Photograph courtesy of Tom Murray.

FAMILY COREIDAE—LEAFFOOTED BUGS

Leaffooted bugs are among the largest of the terrestrial true bugs, with the largest North American representative, *Thasus acutangulus* (the mesquite bug), reaching 35–40 mm. In most species, the hindlegs are much longer than the other pairs and often have a broad, flattened area providing the "leaf foot" of their name. To further distinguish them from some other families, their head is narrower and often shorter than the prothorax. They have scent glands on the thorax, and many can produce very powerful odors.

All members of the family feed on plants, and a few are plant pests. The most notorious member of the family in North America is the squash bug, *Anasa tristis*. In much of the United States, it is the most important pest of pumpkins, winter squash, and some other squash-family plants (Cucurbitaceae). Large populations frequently develop by mid-summer and their feeding destroys large areas of stems and leaves, causing plants to wilt and die. Recently, it has been recognized that squash bugs also may transmit bacteria that cause cucurbit yellow vine disease.

FIGURE 13-17
A leaffooted bug, *Acanthocephala declivis*, showing the flattened areas of the hindlegs, a characteristic of many insects in the family Coreidae. Photograph by Whitney Cranshaw/Colorado State University.

Other important leaffooted bugs include the fruit-spotting bugs of Australasia (*Amblypelta* species) that damage a wide variety of fruits and nuts. In the United States, various *Leptoglossus* species are

FIGURE 13-18
Squash bug, *Anasa tristis*, in mixed life stages. This is one of the most important pests of winter squash in much of North America. Photograph by Whitney Cranshaw/Colorado State University.

common and can damage developing seeds and fruits, although they are considered to be crop pests of only minor importance. One of these, the western conifer seed bug (*Leptoglossus occidentalis*), very commonly moves into homes during fall and has become a major nuisance invader in recent years. Although it is harmless, its rather fearsome appearance and odor production are alarming, causing distressed

FIGURE 13-19
Western conifer seed bug, *Leptoglossus occidentalis*. This insect has become an important nuisance invader of homes in fall. Photograph by Whitney Cranshaw/Colorado State University.

homeowners to make frantic calls for help in dealing with "that stinky pine bug."

FAMILY MIRIDAE—PLANT BUGS

Plant bugs are the largest family of true bugs (1,750 known species in North America) and are abundant in both temperate and tropical regions. They exhibit a wide range of habits and can be found associated with mostly all kinds of plants. They are rather small insects (usually about 3–7 mm), and, since most have a drab appearance, they are frequently overlooked. Furthermore, their rather unoriginal common name, "plant bugs," does little to distinguish them.

Despite the name, a few species are predators of other insects, and even more are omnivores that occasionally feed on insects as well as on sap of plants. However, a majority of these are indeed plant feeders, and a small number (1%–2%) are significant crop pests.

Plant feeding by these insects can be quite destructive as a result of their "lacerate-and-flush" feeding habit. The mouthparts may probe deeply into the plant and directly puncture large numbers of cells. The area is then flushed with saliva containing digestive enzymes that can kill cells up to 3.5 mm beyond the reach of the probing stylets. The resulting soup of released plant cell fluids and saliva is then sucked up, leaving a substantial area of destroyed plant tissue.

Plant bugs that most notably cause crop loss in North America are found in the genus *Lygus*, with the tarnished plant bug (*Lygus lineolaris*) as its most prominent member. Lygus bugs prefer to feed on actively growing tissues, producing a wide range of symptoms that develop after the insect leaves the plant. Injured, newly expanding leaves often grow in a twisted and distorted manner, and new shoots may be killed. Feeding upon young flowers and small seedpods often causes so much damage that these structures abort development. Older fruits may survive a few feeding wounds, but the area of the feeding site is killed; the resulting fruit develops with deep punctures, producing a fruit defect known as "catfacing." In Africa and Asia, the genus *Helopeltis* is the most important, causing serious damage to a wide variety of crops including cocoa, tea, and cotton.

(a)   (b)

FIGURE 13-20

(a) A fourlined plant bug, *Poecilocapsus lineatus*. Photograph courtesy of Tom Murray. (b) Feeding injury produced by the fourlined plant bug. The "lacerate-and-flush" feeding manner of plant bugs produces localized areas of dead tissue. Photograph by Whiteny Cranshaw/Colorado State University.

(a)   (b)

FIGURE 13-21

(a) Adult of the tarnished plant bug, *Lygus lineolaris*. (b) Tarnished plant bug nymph. Photographs courtesy of Scott Bauer/USDA ARS.

FIGURE 13-22

Catfacing injuries to pear produced from feeding wounds of a plant bug. Photograph by Whitney Cranshaw/Colorado State University.

FAMILY REDUVIIDAE—ASSASSIN BUGS

Assassin bugs are common insects in yards, gardens, and forests, where they feed on a wide variety of insects. Most are recognizable by having a short, but prominent, beak arising from the front of a narrow head. Many assassin bugs possess bright colors that help advertise their ability to defend through a painful bite. Over 160 known species of assassin bugs occur in North America.

The forelegs of many species are thickened and raptorial, an adaptation to help grasp their prey. This latter feature is particularly well developed among a common group that frequently inhabits flowers, the ambush bugs (*Phymata* spp.).

FIGURE 13-24

Nymph of an assassin bug, *Zelus* sp., feeding on a sawfly. Photograph by Whitney Cranshaw/Colorado State University.

FIGURE 13-23

Ambush bug, *Phymata* sp., feeding on a fly. Photograph courtesy of Susan Ellis.

Well camouflaged by their colors and patterns, ambush bugs wait for passing flies, bees, or other visitors, then grab them and promptly paralyze them with their injected saliva.

Although most assassin bugs stay outdoors, a few are found inside human dwellings on occasion. One of the more curious is the masked hunter (*Reduvius personatus*), sometimes known as the "bed bug hunter." Adults are rather drably colored, dark brown to black, and the nymphs are well camouflaged by the small pieces of debris, such as household lint, that stick to the body. So if you do see a walking dust ball under your bed, look closer; it may prove to be a masked hunter nymph in its disguise.

One subfamily of the assassin bugs (Triatominae) tackles much bigger prey; they feed on the blood of mammals and birds. Sometimes known as "conenose bugs," "bloodsucking conenoses," or "kissing bugs," they are primarily found in the

(a)

(b)

FIGURE 13-25

(a) Adult of the masked hunter, *Reduvius personatus*, a species that may be found feeding on insects in homes. (b) Nymph of the masked hunter. Debris sticks to the body of masked hunter nymphs, masking their appearance. Photographs by Whitney Cranshaw/Colorado State University.

tropics, where some species frequently wander into homes and bite humans at night in a manner similar to bed bugs (see below). A dozen species of conenose bugs are known to occur in the United States, mostly restricted to the southwestern states. Two species (*Triatoma sanguisuga, T. lecticularia*) range more widely and may be found in parts of the Midwest and mid-Atlantic states. Fortunately, these are almost exclusively associated with wild animals and are only rarely found in homes.

Unlike the bed bugs, the conenose bugs can be important vectors of human pathogens. Chagas disease is produced by the infection with *Trypanosoma cruzi*, a trypanosome that all triatomine bugs are thought capable of harboring. (About a half-dozen species are thought to be particularly important in human cases, because of behaviors that tend to bring them into

FIGURE 13-26

A conenose bug, *Triatoma sanguisuga*, feeding on the foot of a mouse. Photograph courtesy of Sturgis McKeever/Georgia Southern University/Bugwood.org.

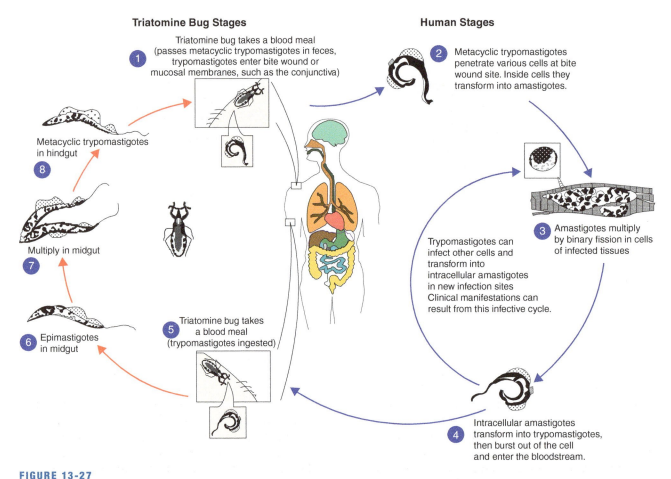

FIGURE 13-27

Life cycle of *Trypanosoma cruzi*, the organism that can produce Chagas disease in humans. Figure courtesy of the Centers for Disease Control and Prevention.

human contact.) *Trypanosoma cruzi* is spread to humans and other hosts in the course of feeding but not directly through the bite. Instead, an infected conenose bug defecates onto the skin of its host. This material may contain infective forms of the *T. cruzi* pathogen, and subsequent scratching rubs the disease organism into wounds or mucous membranes. Pathogen transmission is much like louse-borne typhus discussed in chapter 12.

Chagas disease is estimated to affect up to 12 million people mostly in Central and South America. Although the majority of people do not show serious effects from infection and remain asymptomatic, some victims develop chronic problems that seriously damage the digestive tract or the heart. Despite its prevalence in parts of tropical America, a total of only five human cases

of Chagas disease are known to have originated from conenose bug transmission within the United States. Nonetheless, it is widespread among wild animals, with raccoons, opossums, and dogs most commonly infected.

FAMILY CIMICIDAE—BED BUGS AND THEIR RELATIVES

Despite an almost mythical standing, the bed bug (*Cimex lectularius*) is a very real insect that feeds on human blood. Bed bugs are rather small—about the size of a watermelon seed (6 mm) when full-grown—wingless, mahogany brown insects with a flattened, oval body. Bed bugs can swell to the size of a (blood-filled) small pea when engorged after feeding.

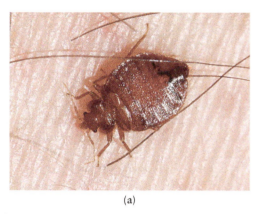 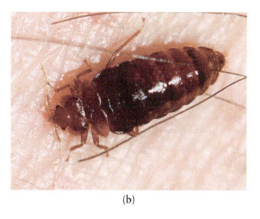

(a)   (b)

FIGURE 13-28
(a) A bed bug settled to begin feeding. (b) Bed bug after feeding for approximately 10 min. Photographs courtesy of Whitney Cranshaw/ Colorado State University.

The association of the bed bug with humans is ancient and probably originated when both were dwelling in caves. The long-standing problems that humans have experienced with the bed bug are also reflected in the very word "bug." Of Celtic/Welsh origin, the word "bug" signified a ghost or goblin, figures of dread that moved quietly in the night. "Bug" later morphed into the English term "bugbear," which meant "bogeyman." Despite this history, for unknown reason, countless children have been put to bed with the rather unsettling advice "good night, sleep tight, and don't let the bed bugs bite."

Bed bugs are active at night, resting in dark cracks associated with bedding and bedroom furniture during the daytime. Peak feeding is around 3–6 AM when they locate their sleeping host from body heat and expelled carbon dioxide. While feeding, the bed bug introduces saliva with anesthetic properties such that bites are painless and undetected by the sleeping victim. Unfortunately, the saliva typically will later induce an allergic reaction resulting in a small blotch that often progresses to a larger whitish, itchy weal. (Due to individual immune responses, the reaction to the bite of a bed bug varies tremendously among humans.) Also, a bed bug may feed at a site for only a few minutes, move and feed again nearby, producing multiple, closely spaced bites. Although bed bug bites can result in uncomfortable itching, it may be some small solace that bed bugs do not transmit any human pathogens while biting.

The female bed bug lays one to five eggs daily over a period of months, tucking them into the tiny cracks associated with the sleeping quarters

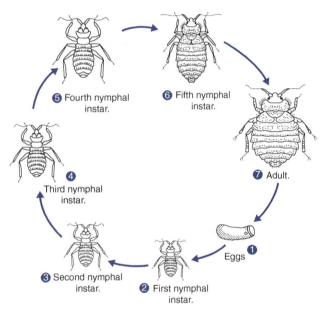

FIGURE 13-29
Life cycle of a bed bug. Illustration provided courtesy of the Centers for Disease Control and Prevention.

FIGURE 13-30
Life stages of a bed bug on a nickel for scale. Photograph by Whitney Cranshaw/Colorado State University.

where she hides during the day. Eggs hatch in about 1–2 weeks and the tiny nymphs begin to search out hosts upon whom they feed upon. They develop through five nymphal stages (instars) before becoming full-grown; at each stage, they must feed at least once. A complete generation can be completed in as little as 5 weeks under optimum conditions of temperature, humidity, and food but usually develops across several months. Bed bugs are quite resistant to starvation and adults can survive up to 1.5 years between meals; nymphs can survive about 3 months without feeding.

Mating by bed bugs takes an unusual form known as **traumatic insemination**. Normal coupling with the genital opening of the female is bypassed. Instead, the aedeagus (penis) of the male, in the form of a stout hook, is used to directly pierce the abdomen of the female. Injected sperm then migrates in the female hemolymph to sperm reservoirs connected to the ovaries. This has costs to the female, and the number of matings is reflected in the number of abdominal scars she carries. In response, the female bed bugs appear to have counter-adapted by modifying an area of the body between the fifth and sixth segments, known as a **spermalege**. Mating punctures into this region are well tolerated and cause little, if any, ill effects on the health of the female; punctures made outside the spermalege can greatly increase mortality of the female. Traumatic insemination is also known to occur with species found in several other families of true bugs including Nabidae, Anthocoridae, and Miridae.

During the day, bed bugs are often found in small groups, and such harborage areas are typically speckled with brownish fecal spots of partially digested blood. The majority of bed bugs are usually found associated with the mattress or bed frames, particularly near the head end. Bed bugs can, nonetheless, be scattered through a room in locations such as at electrical sockets, under the bases of lamps, behind molding, and in drawers. As a result of this scattered hiding, bed bugs become excellent hitchhikers and are readily moved to new sites with furniture or luggage. They may establish wherever there are human hosts for them to feed upon, and the presence of these bed bugs has much more to do with encounters during travel and transport of infested materials than sanitation history.

Incidence of bed bug infestations in North America has resurged in recent years. This followed over a half century when their numbers were dramatically reduced, largely due to development of effective insecticides. Indeed, a great many people grew up during a time when bed bugs had receded to dimming memory. Various explanations have been put forward for the recent dramatic increase in population numbers including introduction of new strains with international travel, the general increase in the mobility of people within North America, inattention to movement of infested materials (e.g., used mattresses and furniture), and changes in indoor insecticide uses.

The family Cimicidae is represented with 15 known species in the United States and 92 species worldwide. Although most of the other known species primarily feed on birds or bats, a few of these also bite humans on occasion. Frequently, barn swallow nests also house swallow bugs (*Oeciacus vicarius*) which will bite humans, particularly in spring when the hungry insects resume their activity in anticipation of the return of nesting birds. In the western United States, *Hesperocimex coloradensis* is associated with nests of owls, woodpeckers, and martins. Nesting bats support bat bugs (*Cimex pilosellus*, *C. adjunctus*) that sometimes invade living areas of buildings where bats are allowed to roost.

FIGURE 13-31
A western bat bug, *Cimex pilosellus*. Bats are the only host on which this insect can reproduce, although they may occasionally bite humans when they disperse from a site where bats were roosting. Photograph by Whitney Cranshaw/Colorado State University.

Suborder Auchenorrhyncha—Cicadas and Hoppers

The suborder Auchenorrhyncha includes several families of insects that all feed on plant fluids. The beak containing the mouthparts arises from the back of the head, just behind the plane of the eyes. The front of the head is often enlarged and is packed with powerful muscles (cibarial pump) that allow certain species to withdraw plant sap rapidly and even, in some species, extract fluid from the xylem of plants.

FAMILY CICADIDAE—CICADAS

Cicadas include the heaviest, longest-lived, and loudest species in the order Hemiptera. Much of their life remains unseen since cicada nymphs develop on the roots of trees and shrubs, sucking fluid from the nutrient-poor xylem and growing slowly. Life cycles vary, but most species likely take between 2 and 8 years to become fully grown. At the end of the last nymphal instar, they emerge from the soil and crawl up a vertical surface to which they cling while molting to the winged adult stage. Initially, the newly emerged adults are pale colored, but they soon begin to darken as the exoskeleton sclerotizes and hardens.

Adult females deposit their eggs into twigs of plants using a long, sharp ovipositor. This frequently produces noticeable plant injuries; the splintered twigs are weakened and may subsequently break, a symptom called "flagging." After about a month, the eggs hatch and the first instar nymphs drop to the ground, burrow into the soil, and seek roots.

In most of the 100-odd cicada species that occur in North America, at least some emerge every year. Their annual presence leads them to often be known as "annual cicadas." The various large "dog-day cicadas" (*Tibicen* spp.), which produce droning songs that often define evenings during the "dog-days" of summer, are often the most commonly seen (and heard) of the annual cicadas in the United States.

FIGURE 13-32
A dog-day cicada, *Tibicen dealbatus*. The adult stage has just recently emerged from the last nymphal stage, which developed underground. Photograph courtesy of David Leatherman.

FIGURE 13-33
Periodical cicada female ovipositing in a branch. Photograph courtesy of Pennsylvania Department of Conservation/Bugwood.org.

FIGURE 13-34
Periodical cicada eggs inserted into a small branch. Photograph courtesy of Pennsylvania Department of Conservation/Bugwood.org.

In many places of the eastern and central United States, it is the **periodical cicadas** (*Magicicada* spp.) that are best known, either through legend or a personal experience. Unlike other cicadas, these insects display synchronized emergences at long time intervals. For example, the 17-year cicadas emerge at a given site every 17 years. Other species have a 13-year life cycle and will appear every 13th year. In the intervening time, about 16 years and 10 months for the 17-year cicada, the insects are not seen aboveground and remain as nymphs feeding on the roots of their host plants. Seven species of periodical cicadas are found in the eastern United States: three with a 17-year life cycle (*Magicicada septendecim*, *M. cassini*, *M. septendecula*) and four with a 13-year life cycle (*Magicicada tredecim*, *M. tredecassini*, *M. tredecula*, *M. neotredecim*).

Their synchronized emergence has long been recognized, is well documented, and is predictable. Known as **broods**, synchronized emerging populations have been conventionally numbered I–XVII for the 17-year species and XVIII–XXX for the 13-year species. Each has a distinct geographic distribution (table 13-1), with the largest being Brood X, found through 15 states. The smallest brood, Brood VII, is restricted to a small area of upstate New York. As there are only 12 recognized broods of the 17-year species, there are years when no broods of these species emerge. (Brood XI, formerly known from the Connecticut River Valley of New England, apparently went extinct over 100 years ago.) Thirteen-year periodical cicada species are present as

TABLE 13-1 A chronology of recently completed and predicted future emergence of periodical cicadas.

17-YEAR BROODS	RECENT AND EXPECTED FUTURE EMERGENCES			GENERAL REGION OF EMERGENCE
I	2012	2029	2046	VA, WV
II	1996	2013	2030	CT, MD, NC, NJ, NY, PA, VA
III	1997	2014	2031	IA, IL, MO
IV	1998	2015	2032	IA, KS, MO, NE, OK, TX
V	1999	2016	2033	MD, OH, PA, VA, WV
VI	2000	2017	2034	GA, NC, SC
VII	2001	2018	2035	NY
VIII	2002	2019	2036	OH, PA, WV
IX	2003	2020	2037	NC, VA, WV
X	2004	2021	2038	DE, GA, IL, IN, KY, MD, MI, NC, NJ, NY, OH, PA, TN, VA, WV
XIII	2007	2024	2041	IA, IL, IN, MI, WI
XIV	2008	2025	2042	KY, GA, IN, MA, MD, NC, NJ, NY, OH, PA, TN, VA, WV
13-Year Broods				
XIX	1998	2011	2034	AL, AR, GA, IN, IL, KY, LA, MD, MO, MS, NC, OK, SC, TN, TX, VA
XXII	2001	2014	2027	LA, MS
XXIII	2002	2015	2028	AR, IL, IN, KY, LA, MO, MS, TN

FIGURE 13-35

A dog-day cicada, *Tibicen dealbatus*. Dog-day cicadas make the loud droning songs of the midsummer "dog days." Photograph by Whitney Cranshaw/Colorado State University.

FIGURE 13-36

A cactus dodger, *Cacama valvata*. This cicada is capable of producing a very loud startling sound when disturbed. Photograph by Whitney Cranshaw/Colorado State University.

adults in the United States in only 3 of 13 years. (Brood XXII apparently also went extinct.)

Emergence events of periodical cicadas are among the most spectacular natural phenomena. Thousands of individuals may emerge from the base of a single tree; millions per hectare may be active aboveground. The humpbacked nymphs emerge from holes during the night and nymphal "shells" (cast off exoskeletons) may cover areas underneath host plants. Suddenly, the large flying adults can be seen everywhere, and the air is full of the buzzing calls, which can be deafening. All this activity occurs during just a few weeks in spring (between late April and late June)—and then it is all over—not to be repeated again for nearly another human generation.

It is not known how long periodical cicadas have been present in North America, but they were soon

noted by the early European colonists. Among the accounts from the 1600s (quoted from Howard Ensign Evans' *Life on a Little Known Planet*) were the following:

> ....there was a numerous company of Flies, which were like for bigness unto Wasps or Bumble-Bees, they came out of little holes in the ground, and did eat up the green things, and made such a constant yelling noise as made all the woods ring of them, and ready to deaf the hearers;....

> ....there was such a swarm of certain sort of insect in that English colony, that for the space of 200 miles they poyson'd and destroyed all the trees of that country. There being found innumberable little holes in the ground, out of which those insects broke forth in the form of maggots, which turned into flyes that had a kind of taile or sting, which they struck into the tree, and thereby envenomed and killed it...

These early settlers were ill-prepared for the new life forms that North America held and certainly did not have the aid of entomology identification guides. Pretty much all they had was the Bible and, in trying to make sense of the cicada emergence, the closest parallel they could find were references to the plague of locusts. As a result, these insects have since been inappropriately referred to as "locusts," or more specifically "17-year locusts." Of course the term locust is only correctly used for very different insects, certain migratory grasshoppers that may become abundant, and these are discussed with the order Orthoptera in chapter 9.

Their sound production abilities are particularly remarkable, and cicadas include the noisiest and loudest of all insects. Although cicadas may make sounds by various means, the most distinctive and loudest songs result from unique modifications of the abdomen. The most important is a structure known as the **tymbal**, which is domed and ribbed on the ventral side of the abdomen. Rapid contraction of the tymbal by special muscles can buckle the ribs and produce a series of loud clicks. The body then resonates and amplifies these sounds in hollow cavities that can comprise up to 70% of the abdomen of a male cicada. As with most sound-producing species, it is only the males that produce the loud sounds used to attract female mates.

FIGURE 13-37
Mass of the periodical cicadas following their periodic (17 years) emergence. Photograph courtesy of Pennsylvania Department of Natural Resources/Bugwood.org.

FIGURE 13-38
Branch dieback (flagging) produced by twig wounding of periodical cicadas when ovipositing. Photograph courtesy of Pennsylvania Department of Natural Resources/Bugwood.org.

Cicadas may make several different types of sound signals. A typical sequence begins with the male calling song, which is perceived by both sexes within hearing range. As potential mates approach the male, he switches to a courting song. In turn, the female may respond with a subtle wing flicking or other behavior indicating receptivity. Additional courting songs may be produced if courtship successfully progresses. Cicadas can also make alarm calls—often extremely sharp and loud—when startled or approached by potential predators.

Despite apparently making their general presence known to potential predators, the loud noises cicada's produce may actually repel potential predators by saturating the ear with noise and thus making them difficult to locate. The loud calls also have the advantage of allowing cicadas to attract mates over long distances or through heavy vegetation.

There are several candidates for noisiest cicada, with contenders from North America, Africa, Australia, and Southeast Asia. Many of these species can produce signals in excess of 105 decibels (dB), perceived at a distance of 50 cm. (A loud car horn produces sound in the range of 110 dB; a jet engine emits noises in the range of 120 dB or more.) Clearly, these animals can hold their own in the world of noise. On the other hand, the songs of a great many cicadas are subtle in human perception, producing faint clicking or rustling sounds. Still other cicadas produce songs that are undetectable being pitched above the range of the human ear.

FAMILY CERCOPIDAE—SPITTLEBUGS OR FROGHOPPERS

Spittlebugs are one of the very few insects (as well as cicadas and sharpshooter leafhoppers) that can pull fluid from the water-bearing xylem vessels of a plant, a difficult task requiring strong pumping muscles. As the fluid in the plant xylem is nutrient poor, they must ingest very large volumes of fluid. Spittlebugs

FIGURE 13-39
The spittlebug nymphs in spittle mass. Photograph courtesy of David Cappaert/Michigan State University/ Bugwood.org.

have been measured pumping 150–300 times their weight in xylem fluid per hour.

After removing the nutrients from this liquid, a spittlebug nymph uses its excreted waste for a unique purpose, that is, the production of a spittle mass. By mixing the waste fluid with a sticky substance from special abdominal glands and air, a stable protective foam is produced. This foam or "spittle" flows over the insect's body, providing it with protection from many predators and a high-humidity shelter.

Spittlebugs can be found on many kinds of plants but appear to be more common on those that have higher levels of amino acids in the nutrient-poor xylem fluids. The stubby-bodied, winged adults, known as froghoppers, do not produce spittle masses.

FAMILY CICADELLIDAE—LEAFHOPPERS

The leafhoppers are a very large and economically important group of insects with over 2,500 known species in North America. Most are small, typically only a few millimeters long, generally wedge-shaped, and rather drably colored—with some notable exceptions. Leafhoppers are active insects with

nymphs capable of walking rapidly, often in a crab-like fashion. Adults jump from plants to launch themselves into flight. Small spines on their legs allow them to be distinguished from related hemipteran "hoppers," such as the planthoppers (Delphacidae family), treehoppers (Membracidae family), and froghoppers/spittlebugs (Cercopidae family).

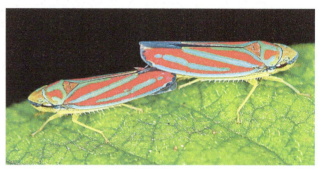

All leafhoppers feed on plant fluids, although their mouthparts may reach and withdraw fluids from different locations within a plant causing a range of different symptoms. Most feed on the sugary sap found in the plant phloem, much like aphids (see following), and they may similarly produce honeydew as a waste product. Usually, injuries from phloem-feeding leafhoppers are minor; but a particularly important species, the potato leafhopper (*Empoasca fabae*), destroys cells of the phloem as it feeds. Feeding by the potato leafhopper disrupts the movement of sap, reduces photosynthetic ability throughout the leaf, and produces a progressive death of leaf tissue along the leaf margins known as "hopperburn." It is a serious pest of potato, alfalfa, beans, and many nursery plants in the midwestern United States.

Other leafhoppers extract the cell contents from individual cells underlying the leaf surface, producing small whitish wounds (**stippling**). Still other leafhoppers, including those known as sharpshooters, have large extensions of the head packed with muscles that allow them to extract the watery fluids found in the xylem vessels of plants. The common name "sharp-shooter" is often given to the xylem-feeding leaf-hoppers in reference to their habit of using their abdomen and legs to flick away the excreted watery waste droplets.

(a)   (b)

FIGURE 13-42

(a) Adult and nymph of the potato leafhopper, *Empoasca fabae*. (b) Damage to potato produced by the potato leafhopper. The field in foreground has been killed due to "hopperburn" produced by feeding injuries produced by the potato leafhopper. Photographs courtesy of Ted Radcliffe/University of Minnesota.

Many leafhoppers are important as vectors of plant pathogens, and, worldwide, they are known to transmit about 80 different plant disease organisms. In Asia, leafhoppers (along with planthoppers) are extremely important in the spread of key virus diseases affecting rice. In the United States, the most important virus disease involving leafhoppers is beet curly top, spread by the beet leafhopper (*Circulifer tenellus*), a species native to Europe. Periodically, beet curly top causes devastating outbreaks among peppers, tomatoes, sugar beets, and other crops grown in the western United States.

Even more important is the role leafhoppers have in transmitting certain types of bacterial diseases. The plant disease known as aster yellows is a very common problem in much of the United States, where it can affect lettuce, carrots, celery, potatoes, and scores of garden flower species. It is produced by a type of bacterium known as a phytoplasma that grows in the phloem of the plant and is spread solely by leafhoppers, particularly the aster leafhopper (*Macrosteles quadrilineatus*).

Perhaps even more economically significant are the diseases caused by the bacterium *Xylella fastidiosa*. Depending on the plant species and strain of the bacterium, *X. fastidiosa* causes Pierce's disease of grapes, almond leaf scotch, alfalfa dwarf, phony peach disease, oleander leaf scorch, citrus variegated chlorosis, and a host of hardwood scorch diseases. Pierce's disease is of special concern as it is fatal to several *Vitis* species and varieties of grapes (including vines that produce grapes for wine, table grapes, raisins, and juice).

(a)   (b)   (c)

FIGURE 13-43

(a) Adult beet leafhopper, *Circulifer tenellus*. This insect is primarily important because it is the vector of several viral diseases affecting crops, particularly curly top virus. (b) A beet leafhopper nymph just hatched from the egg. Eggs are inserted into plant tissue. (c) Late instar nymph of the beet leafhopper. Photographs courtesy of Ken Gray/Oregon State University.

FIGURE 13-44

The glassy-winged sharpshooter, a leafhopper that is an efficient vector of the bacterium *Xylella fastidiosa*, which can produce several diseases in plants. Photograph by Jack Kelly Clark and provided courtesy of the University of California IPM Program.

The disease has long been present in California grape-growing areas but usually only occurred at a relatively low rate of incidence (there have been historical exceptions to this). The situation changed dramatically in the early 1990s when the

FIGURE 13-45

Wine grapes showing symptoms of Pierce's disease, a serious disease of grapes. Incidence of this disease can rise sharply when the glassy-winged sharpshooter (*Homalodisca vitripennis*) establishes, as recently occurred in parts of southern California. Photograph by Jack Kelly Clark and provided courtesy of the University of California IPM Program.

glassy-winged sharpshooter (*Homalodisca vitripennis*) became established in southern California and triggered at least two epidemics of Pierce's disease. Due to it polyphagous nature, strong dispersal abilities, and its ability to transmit the bacteria, the glassy-winged sharpshooter has become a serious threat to a multibillion dollar California wine grape industry. Currently, intensive efforts are being made to slow the spread of the insect and prevent its establishment within the large wine-producing areas in the central and northern portions of California.

Suborder Sternorrhyncha— Aphids, Psyllids, Whiteflies, and Scales

Hemipterans found within the suborder Sternorrhyncha are all small insects that suck plant fluids, usually those of the phloem tissues. The beak containing the mouthparts is also small and arises so far back in the head that it often appears near the base of the front pair of legs. During most of their life cycle, they are immobile insects that move little, but some have winged adult forms capable of long distance, wind-assisted migration.

In the course of feeding upon the sugar-rich phloem of plants, many of these insects excrete a

FIGURE 13-46

Brown soft scale, *Coccus hesperidum*. The fluid that surrounds the insect is honeydew, a sweet sticky fluid excreted by soft scales and some other phloem-feeding Hemiptera such as aphids, whiteflies, and mealybugs. Photograph by Whitney Cranshaw/ Colorado State University.

sweet, sticky waste product known as **honeydew**. Often, people first encounter honeydew when their vehicle has been parked under some aphid-infested tree, and they return to find the windshield sparkled with tiny sticky honeydew droplets. A common assumption is that the tree is dripping "sap," when in actuality the offending substance is sap that has traveled through the digestive system of an aphid and emerged transformed as a honeydew droplet. In addition to being sticky, honeydew is attractive to many insects, including flies, ants and various wasps and bees. Honeydew also serves to support the growth of common fungi known as sooty molds that darken any surface on which they grow. Other species have evolved the ability to transform excess sugars found in their food into waxes that they use for protective coverings or are excreted as pellets known as **lerps**.

Some Sternorrhyncha may display complex life histories. In extreme cases, these may involve numerous generations that require one, sometimes two, years to fully complete and involve stages of different form and habit.

FAMILY APHIDIDAE—APHIDS

In temperate climates, aphids are usually, by far, the most common sucking insects associated with garden plants. Over 1,350 known species occur in North America, and there are few plants that do not support at least one kind of aphid or another. Aphids are called "plant lice" in Great Britain, a term that sometimes enters the North American garden literature.

Physically, aphids are small insects, typically about 3–5 mm in length, and most have a pear-shaped body form. A pair of tailpipe-like **cornicles** project from the abdomen through which aphids release defensive chemicals and alarm pheromones.

FIGURE 13-47
Dark staining of a sidewalk underneath a linden tree. The staining is produced by sooty mold that grows on the honeydew of the linden aphids that regularly infest this tree. Photograph by Whitney Cranshaw/Colorado State University.

FIGURE 13-48
Asian citrus psyllid nymphs excreting strings of lerps. Many psyllids produce lerps as a waste material, which are in the form of solid pellets or strings. Photograph by Jack Kelly Clark and provided courtesy of the University of California IPM Program.

FIGURE 13-49
Colony of spirea aphids, *Aphis citricola*, showing a range of life stages. Photograph by Whitney Cranshaw/Colorado State University.

Aphids can come in a wide range of colors and many have distinctive body patterning. Some, known as "woolly aphids," secrete waxy threads through pores in the body. These threads may completely cover the insect giving it a woolly appearance.

FIGURE 13-51
A mother aphid giving live birth to a daughter. Photograph by Whitney Cranshaw/Colorado State University.

FIGURE 13-50
A colony of leafcurl ash aphids, *Meliarhizophagus fraxinifolii*. These aphids live within a leaf curl that they induce on emerging ash leaves. They produce a considerable amount of wax, which is protective and also helps to prevent honeydew from coating their body. Photograph by Whitney Cranshaw/Colorado State University.

Reproduction in aphids is notable for the nearly complete absence of males. For some species, males are totally unknown, while species that do produce males only produce them during one generation per year. Instead, routine reproduction occurs through **parthenogenesis**, in which the eggs hatch internally within the mother, resulting in live birth. All such are genetically identical to each other and to their mother. Furthermore, at birth, the newborn daughters have already begun to mature their own young, arriving pregnant at birth. (A female aphid that has given birth is thus not only a new mother but a new expectant grandmother as well.) The ability to reproduce in this matter provides aphids with a tremendous reproductive potential; a population of several dozen may arise in just a few weeks following the arrival of a single female.

Aphids also have the ability to produce either winged or wingless adult forms known as **alatae** or **apterae**, respectively. The ability to develop wings or remain wingless is due to hormonal influences, not genetics. Most often, the wingless apterae remain on the host plant colonized by their mother. Here, they may produce up to five daughters per day during their relatively brief life of a few weeks. Alatae may

FIGURE 13-52
A winged form of green peach aphid, *Myzus persicae*, surrounded by wingless forms of various age. Photograph by Whitney Cranshaw/Colorado State University.

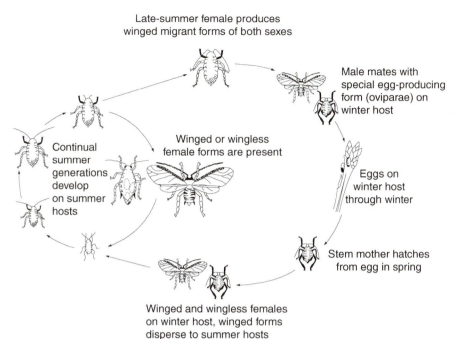

Late-summer female produces
winged migrant forms of both sexes

Male mates with
special egg-producing
form (oviparae) on
winter host

Continual
summer
generations
develop
on summer
hosts

Winged or wingless
female forms are present

Eggs on
winter host
through winter

Stem mother hatches
from egg in spring

Winged and wingless females
on winter host, winged forms
disperse to summer hosts

FIGURE 13-53
An aphid's generalized life cycle that is holocyclic, involving both summer and winter host plants. Drawing by Loretta Mannix.

also occur on this host plant selected by the mother, but they are relatively few in number when the overall population density is low. The proportion of alatae tends to increase within the population when there is some need to leave the plant. Overcrowding and a decline of food quality are common environmental cues that shift aphids to produce more winged adult forms.

Winged stages also occur during times of the year when aphids are required to shift to new species of host plants, such that it commonly occurs late in the growing season. This alternation of host plants is yet another unusual aspect in the complex life cycle of many aphids. Host alternation arises largely as a means to cope with periods of freezing temperatures that make host plants unavailable through winter.

Aphids have responded by producing eggs that survive through this harsh period. The eggs are laid in fall, usually near the buds of a perennial plant so that the nymphs that hatch from them can be near food at bud break and leaf emergence the following spring. Often, the overwintering plant (winter host) where these eggs are laid is very different from the plants the aphid will feed upon in summer (summer host). Aphids with this complicated life cycle, involving alternation of host plants, are said to have a holocyclic life cycle (figure 13-53).

FIGURE 13-54
Overwintered eggs of a maple aphid, *Periphyllus* sp., laid near a bud. Photograph courtesy of Ken Gray/Oregon State University.

Each species of aphid is highly particular in its choice of winter host. For example, in late summer and early fall, the green peach aphid (*Myzus persicae*) lays its eggs on only a couple kinds of *Prunus* spp. trees, such as peach and apricot. Just prior to this, the sole male form is produced, along with a unique sexual form of the female. Eggs hatch the following spring, coincident with budbreak, and a few generations of apterae are produced on the leaves of these trees. Later, alatae are produced so that all leave the winter host plant and move to more succulent annual plants, including many important vegetable crops.

FIGURE 13-55
Green peach aphids, *Myzus persicae*, just prior to leaving their winter host, peach. Several generations were completed on this plant, following hatch of the overwintered eggs, and this species causes the emergent leaves to curl. Winged forms produced will leave peach and seek their summer hosts, which include many vegetable crops and weeds. Photograph by Whitney Cranshaw/ Colorado State University.

Many aphids are important pests damaging plants in a variety of ways. If very high numbers are sustained on a plant for long periods of time, they can cause wilting and slow growth, particularly if plants are under drought stress. Aphids may also distort the new growth of some plants. Many plant viruses are vectored by aphids.

In other situations, aphids become important not because of plant injuries but because of their excreted honeydew waste product. Aphids are the most abundant single source of honeydew in most of the United States and Canada. Some other Hemiptera with similar feeding habits (e.g., whiteflies, soft scales, mealybugs) may be more important honeydew producers locally on certain plants.

Although aphid numbers may increase rapidly, high populations outdoors infrequently persist for long periods as they are prey to numerous natural enemies. Aphids possess few defenses and are soft-bodied so they serve as an important food resource not only for various insect predators (e.g., lady beetles, green lacewings, syrphid flies, certain parasitic wasps) but also for many songbirds and reptiles. In addition, the honeydew that is continuously excreted by feeding aphids is a source of sweet food used and relied upon by a great many insects, including many bees and ants.

FIGURE 13-56
This aphid colony shows typical activity that might be seen during late spring and summer. All stages of this bright yellow species, *Aphis lutescens*, are present as are some shed exoskel-etons following molts. Two lady beetle larvae (lower right and along leaf edge) are feeding on some aphids. An "aphid mummy" is present in the lower left, indicating that this aphid has been parasitized by a small wasp. Photograph courtesy of David Cappaert/Michigan State University/Bugwood.org.

● Ants and Aphids—An Order of Benevolent Sisterhood

Aphids and ants often are found in relationships that are of mutual benefit to both. Aphids, through their plant-feeding habit and excretion of abundant amounts of honeydew, provide a source of sweet material that is a favored food of many ants. In return, the ants provide protection, attacking potential aphid predators such as lady beetles and green lacewings that are important aphid natural enemies.

Often the relationship is fairly casual; an aphid colony is discovered by the ants and is subsequently provided protection. In some cases, the arrangement is more elaborate, and aphids will be physically

(continued)

FIGURE 13-57
A carpenter ant tending a colony of aphids on an aspen twig. The ants will collect the honeydew droplets as they are excreted by aphids and return it to their colony where it is a valuable source of sugars. In return, the ant will protect the aphids from potential predators. Photograph by Whitney Cranshaw/Colorado State University.

moved from plant to plant by the ants, in a manner resembling herding cows. The association is even more advanced with the corn root aphid (*Aphis middletonii*) and sweet-loving cornfield ants (*Lasius* spp.) found in the midwestern United States. These aphids develop belowground and feed on plant roots. At the end of the season, as the host plants die and

corn root aphids produce overwintering eggs, the cornfield ants carry the eggs into their colony. In spring, the aphid eggs hatch and the ants carry the young aphids to the roots of smartweed (*Polygonum* spp.) and other plant species when plant growth is renewed in spring. Later in the year, the ants move the aphids to the roots of corn.

Grape Phylloxera—A Hard Lesson on International Trade

In the 1860s, the massive French wine industry faced destruction from the work of a small insect, the grape phylloxera (*Daktulosphaira vitifoliae*) (Hemiptera: Phylloxeridae). Native to North America, grape phylloxera was accidentally introduced in

Europe and first noticed in France in 1863. Subsequently, it spread throughout the country and later to Germany, Italy, and Spain. By 1875, wine production in France had plummeted by over two-thirds.

FIGURE 13-58
Grape phylloxera, *Daktulosphaira vitifoliae*, on grape roots. Infestations can seriously damage the roots of wine grapes. Photograph by Jack Kelly Clark and provided courtesy of the University of California IPM Program.

FIGURE 13-59
Grape vines in California decline due to root infestation by grape phylloxera. Photograph by Jack Kelly Clark and provided courtesy of the University of California IPM Program.

(continued)

The grape phylloxera is an innocuous-looking, small, oval-shaped insect related to aphids. Some forms of this insect develop on roots of grape vines, producing enlarged nodules or galls. These injuries, plus root-rotting fungi that colonize the galls, destroy the roots and ultimately often cause the plants to die.

Desperate for help, France invited the pioneering American entomologist Charles Valentine Riley to visit and make recommendations. After assessing the severity of the problem, he returned to the United States full of energy and ideas, and he dove into research on this unusual insect.

Because so little was known about grape phylloxera, and there was essentially no US grape industry that was threatened by it, Riley had his job cut out for him. One of the first things he had to do was determine the life cycle of the insect, which is extraordinarily complex. The complete life cycle of the grape phylloxera produces four different adult forms and involves multiple generations over a period of 2 years. Some of these feed on leaves, producing raised galls on grape leaves; other morphs develop on roots. Essentially, all grape phylloxeras are females that reproduce parthenogenetically; males and sexual females are produced during only one generation at the end of the 2-year life cycle. (This life cycle is short-circuited in Europe and California where only all female root-feeding forms are present. Leaf gall-producing forms still predominate in the eastern half of the United States, particularly on French-American hybrid grapes.)

Additionally, Riley noted that the grape phylloxera does not cause serious injury to the roots of any of the 16 species of North American grapes. The insect may be present but the plant tolerates it without the resulting root injury. Thus, he proposed to graft the French wine grapes, *Vitis vinifera*, onto rootstocks of the phylloxera-resistant North American grape, *Vitis labrusca*. This was only partially successful because the rootstocks were not well adapted to the alkaline, chalky soils of France. Rootstocks produced by hybridizing the French and American grapes solved that particular problem. With solution in hand, there was a massive effort to replant all of the hundreds of millions of European grape vines onto these new, resistant rootstocks. Begun in 1880, this huge task was largely complete within a decade and the French wine industry was saved.

Although grape phylloxera provides a classic example of how the use of resistant plants can be effectively used in pest management, it also provides examples for caution. Not all rootstocks used are equally resistant to the insect, and in 1915 a strain of phylloxera that had evolved to attack the presumably resistant, and by then commonly grown, rootstock used in Spain appeared. This resulted in the need to again remove and replant vines, using yet again rootstock of a different genetic origin. More recently, grape phylloxera arose during the early 1980s in California, again associated with rootstocks with an incomplete resistance that the insect was able to overcome.

There are also lessons to be learned from the hazards of the unrestricted movement of plant materials that took place between the United States and France during the 1800s. It is believed that grape phylloxera was originally introduced into France on grapes brought over to help the French breed and develop new varieties of grapes resistant to a new and devastating fungal disease, called powdery mildew. (Powdery mildew apparently had been accidentally introduced earlier on grapes that also originated from North America.) Subsequently, as the grape breeding efforts between the two countries increased in their efforts to meet the grape phylloxera crisis, a second fungal disease, downy mildew, was introduced. During a relatively brief period, three new pests of grapes from North American became permanently established in Europe: powdery mildew, grape phylloxera, and downy mildew, all of which to this day must be managed by European growers.

FIGURE 13-60
On wild grapes in eastern North America, grape phylloxera has a complex life cycle that also includes stages that make galls on leaves, in addition to a root-infesting stage. Photograph courtesy of Jim Kalisch/University of Nebraska.

FAMILIES PSYLLIDAE AND TRIOZIDAE—PSYLLIDS

Psyllids are small, plant-feeding insects that rarely attract much attention. Adults somewhat resemble miniature cicadas but have the ability to jump, lending them the common name "jumping plant lice." Immature forms have a substantially different form and most are flattened and oval-shaped.

FIGURE 13-61
Bluegum psyllid, *Ctenarytaina eucalypti*, adult, and nymph. Photograph by Jack Kelly Clark and provided courtesy of the University of California IPM Program.

When psyllids are noticed, it is usually their excrement that is first observed and not the insect. Although some psyllids produce clear droplets of honeydew in the same manner as aphids, soft scales, and other phloem-feeding hemipterans, many convert their waste into wax-coated **lerps**. In some species, the lerps may be excreted in the form of long strands of material that coil and cover the insects; in other psyllids, small pellets that resemble granulated sugar are produced.

Unusual plant growth disorders can be produced by psyllid feeding through the effects of the insect's saliva on plant tissues. For example, the native hackberry and sugarberry trees (*Celtis* spp.) host several different *Pachypsylla* species of psyllids that induce various lumpy galls to form on leaves, buds, and branches. Other species of plants host psyllids that produce symptoms of leaf pitting, curling, and other distortions.

Psyllid saliva may also produce more systemic plant symptoms. One of the most important insect pests of pears is the pear psylla (*Cacopsylla pyricola*). It produces large amounts of honeydew and, at high

FIGURE 13-62
Colony of acacia psyllids, *Acizzia uncatoides*. This species produces long waxy strands of lerps as a waste material. Photograph by Jack Kelly Clark and provided courtesy of the University of California IPM Program.

FIGURE 13-63
Nipplegalls on hackberry produced by the hackberry nipplegall psyllid, *Pachypsylla celtidismamma*. The feeding nymphs induce this leaf gall to form on the newly expanding leaves and live within the gall until late summer or early fall, when the adults emerge and disperse. Photograph courtesy of Jim Kalisch/University of Nebraska.

densities, they can cause leaves to drop and plant growth to greatly slow, a condition known as "psylla shock." Arguably, the insect with the most phytotoxic saliva of all is the potato/tomato psyllid (*Bactericera cockerelli*). The saliva that this insect introduces can disrupt development of plants both above- and belowground, and the insect causes tremendous crop losses in parts of the western United States during its periodic outbreaks. The various symptoms produced by this insect are often referred to as "psyllid yellows."

Some psyllids also transmit plant pathogens, particularly bacteria known as phytoplasmas.

FIGURE 13-64
Nymph of a pear psylla, *Cacopsylla pyricola*. The nymphs are often nearly covered by the honeydew they excrete. Photograph by Whitney Cranshaw/Colorado State University.

FIGURE 13-65
Adults, nymphs, cast skins, and lerps associated with a colony of the potato/tomato psyllid, *Bactericera cockerelli*. Photograph by Whitney Cranshaw/Colorado State University.

Potato psyllid has recently been recognized to transmit a pathogen (*Liberibacter* species) that produces the disease known as "zebra chip," which has devastating effects on potatoes grown for chipping. Of particular concern at present is the Asian citrus psyllid (*Diaphorina citri*), accidentally introduced into Florida sometime prior to 1998 and since found in Texas and southern California. This insect is capable of transmitting *Liberibacter asiaticus*, a phytoplasma that produces citrus greening disease (Huanglongbing disease). Citrus greening disease has destroyed citrus production in many Asian countries and now poses a serious threat to these crops in the United States and Brazil.

FIGURE 13-66
The Asian citrus psyllid, *Diaphorina citri*, is considered to be a major threat to the US citrus industry since it is an efficient vector of the pathogen that produces citrus greening disease. Photograph courtesy of Michael Rogers/University of Florida.

FAMILY ALEYRODIDAE—WHITEFLIES

Whiteflies are small insects, typically about 2–3 mm in size, with adults that possess wings covered with whitish powdery wax. They are most abundant in the subtropics and tropics since few whitefly species can survive areas where freezing temperatures periodically kill their host plants. (In the northern half of the United States and Canada, the greenhouse whitefly, *Trialeurodes vaporariorum*, is the overwhelmingly most abundant whitefly species. In those areas, all outdoor infestations annually

FIGURE 13-67
Different life stages of the greenhouse whitefly, *Trialeurodes vaporariorum*. This is the common species that infests houseplants and greenhouse crops in North America. Photograph by Whitney Cranshaw/Colorado State University.

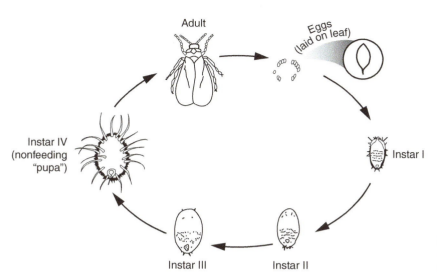

Adult

Eggs
(laid on leaf)

Instar I

Instar II

Instar III

Instar IV
(nonfeeding
"pupa")

FIGURE 13-68
Generalized life cycle of a whitefly. Drawing
by Loretta Mannix.

originate from infested plants maintained indoors in greenhouses during winter.)

Whitefly development follows the general form of simple metamorphosis but has several unique aspects. Following egg hatch, the first nymphal instar (Instar I or "crawler") is active and moves about the plant until it settles, inserts its stylets into the plant, and begins to feed. Following the next molt, the legs are lost and the insect remains in place while it continues to feed during Instars II and III. The final juvenile stage (Instar IV) does not feed, and during this time wings and other adult features begin to form. Finally, an adult emerges through a weakened area on the dorsal surface. Because of this peculiarity in the simple metamorphosis pattern, the first three instars are typically referred as nymphs and the fourth as a "pupa," the latter term normally reserved for immature forms of insects that have complete metamorphosis.

Some species of whiteflies are very important crop pests. In high population densities, they can withdraw so much fluid from their host plants that growth is retarded, wilting may occur, and leaves yellow. Copious amounts of honeydew also may be produced. Often associated with honeydew are gray or black sooty mold fungi that grow wherever honeydew persists.

The saliva of whiteflies, introduced during feeding, may also have direct adverse effects on plant growth, such as the changes in leaf color that occur on many vegetables when supporting heavy infestations of the silverleaf whitefly (*Bemisia tabaci*

FIGURE 13-69
Adults and nymphal skins of the silverleaf whitefly, *Bemisia tabaci* (Strain B). Photograph courtesy of Scott Bauer/USDA ARS.

Strain B). A few species of whiteflies, notably the sweetpotato whitefly (*Bemisia tabaci*), also can transmit plant viruses. Currently, there are over 60 Begomoviruses (Geminiviridae family) known to be transmitted to plants by whiteflies; most of these viruses have only been identified recently. Finally, whiteflies garner attention and great respect as plant pests because they are difficult to control and have often developed resistance to insecticides.

● Insecticide Resistance

To scientists that work on problems of managing insect pests, there is probably no clearer example of the adaptability of insects than their ability to evolve resistance to insecticides. When this occurs, insects can no longer be controlled with an insecticide that was formerly effective. Furthermore, when such a change does occur, it is almost always permanent, so that the insecticide will no longer be a useful means of control. The occurrence of insects (and mites) that have become resistant to pesticides has increased tremendously over the past 40 years. To date, over 540 species of insects and mites are now resistant to at least one kind of pesticide.

Insecticide resistance develops because of genetic changes in an insect population in response to insecticide exposure. Often, originally a small percentage of individuals are present that have some genetic trait or traits that allow them to somehow better survive the effects of exposure to the insecticide. Most susceptible individuals are killed off, while those with some ability to better resist its effects tend to survive. The genes for this resistance trait are then passed on to their offspring. Continued exposure to the insecticide will continue to concentrate the occurrence of genes controlling traits that provide resistance, a process known as **selective pressure**. Ultimately, the occurrence of genes for resistance becomes so widespread in an insect population that the insecticide loses its effectiveness. Also, when insects become resistant to one kind of insecticide, they usually become resistant to others that have a similar **mode of action**, producing the phenomenon known as **cross-resistance**.

There are many genetic traits that can increase insecticide resistance. Often these involve biochemical differences that provide increased activity of certain enzymes to more rapidly break down and inactivate the insecticide. Resistant insects may have nerve receptors that are less disrupted by the insecticide (target-site insensitivity), have features of the exoskeleton that decrease penetration, or have internal functions that allow an insecticide to be more rapidly excreted. Some insects have even developed behavioral changes so that they avoid surfaces that are treated with certain insecticides. Regardless of the resistance mechanism, insecticide resistance develops because the selective pressure from insecticide exposure produces an evolutionary change in the occurrence of genes that

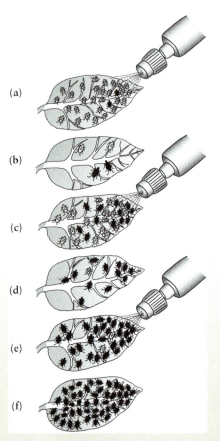

(a)

(b)

(c)

(d)

(e)

(f)

FIGURE 13-70

An illustration about how insecticide resistance characteristics can come to predominate in insect populations. In this example, the original population has a very low number of individuals (indicated in dark) with inherited characteristics that allow resistance to the effects of the insecticide. Following insecticide applications, the susceptible insects (indicated in light) are killed, leaving those with resistance, which then reproduce and pass on the resistance characteristics to their progeny. Following repeated exposure to the insecticide, insects with resistance characteristics come to predominate and the insecticide is no longer effective. Drawings by Matt Leatherman.

provide the insect resistance. It is an excellent example of artificial (human-caused) selection (as opposed to natural selection).

Development of insecticide resistance is a problem that can seriously undermine the ability to control critically important pest insects. Management strategies

(continued)

to prevent or retard insecticide resistance primarily seek to minimize the use of insecticides such that resistance genes are not concentrated in the population (i.e., selective pressures are reduced). Most fundamentally, this involves coordinated use of a variety of alternative control methods, such as pest-resistant plants, trapping, or methods that improve the activities of biological control organisms. Furthermore, monitoring of pests can allow detection of potential problems so that insecticides need only be used when the potential for damage is present. Finally, insecticides that work differently to kill insects (have different modes of action) can be used in sequence to help delay the development of resistance to any one type of insecticide.

SUPERFAMILY COCCOIDEA—SCALE INSECTS

The insects described as scales are found in many families (16) and presently include over 740 known species in North America. Many of the economically most important species are accidental introductions from other parts of the world.

A common characteristic of the scales is that they secrete waxy material over their body that covers and may substantially protect the insect. The nature of this waxy cover is relatively distinctive among the soft scales (Coccidae family) and the armored scales (Diaspididae family). Other families of scale insects produce waxy covers that appear as a tangle of waxy threads. Mealybugs (Pseudococcidae family) and cochineal insects (Dactylopiidae family) are examples of the latter.

FIGURE 13-71
A Boisduval scale, *Diaspis boisduvalii*, with the cover removed to show the adult scale insect and eggs that are under the protective cover. Photograph courtesy of Jim Kalisch/University of Nebraska.

All scales feed on plant sap and are among the most sessile (nonmoving) insects associated with plants. The primary time when scales are mobile is the brief period just after egg hatch when the minute first instar nymphs (Instar I), also known as crawlers, actively move about the plant seeking a location to settle and feed. (This behavior is similar to whitefly crawlers.) Additionally, nymphs may be transported to considerable distances by hitching a ride on the legs of larger insect when the two stumble upon one another. After the first molt, a week or so after egg hatch, most scale insects move little, if at all, for the rest of their life.

Adult female scales are wingless, often essentially legless, and lay their eggs under their waxy covers. Males, if they occur (many scale insect species are parthenogenetic), are tiny insects with a single pair of mesothoracic wings.

Family Diaspididae—armored scales. Within North America, the armored scales are the largest in number (about 310 species) and are among the most important in terms of causing plant injuries. Many of the most damaging species are native to Europe or Asia, and a great many species of armored scales have been accidentally spread around the world in the course of human commerce. (These introductions continue today.) The armored scales are only active during the first instar "crawler stage" and will either successfully settle on a plant or die within about a week following egg hatch. Legs are lost after the first molt. They produce a thick hard wax cover of a characteristic elongate or oval shape. The wax cover completely encloses the insect's body except for the mouthparts protruding into the host plant. No honeydew is produced, since they feed on individual plant cells immediately under the location where they rest, rather than sap from the phloem. Armored scale feeding may cause some localized cell death and at high population densities can induce dieback of branches or leaf drop. Among the more important armored scales presently occurring in the United States are oystershell scale (*Lepidosaphes ulmi*), elongate hemlock scale (*Fiorinia externa*), cycad aulacaspis scale (*Aulacaspis yasumatsui*), California red scale

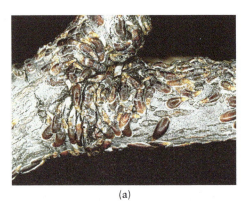

(a)   (b)

FIGURE 13-72

Two armored scales (Diaspididae family): (a) oystershell scale, *Lepidosaphes ulmi*, is one of the most destructive scale insects on hardwood trees and shrubs in North America. Photograph courtesy of Jim Kalisch/University of Nebraska. (b) Black pineleaf scale, *Dynaspidiotus californica*, develops on the needles of pines and is a sporadic pest. Photograph by Whitney Cranshaw/Colorado State University.

(*Aonidiella aurantii*), pine needle scale (*Chionaspis pinifoliae*), euonymus scale (*Unaspis euonymi*), and San Jose scale (*Quadraspidiotus perniciosus*).

Family Coccidae—soft scales. Soft scales are often the most commonly noticed scales; they are rather large (4–8 mm), usually globular in form. At least 105 species occur in North America, and they occur on a wide variety of trees and shrubs. Among the more important soft scales on ornamental plants are cottony maple scale (*Pulvinaria innumerabilis*), tulip-tree scale (*Toumeyella liriodendri*), European fruit lecanium (*Parthenolecanium corni*), and various wax scales (*Ceroplastes* spp.). The most common scales on houseplants are in this family, such as the brown soft scale (*Coccus hesperidum*), citricola scale (*Coccus pseudomagnoliarum*), and hemispherical scale (*Saissetia coffeae*).

Soft scales do retain mobility through most of their lives. Typical life cycles of outdoor species take a year to complete with summer stages on leaves or needles and a return to twigs in the fall where they are found through the next spring. Individual soft scales produce hundreds of eggs that hatch over an extended period of weeks. Unlike armored scales, the soft scales excrete abundant amounts of honeydew as they feed.

(a)   (b)

FIGURE 13-73

Two soft scales (Coccidae family): (a) calico scale, *Eulecanium cerasorum*, adults and crawlers. Photograph courtesy of Dan Potter/University of Kentucky; (b) cottony maple scale, *Pulvinaria innumerabilis*, with large egg sacs. Photograph courtesy of Frank Peairs/Colorado State University.

Family Pseudococcidae—mealybugs. The mealybugs are named because of the mealy textured wax secretions that typically cover their body. They constitute a large family, particularly common in tropical and subtropical areas. In North America, some 240 species occur, many of which are accidentally introduced. Many are serious plant pests, notably the citrus mealybug (*Planococcus citri*), longtailed mealybug (*Pseudococcus longispinus*), and pink hibiscus mealybug (*Maconellicoccus hirsutus*). In addition to direct injuries produced during feeding, some mealybugs are very important as vectors of the various swollen shoot viruses that devastate cacao production in western Africa.

The manna mealybug, *Trabutina mannipara,* is a common species in the Middle East. It excretes large amounts of honeydew that often dries as thick crusts on twigs and branches of tamarisk (salt cedar). This has been collected since biblical times and continues to be used as a sweet food source in rural areas of Iraq and Arabia. This mealybug has been considered for introduction into the United States as a possible biological control organism for the control of tamarisk, which has become a serious invasive plant in the western states.

Family Dactylopiidae—cochineal insects. The cochineal scales are large scales covered with wax that gives them a cottony appearance similar to mealybugs.

(a)   (b)

FIGURE 13-74

Two mealybugs (Pseudococcidae family): (a) citrus mealybug, *Planococcus citri*; (b) longtailed mealybug, *Pseudococcus longispinus*. Photograph of citrus mealybug by Whitney Cranshaw/Colorado State University. Photograph of longtailed mealybug courtesy of David Shetlar/The Ohio State University.

They feed on cacti pads, and three species are known from the United States. All produce hemolymph that is highly pigmented with carminic acid, and a Mexican species (*Dactylopius coccus*) has been extensively used as a source of red (carmine) dye.

The cochineal scale was long cultivated in pre-Hispanic Mexico. Cortez introduced the product (dried scales) to Europe, where it was highly prized for the intensely bright red dyes made from it; after gold, cochineal was the most valuable product exported from the region. Interestingly, it was almost 150 years before Europeans recognized its insect origin. (It was previously thought to be a type of seed.) Until synthetic aniline dyes became available in the late 1800s, cochineal was the primary source for

FIGURE 13-75

Cochineal scales being reared on nopal cactus pads. When full grown, the insects will be brushed off, dried, and crushed to produce the carmine dye from their blood. Photograph by Whitney Cranshaw/Colorado State University.

producing brilliant red-colored fabrics for over 300 years. During the American Revolution, its vivid scarlet color made appearances both in the red of the American (and British) flag and the "redcoats" of the British soldiers. Cochineal culture was introduced into other areas of the world, particularly the Canary Islands. Cochineal is produced by delicately scraping the insects off the pads of prickly pear (*Opuntia* spp.) on which they develop. They are then boiled to remove the wax, dried, and ultimately ground to a fine powder. Up to 10% of the hemolymph of the cochineal scale consists of carminic acid, a very bitter material that is the source for the color carmine. Over 155,000 scales are processed to produce a kilogram of finished cochineal. During dyeing, it is boiled with the fabric and mixed with a mordant material (e.g., oxalic acid, urine) to affix it to the fabric and intensify the color.

Production of cochineal continues today although at a much lower level than was the case a century ago. It is still primarily used in fabric dyes but also is a source of natural red used in coloring

FIGURE 13-76
Yarn dyed with cochineal extract. Both the red and orange are from cochineal dye, the color differences related to fixatives used in the dye process. The blue yarn is from an indigo dye. Photograph by Whitney Cranshaw/Colorado State University.

cosmetics and some food items, such as red-colored fruit drinks. The presence of cochineal is indicated on food and cosmetic labels by the signal words "carmine," "cochineal," "cochineal dye," or "cochineal extract." Use in food items has decreased in recent years within the United States due largely to strongly publicized objections by some religious groups and vegans with prohibitions against eating certain arthropods.

Another scale insect with a substantial history of human use in fabric dyeing is *Kermes ilicis*, from which the word crimson is derived. This scale is a member of the gall-like scale family Kermesidae and it develops on European live oak. It was widely traded throughout the Middle East by the Phoenicians over 2,500 years ago.

Family Kerriidae—lac scales. The lac scales produce a covering of resinous material and live within resinous chambers that they construct. Most are tropical or subtropical in distribution although seven species in the genus *Tachardiella* occur on desert plants in the southwestern United States.

One Asian species is cultivated, and the resin it produces provides the source of a material known as shellac. The common lac insect, *Kerria* (*Laccifer*) *lacca*, develops on the twigs of kusum, palas, ber, and other trees native to southern Asia. It can build to tremendous numbers forming thick crusts of resin. These crusts are then scraped from the trees to make a product known as "stick lac." Through a process of grinding, washing, heating, and filtering, the stick lac is transformed to

FIGURE 13-77
Lac insects, *Kerria lacca*, on a twig. These are cultured, then scraped off for processing into "shell lac," which is the base for producing shellac used in wood finishing and other surface treatments. Photograph courtesy of Jeffrey W. Lotz/Florida Department of Agriculture and Consumer Services/Bugwood.org.

flaky sheets of "shell lac." About 300,000 individual lac insects are collected in the production of a kilogram of finished shellac. Almost all production currently occurs in northeastern India and Thailand.

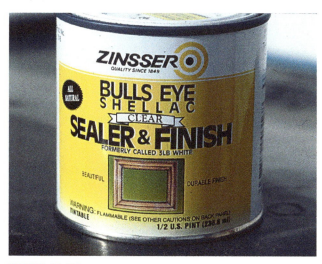

FIGURE 13-78
A shellac product used for wood finishing. Photograph by Whitney Cranshaw/Colorado State University.

The shellac resins are a natural polymer (natural plastic) and have features that make them useful for a variety of uses. These include thermoplasticity, allowing molding, good film-forming capacity, high resistance to organic solvents, excellent adhesion, and good electrical properties. Furthermore, shellac is nonpoisonous and odorless. Prior to the development of vinyl in 1938, the largest single use was for production of phonographic records, and it was still commonly used for this purpose until the early 1950s. It also had extensive use as a wood sealant and polish although synthetic alternatives such as polyurethane have eliminated much of this market. Shellac continues to be used as a natural glaze on pills, candy, and fruit.

Order Thysanoptera— The Thrips

Entomology etymology. The order name means "fringe" (thysanos) "wings" (ptera) in Greek, referring to the distinctive wing type of these insects. Collectively, these insects may be referred to as thysanopterans.

However, the common name "thrips" has a less satisfactory origin, derived from a Latin word translating to "woodworm" or "wood louse." This poorly describes the habits of these insects, which are found in all manner of habitats. Furthermore, "thrips" is also a name that causes frequent confusion, since the name is both singular and plural. There is no such thing as a "thrip." It is one thrips, two thrips … a million thrips.

FIGURE 13-79
A western flower thrips, *Frankliniella occidentalis*. In this specimen, the wings are spread to show the fringed wings characteristic of the order Thysanoptera. Photograph courtesy of Jack Reed/Mississippi State University/ Bugwood.org.

Although small in size, the thrips display so many peculiarities of feature and habit that they are clearly a unique group of insects. To begin, their rodlike wings are fringed with long setae or hairs, greatly extending the surface area of the narrow wings. Although fringe wings are found in a few other tiny insects (fairyflies of the hymenopteran family Mymaridae and the feather-winged beetles of the Ptiliidae), it is a defining characteristic from which the order name is derived. Over 5,500 known species of thrips occur worldwide, with about 700 found within the United States.

The mouthparts of thrips are modified in ways that allow them to suck fluids but achieve this with an arrangement far different from the Hemiptera. In side view, a small cone appears to project from the base of the head; this cone is made up of the labium, labrum, and parts of the maxillae. Confined within the cone are stylets comprised of the maxillae and a single mandible, the left one. (In this asymmetrical arrangement, the right mandible is highly reduced and nonfunctional.) The mandible is used to puncture the surface of its food, usually a plant, and the thinner pair of maxillary stylets probes further to penetrate underlying cells. The released fluids and cell contents produced by the wounding are then drawn into the cone and ingested. Thrips feeding has sometimes been described as "rasping-sucking," but "puncture, poke, and suck" far better describes the thrips feeding process.

FIGURE 13-80
Head of a dandelion thrips, *Tenothrips frici*. The mouthparts are in the form of a cone, within which are a single, spiked mandible and a pair of lancelike maxillae. Photograph courtesy of John Dooley/USDA APHIS PPQ/Bugwood.org.

Their slender body and small size, typically less than 2–3 mm, allow thrips to get into all manner of places. They are further aided in their explorations by small bladders on the pretarsi of the legs that balloon out upon contact with a surface. These allow them to adhere to and crawl upon smooth surfaces that would defy most any other insect.

FIGURE 13-81
Side view of a female western flower thrips, *Frankliniella occidentalis*. The stout ovipositor on the hind end is used to insert eggs into plant tissue. Photograph courtesy of Alton N. Sparks, Jr./University of Georgia/Bugwood.org.

Thrips undergo a type of hemimetabolous metamorphosis but have peculiarities in this regard as well. Eggs are often inserted into the surface of plants and subsequently there are two active feeding stages, or larvae, Instars I and II. Following this, they

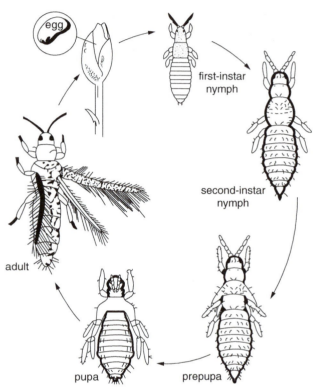

FIGURE 13-82
Generalized life cycle of a thrips. Line drawing by Loretta Mannix.

FIGURE 13-83
Adult and Instar II nymph of the onion thrips, *Thrips tabaci*. Photograph courtesy of Alton N. Sparks, Jr./University of Georgia/Bugwood.org.

usually drop to the ground or find a small crack to hide in and subsequently undergo two nonfeeding stages, Instars III and IV. During this time, the developing thrips are immobile but undergo gradual physical changes that allow transition to the ultimate adult form. Although these later nymphal stages are sometimes known as a prepupa (Instar III) or pupa (Instar IV), these terms are misleading as they little resemble the pupal stages of insects that undergo complete metamorphosis.

Within the order Thysanoptera, a wide range of foods are used. Most thrips are plant feeders and are ubiquitous inhabitants of almost any crop. (In North America, thrips are undoubtedly the most common insects incidentally ingested by people. Most visits that pick up some greens at the salad bar will likely have a few of thrips as incidental garnish.) Their feeding characteristically produces small silvery scars where the plant cell contents have been removed. Often their feeding activities result in minor injuries, largely reflecting their small size, but heavy infestations of thrips can reduce the photosynthetic ability of the plant and produce disfiguring scars on flowers, seedpods, or fruit. Thrips are also important in the transmission of some plant viruses responsible for serious diseases.

● Tospoviruses—Emerging Diseases of the Plant World

An ebb and flow of human diseases has been a constant occurrence for at least as long as people have been on the planet, and often these diseases have had transforming effects on human history. Plants, like people, also suffer from disease—produced from infection by many of the same kinds of potential pathogens that can plague humans—bacteria, viruses, fungi, and protozoa. Among those that have increased most dramatically over the past couple of decades are the tospoviruses (Bunyaviridae family) transmitted from plant to plant by thrips.

The most important is tomato spotted wilt (TMSV), a virus that has the most extraordinary range of host plants. For a long time, it was the only known thrips-transmitted plant virus, but recent advances in detection have now allowed description of 19 new viruses that these insects transmit to plants, including most notably impatiens necrotic spot virus (INSV) and iris yellow spot virus (IYSV). To date, there are at least 110 families of plants and 850 plant species that are known to be susceptible to infections with at least one of these three viruses, with tomato spotted wilt virus and impatiens necrotic spot virus having the best documented range of host plants.

The ability of a thrips to transmit one of these viruses requires a whole sequence of steps be successfully completed. After the thrips feeds on an infected plant and ingests some virus particles, the virus must then migrate through the midgut lining and into the body cavity of the insect. From the body cavity, the virus particles must travel to the salivary glands. Along the way, the virus must also replicate within tissues of the thrips.

Unique among insects that transmit plant viruses, the only time this all comes together with thrips is in the first juvenile stage (Instar I). This is because in the very young thrips, the head is filled with muscles used to pump sap. This forces the brain and the salivary glands into the thorax where they are in very close contact with the midgut, thus allowing the virus to make the crossover into the body cavity and salivary glands. After the next molt, this arrangement of organs changes such that the brain and salivary glands move back into the head and are no longer in proximity to the midgut.

Older thrips nymphs and adults never have the bodily arrangement that allows them to acquire a plant virus and later transmit it; however, those Instar I nymphs that do feed on an infected plant can later transmit the virus for the remainder of their lives. Very few species of thrips are capable of all this and almost all plant viruses transmitted by thrips involve either the western flower thrips (*Frankliniella occidentalis*), the tobacco thrips (*Frankliniella fusca*), or the onion thrips (*Thrips tabaci*).

(*continued*)

FIGURE 13-84
Symptoms of infection with tomato spotted wilt virus, a serious plant disease that is thrips transmitted: (a) stunting of infected bell pepper plants, which later died prematurely; (b) ringspot symptoms on tomato fruit. Photographs by Whitney Cranshaw/Colorado State University.

FIGURE 13-85
Adult and nymph of the greenhouse thrips, *Heliothrips haemorrhoidalis*. This species excretes a conspicuous tarry fluid that spots infested leaves. Photograph by Jack Kelly Clark and provided courtesy of the University of California IPM Project.

Many thrips feed on pollen, while others feed on the mycelia of fungi. Still others are predators of mites, other thrips, small insects, and insect eggs.

FIGURE 13-86
Scarring injuries on the skin of citrus produced by thrips feeding wounds. Photograph by Jack Kelly Clark and provided courtesy of the University of California IPM Project.

Some of the most economically important thrips species are highly polyphagous and will feed on a wide range of foods. Still other members of the order display a more restricted feeding habit, limited to a genus or two of plants (oligophagy) or sometimes a single species (monophagy).

Thrips feeding on new growth commonly causes leaf curling and some species induce such substantial changes in plant growth that they are considered to be gall producers. Gall-making thrips are particularly common in arid areas of Australia where such growths appear to provide useful shelter from the heat. Many of these leaf-curling and gall-making thrips also display truly social

FIGURE 13-87
Colony of Cuban laurel thrips, *Gynaikothrips ficorum*, showing a range of life stages, including a mass of eggs. Cuban laurel thrips produce leaf curls on susceptible hosts, notably certain *Ficus* spp. Photograph courtesy of David Shetlar/The Ohio State University.

behaviors. At least six species in the genera *Oncothrips* and *Kladothrips* have colonies established by a founding female who initiates the gall, sometimes with the assistance of a male. Emerging from the first eggs are individuals of very different form and function, the soldiers. Soldier-form thrips have reduced wings and stout forelegs, allowing them to defend the colony. As these insects have sucking mouthparts and feed individually, communal feeding of the young by the adults does not occur as it does with more familiar social insects (termites, ants, some bees and wasps). Communal protection does occur in the form of the specialized soldier caste and communally feeding activity produces the plant distortions that house and protect the small colony.

Social behavior in thrips is also found in the American tropics where the large, lichen-feeding thrips *Anactinothrips gustaviae* forms large groups that communally guard the eggs and young nymphs.

Insect Bruisers and Their Lacewinged Cousins

Beetles, Lacewings, Dobsonflies, and Relatives

Because of the tremendous species richness and diversity of habits it has been argued that the beetles (Coleoptera order) are the most successful life-forms on earth. (Others make the argument that the title of "Most Successful" should go to the bees, wasps, and ants of the order Hymenoptera.) Their hardened forewings are highly protective and are a feature that contributes greatly to their success as a life-form. Interestingly, their closest relatives among the insects are a group of insects often noted for their wing delicacy—the gauzy-winged insects of the order Neuroptera.

FIGURE 14-1
A carrion beetle with hindwings unfolding from under the forewing covers in preparation for flight. Photograph courtesy of Joseph Berger/Bugwood.org.

Order Coleoptera—The Beetles

Entomology etymology. The order name is derived from the Greek words for sheath (coleos) and wing (pteron), a reference to the hardened forewings of beetles covering and protecting the hindwings and abdomen. Collectively, members of the order may be referred to as coleopterans.

Beetles are the most diverse life-forms on Earth with an estimated 350,000 + described species. Over a third of all known insect species are some type of beetle, and they are found in essentially all terrestrial and aquatic fresh water environments. Their behaviors are similarly diverse, with species that include predators, parasitoids, scavengers, plant feeders, and fungivores. Within the order, there is a 500-fold range in body length, from tiny featherwing beetles (Ptiliidae family) only 0.3 mm in length to the South American long-horned beetle *Titanus giganteus* that may be over 165 mm. Differences in range of weight are orders of magnitude greater.

Perhaps the most famous comment on the immense diversity displayed by the Coleoptera was by the British geneticist J.B.S. Haldane (1892–1954)

FIGURE 14-3

Titanus giganteus, a longhorned beetle native to South America, is one of the largest beetles, reaching body lengths of over 165 mm. Photograph by Whitney Cranshaw/Colorado State University.

who, in response to a query by theologians on the lessons learned from his career studying life's creations, stated that the Creator has "an inordinate fondness for beetles." Much of the success of the beetles is due to the hardened forewing, the **elytra**. The elytra well protects the underlying hindwing and, with this adaptation, adult beetles can burrow into soil, tunnel into wood, and exploit other sites while still retaining their ability to fly.

Beetle larvae are often known as grubs, particularly the less mobile species that feed within plants or on roots. Like the adults, they have chewing mouthparts, and most have a distinct head area. There are three segmented legs on the thorax but no prolegs on the abdomen, while the thoracic legs are much reduced in weevils and wood borers. Many species will spin a silken cocoon in which they pupate. Within Coleoptera, silk is produced in modified Malpighian tubules and excreted from the anus.

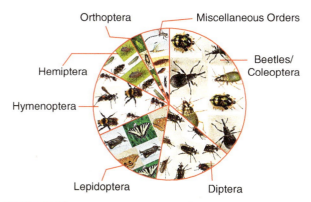

Orthoptera   Miscellaneous Orders

Hemiptera

Hymenoptera

Lepidoptera   Diptera

Beetles/ Coleoptera

FIGURE 14-2

The relative number of species among the insect orders. There are over 360,000 described species of Coleoptera, which comprise over 1/3 of all presently described insects. Diptera, Hymenoptera, and Lepidoptera are also particularly species-rich orders. However, although there are nearly one million described insect species, it is estimated that this represents only about 1/5 of the world's present insect fauna, the overwhelming majority remaining undescribed. Photographs courtesy of Tom Murray.

● The Naming of the Beatles

When the little known band the Quarrymen, out of Liverpool, England, were considering a change in direction and new name, band member Stuart Sutcliffe suggested The Beetles as a tribute to Buddy Holly and the Crickets, a pioneering rock and roll band that had recently met a tragic end in an airplane crash, which killed Mr. Holly. For a brief period, this morphed into the band name "Silver Beetles" before they ultimately settled on "The Beatles."

Exactly who originally made this final suggestion will likely always be up for debate as two other band members, John Lennon and Paul McCartney, are variously given credit.

Although the order Coleoptera is far and away the most diverse insect order, based on the number of currently described species, the Hymenoptera may ultimately surpass the beetles as new species become cataloged. Similarly, though the Beatles have been

(continued)

FIGURE 14-4
The all time best-selling rock and roll group, the Beatles, chose their insect-themed name as a tribute to the pioneering Buddy Holly and the Crickets, shortly after Buddy Holly died in a plane crash. Photograph by Whitney Cranshaw/Colorado State University.

named the #1 most influential rock and roll band of all time, beetles are not the most common insect order embraced by rock and roll artists. Joseph Coelho of Quincy University recently did a review of insect imagery appearing on album cover art for the period 1955–2004. During this period, he cataloged 392 albums that depicted insects, with the following orders making the top five (percent of total):

Lepidoptera—butterflies and moths (36%)
Hymenoptera—bees, wasps, ants, and sawflies (17%)
Coleoptera—beetles (11%)
Diptera—flies (9%)
Odonata—dragon- and damselflies (8%)

It is clear that other insects and arthropods (e.g., spiders and scorpions) also have a long history of inspiring pop bands. From Buddy Holly and the Crickets, through the Beatles and Iron Butterfly in the 1960s, to Adam and the Ants and more recently Band of Bees, insects remain a lasting source of inspiration among musicians.

FAMILY CARABIDAE—GROUND BEETLES

Ground beetles (carabids) are insects (2,600 North American species) perhaps most often seen when turning over a rock or board, exposing a shiny black beetle that scurries for cover. They are active insects with legs adapted for running. Viewed from above, the hindlegs appear to arise unusually far back, on the abdomen, but they are attached to a plate of the metathorax extending rearward and covering the first segments of the abdomen. Most are dark, shiny, and somewhat flattened, with grooved elytra. The majority are predators, with strong jaws that allow them to capture and feed on various insects and other invertebrates. A few of them are seed feeders. The predatory species are considered very valuable in agricultural systems as they feed upon pest species of insects such as cutworms and other pest caterpillars.

Although the great majority of ground beetles are basically black in color, modifications do exist. Some are brilliantly colored, notably the "caterpillar hunter," *Calosoma sycophanta*, a large species with iridescent green, blue, and red hues. Ground beetles that specialize in snails have very narrow heads and mouthparts that they use as hooks or spoons to extract the flesh of their prey.

FIGURE 14-5
Calosoma scrutator, a largely brightly colored ground beetle that feeds on caterpillars. It is one of the few ground beetles adapted to climbing trees and was introduced into North America to prey on gypsy moth. Photograph courtesy of Jim Kalisch/University of Nebraska.

Likely the most spectacular adaptation that has been found among ground beetles, so far, is the defensive capabilities of the bombardier beetles (*Brachinus* spp.). These beetles possess specialized reactor glands in their abdomen that allow them to almost instantaneously create a chemical reaction

FIGURE 14-6
A ground beetle larva feeding on a caterpillar. Photograph courtesy of Jim Kalisch/University of Nebraska.

FIGURE 14-7
Harpalus sp. ground beetles massed near an evening light. Photograph courtesy of Jim Kalisch/University of Nebraska.

that produces caustic chemicals (benzoquinones) that are immediately expelled at near-boiling temperatures. These chemicals are released as a pulsed stream, spurting up to 500 times per second, and can be made in directed shots through a turret at the tip of the abdomen. Such massive defense serves them very well to protect them against predation by ants and even large vertebrates such as toads.

The family Carabidae also includes the tiger beetles (subfamily Cicindellinae). Many tiger beetles have brilliant colors and patterns, often with metallic and iridescent hues. They run rapidly, and most can quickly escape by flying, which makes them difficult to capture. Due to their elusiveness and their vivid coloration, these beetles are among the most avidly sought insects by collectors.

FIGURE 14-8
A bombardier beetle, *Brachinus* sp. Photograph courtesy of Jillian Cowles.

FIGURE 14-9
A brightly colored species of tiger beetle, *Cicindela sexguttata*. This beetle has recently emerged from the soil where it had pupated. Photograph courtesy of Jim Kalisch/University of Nebraska.

Larvae of tiger beetles live within vertical tunnels dug deeply into soil. They are ambush hunters that quickly seize passing insects with their large curved jaws. Most occur near water, but many are very choosy about the exact site where they will burrow and develop. Some occur in certain types of sandy soils, others in salt flats with a saline crust. Because of this finicky requirement, they are often very restricted in range, and four US species are currently listed as Endangered or Threatened because of the threat of habitat destruction.

FIGURE 14-10
Larva of a tiger beetle, *Cicindela punctulata*, removed from its burrow. Photograph by Whitney Cranshaw/Colorado State University.

FAMILY SILPHIDAE—CARRION BEETLES

Carrion beetles are a small family (175 described species worldwide) with some colorful representatives. Most are fairly large beetles, over 10 mm, and they have conspicuously clubbed antennae. Although a few members of the family prey on snails or caterpillars, one is most likely to find a carrion beetle near a dead animal.

Carrion beetles pursue one of two main life history strategies. The majority of genera (e.g., *Heterosilpha*, *Thanatophilus*, *Oiceoptoma*, and *Necrophila*) have gray-black adults with broad flattened bodies, and these tend to be attracted to larger carcasses. They produce dark-colored, active, free-living larvae found foraging among the tissues of the carrion on which they feed.

FIGURE 14-11
Adults and larva of the carrion beetle *Oiceoptoma noveboracense*. Photograph courtesy of Susan Ellis/Bugwood.org.

Those of the genus *Nicrophorus* have a more elongate body and often are brightly marked with orange or red. These are attracted to smaller carcasses, about the size of a mouse. (Chicken legs can be used successfully to attract *Nicrophorus* carrion beetles.) The first beetle to find a carcass guards it aggressively until an acceptable member of the opposite sex arrives. They then take on the joint task of burying the carcass, working together, and ultimately interring it within a chamber about an inch below the surface. Because of this behavior, they are known as sexton beetles or burying beetles.

The beetle pair then works to prepare the carcass, stripping off the fur and rolling the flesh into balls that they cover with preservative secretions. In addition,

FIGURE 14-12
The American burying beetle, *Nicrophorus americanus*, an endangered species of carrion beetle. Photograph courtesy of Jim Kalisch/University of Nebraska.

they always carry with them tiny mites (*Poecilochirus necrophori*), which, once within the burial chamber, move off the beetles' bodies and patrol over the food, consuming fly larvae and other insects that attempt to establish on the prepared carcass.

FIGURE 14-13
A carrion beetle, *Nicrophorus sayi*. Mites are present on the prothorax and clustered on some of the legs of the beetle. Photograph courtesy of Joseph Berger/Bugwood.org.

A small side chamber is constructed and the female lays her eggs there. These soon hatch and the parents remain with the young, feeding them a nutritious diet of partially digested flesh that allows them to grow very rapidly. They also notify the young of food, regularly "calling them to dinner" with stridulation sounds the adults make by rubbing the inner surface of the elytra along ridges on the abdomen.

FAMILY STAPHYLINIDAE—ROVE BEETLES

Rove beetles have such a distinctive appearance that most people would likely not recognize them as beetles at all. Although there is a considerable variation in form and color, most have an elongate body with very short elytra that barely reach the abdomen. Under the elytra are the hindwings, which can unfold for flight. Rove beetles are also quite flexible and can curl and twist their body. Despite their general lack of popular recognition, the rove beetles are a huge family of insects, with over 46,000 known species. In recent years, some 400 new species have been described annually, and it is likely that they could eventually surpass the Curculionidae (weevils) as the largest insect family once all species have been described.

FIGURE 14-14
Rove beetles feeding on a fly larva. Photograph courtesy of Jim Kalisch/University of Nebraska.

Rove beetles occur almost everywhere except in the driest of sites. Some feed on decaying vegetation, but the majority prey on other insects; they can be very important in the control of many pest groups, particularly flies. A few have developed physical and chemical adaptations that allow them to be taken into and accepted by ant and termite nests, where they then respond to the hospitality by eating the eggs and young of their hosts. Larvae and adults usually share the same feeding habits.

Rove beetles of the genus *Paederus* defend themselves using a very potent toxin in their blood, pederin. When released onto skin, it can produce large areas of painful blistering and has even been used to burn off warts. Occasionally, local species may become very abundant and attracted to artificial lights. Such abundant occurrences of *Paederus* species

FIGURE 14-15
A rove beetle, *Platydracus maculosus*, showing the hindwings unfolded from under the wing covers of the forewings. Photograph courtesy of Susan Ellis.

FIGURE 14-16
A rove beetle larva. Photograph courtesy of Ken Gray/Oregon State University.

are referred to as "Nairobi fly" outbreaks (involving *Paederus crebinpunctatus* or *P. sabaeus*) in West Africa. In Malaysia, such outbreaks of these beetles are known as the "semut semai" or "semut kayap."

● Ancient Egypt—When Scarab Beetles Were Gods

Scarab beetles have been tremendously important throughout the long period of the Ancient Egyptian civilizations. The activity of a common dung rolling scarab of the Nile valley, *Scarabaeus sacer*, was likened to the daily passage of the sun across the sky. Their spiny legs furthered this association, as rays of the sun. Furthermore, the appearance of the beetles as they emerged from the earth seemed to be sudden and spontaneous, linking them as strong symbols of resurrection and reincarnation.

Embodying this was Khepri (Kheper, Kephir) a creator god and solar deity known as "He who is coming into being." Khepri was pictured as either a scarab or a god with human body and the head of a scarab. Every

day, Khepri pushed the orb of the sun across the skies from the east to the west, where he sunk into the underworld at dusk and emerged reborn at dawn.

Depictions of scarabs were powerful symbols of life's victory over death and were as symbolically important in Ancient Egypt as the cross was later to become to Christians. Winged and heart scarabs were routinely entombed among the wrappings of mummies to ensure the safe passage to the afterlife. Other scarabs took on more general roles such as ensuring health and good luck or for commemoration of events and pharaohs. The flat underside of each scarab was often inscribed with mottos, wishes, names, or other messages reflecting its purpose.

FIGURE 14-17
Scarab form cartouches recovered from Egyptian tombs. Photograph by Whitney Cranshaw/Colorado State University.

FAMILY SCARABAEIDAE—SCARAB BEETLES

The scarab beetles are heavy-bodied beetles with a convex shape. Variously known as "dung beetles," "May/June beetles," "rhinoceros beetles," "Hercules beetles," or "chafers," the adults of many species are familiar animals, a few are important pests of plants, and others are scavengers of decaying plant matter or animal dung. The larvae are known as "white grubs" and they have a distinctive C-shaped body form. Among the 1,400 described species in North America, size ranges considerably from about 5 to 65 mm.

Several of the big tropical species of scarab beetles are frequently considered to be among the largest of all insects, at least by some measures. The University of Florida web site *Book of Insect Records* provides an excellent review of the topic by David Williams, including various difficulties in establishing what measurements are appropriate for judging large or heavy insects. Among the five proposed finalists for the title of largest insects, four are scarab beetles:

Megasoma actaeon—native to Ecuador, Peru, Brazil, and Colombia;
Megasoma elephas (elephant beetle)—found in lowland rainforests from Mexico to Venezuela;
Goliathus regius—occurring in the coastal countries of Ghana and Ivory Coast of West Africa;
Goliathus goliatus (Goliath beetle)—present along a wide swath across Equatorial Africa.

FIGURE 14-18
Three species of turfgrass-damaging white grubs, the larval stage of scarab beetles: (l–r) Japanese beetle, European chafer, and a May/June beetle (*Phyllophaga* sp.). Photograph courtesy of David Cappaert/Michigan State University/Bugwood.org.

FIGURE 14-19
A May/June beetle, *Phyllophaga futilis*. Photograph courtesy of Jim Kalisch/University of Nebraska.

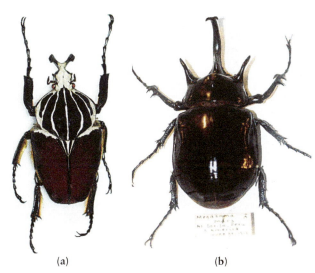

(a)   (b)

FIGURE 14-20
(a) A Goliath beetle, *Goliathus giganteus*. (b) *Megasoma mars*, one of the large South American scarab beetles known as "elephant beetles." Photograph by Whitney Cranshaw/Colorado State University.

Among the best known of the scarab beetles are the dung feeders (**coprophages**) of the subfamily Scarabaeinae. These beetles are very important in the recycling of animal waste. Adults are quickly attracted to the fresh manure and use it to feed their young. The behavioral approach subsequently used to exploit this food varies among species. Some lay eggs directly into dung pads, and their white grub larvae develop within. (This method is more widely used by scarabs in other subfamilies.) Adults of other species tunnel beneath the dung pad, excavating underground chambers that they then provision with manure pieces for the young.

● Australia's Problem with Cattle Manure

The importance of insects in the cycling of organic waste, and dung-feeding scarabs in particular, was made clear from the Australian experience with cattle production. Prior to the introduction of cattle and sheep, animals of similar size never roamed the continent. The native marsupials produced tough, fibrous droppings and native species of dung-feeding scarabs had evolved feeding and digestive adaptations to handle this material.

Cattle droppings, produced by the hundreds of millions daily following the introduction of livestock, overwhelmed these native beetles. Other insects thrived on the fresh cattle dropping, such as the notorious bush fly (*Musca vetustissima*), producing terrible nuisance problems for humans and livestock. Blood-sucking buffalo flies (*Haematobia irritans exigua*) also did well on this new food source.

These flies only consumed a relatively small portion of the available manure. Most of the manure was left to dry, ultimately breaking down by the slow action of termites and weathering. Large areas of the continent were on the way to being covered with cow pies, accompanied by nuisance flies in plague proportion.

FIGURE 14-22
Onthophagus gazella, a dung-feeding scarab that has been introduced in several regions to help manage cattle manure. Photograph courtesy of Bart Drees/Texas A&M University.

By 1963, the situation was so severe that actions were begun to introduce dung-feeding scarabs from other countries to help mitigate the problem. Introductions began a few years later and by 1991, 22 species brought from Europe, Africa, or Asia had established in at least one area of the country or another. A few of these species became very abundant, particularly the African species *Onthophagus gazella*, *Euoniticellus intermedius*, and *E. africanus*. These species had dramatic effects on manure accumulation and often incorporated it into soil within 48 h. Problems with bush flies receded greatly and grazing areas recovered.

Scarab beetles have been introduced in other regions of the world where bovine dung has become a problem (The technique was actually first tested successfully, but with less publicity, in Hawaii in the 1920s.) Following the Australian experience, dung-feeding scarabs were brought to mainland areas of the United States (south Texas and Florida) where they have since established, spread, and now assist in control of blood-feeding horn flies (*Haematobia irritans*) that breed in fresh cattle manure.

FIGURE 14-21
Larvae of the bumble flower beetle, *Euphoria inda*, feeding on horse manure. Photograph by Whitney Cranshaw/ Colorado State University.

FIGURE 14-23
A pair of dung rolling scarab beetles, *Canthon imitator*. Photograph by Whitney Cranshaw/Colorado State University.

FIGURE 14-24
A male eastern Hercules beetle. Photograph courtesy of Allen Bridgman/South Carolina Department of Natural Resources/Bugwood.org.

The most interesting to watch, however, are the dung rollers or "tumblebugs." These are dull black beetles that cut off a chunk of the manure and then tamp it into a smooth ball. The ball is then pushed away by the female, often with an assisting male. When they encounter an area of suitably loose soil, the female deposits an egg inside the dung ball and buries it. The young white grub consumes the buried ball as it develops and ultimately emerges to the surface as an adult.

Most of the largest scarab beetles occur in the subfamily Dynastinae, which feed and develop in rotting wood. Adults show strong **sexual dimorphism**, with large hornlike projections on the head or prothorax of the males, which are used in battles for territory. The hornless females typically are larger and heavier. Because of their large horns, various common names are given these insects, including rhinoceros beetles, Hercules beetles, elephant beetles, unicorn beetles, ox beetles, or triceratops beetles. The largest North American scarabs are in this group, which includes the eastern Hercules beetle (*Dynastes tityus*) and western Hercules beetle (*Dynastes granti*). A much larger relative is the Hercules beetle of central and South America (*Dynastes hercules*) with a body length (including the horn of the male) up to 17 cm. The large Asian species *Chalcosoma atlas*, known as the Atlas beetle, is a popular pet in Japan where limited housing space makes its small size a desirable feature.

Very large scarab beetles also occur in the subfamily Cetoniinae. These include the several large "green June beetles" of the genus *Cotinis* that feed on ripe fruit and flowers in the southern half of the

United States. One of the most common species, *Cotinis nitida*, develops on decaying organic matter in lawns and sometimes incidentally causes damage due to its tunneling. The white grub larvae have the peculiar habit of sometimes crawling on their back on the soil surface at night, particularly after rains. Fuzzy "bumble flower beetles" (*Euphoria* species) that somewhat resemble bumble bees can be locally common. Adults visit flowers and oozing plant wounds, while larvae can be common in horse manure and compost piles. The very large African Goliath beetles (*Goliathus* spp.) are also in this latter group of scarab beetles.

FIGURE 14-25
A green June beetle, *Cotinis nitida*. Photograph courtesy of Susan Ellis/Bugwood.org.

Injury to roots of turfgrasses and pasture grasses occurs with several white grubs in the subfamily Melolonthinae. Adults of these are known as May/June Beetles or chafers, and worldwide, they are the most damaging insects to grasses. In the United States, most of the white grubs that damage lawns and golf courses are found within this group of scarabs. These include a mixture of native species, such as the large chocolate-colored May/June beetles (*Phyllophaga* spp.) and masked chafers (*Cyclocephala* spp.), along with various introduced species, such as the rose chafer (*Macrodactylus subspinosus*) and the European chafer (*Rhizotrogus majalis*).

FIGURE 14-26
Diagram of the various life stages of a May/June beetle. Illustration by Art Cushman/USDA Smithsonian Department of Entomology and provided from Bugwood.org.

Plant roots are also damaged by the shining leaf chafers of the subfamily Rutelinae. Most of these are brightly colored, shiny, attractive insects, and some tropical species, known as "jewel scarabs" (*Chrysina* spp.), are used in jewelry and ornaments. One introduced species present in the eastern United States is the Japanese beetle (*Popillia japonica*), a notorious pest that damages roots of grasses while a white grub and damages flowers or leaves of trees and shrubs in the adult stage.

FIGURE 14-27
A mass of Japanese beetles. Photograph courtesy of Jim Kalisch/ University of Nebraska.

FAMILY BUPRESTIDAE—METALLIC WOOD BORERS

The metallic wood borers are a large family (14,600 described species worldwide) that include among their members some of the most brilliantly colored of all insects. Adults of all species have some metallic coloration, particularly on the underside, and maybe copper, green, blue, or metallic black. Iridescent hues reflect off some, produced by precise microscopic sculpturing of the exoskeleton. In much of the world, these are known as "jewel beetles"; many bright-colored scarab beetles are also referred to as jewel beetles.

Metallic wood borers have a characteristic shape that is elongate-ovoid and somewhat resembles a

FIGURE 14-28
A brightly colored metallic wood borer, *Buprestis confluenta*. Photograph courtesy of Jim Kalisch/University of Nebraska.

FIGURE 14-29

A metallic wood borer, *Poecilonota thurea*. In this photo, three males are attempting to mate with the female. Photograph by Whitney Cranshaw/Colorado State University.

FIGURE 14-30

Earrings made from the forewings of a metallic wood borer. Photograph by Whitney Cranshaw/Colorado State University.

bullet. The body may be flattened or cylindrical, smooth or hairy, and there is a wide range in size, from 3 to 100 mm; although most are less than 20 mm. Adult beetles of many species feed on leaves, but some visit flowers for pollen.

The larvae develop as borers in plants and possess a laterally flattened area behind the head, leading to the common name "flatheaded borers." Some mine leaves or stems, while others are wood borers that characteristically make winding galleries through the bark or underlying cambium. Although the great majority of wood boring species limit their attacks to trees or individual branches that are in serious decline, a few are important pest species. In North America, most pest buprestids occur in the genera *Agrilus* (e.g., *Agrilus anxius*, the bronze birch borer) or *Chrysobothris* (e.g., *Chrysobothris femorata*, the flatheaded appletree borer). The recently established emerald ash borer, *Agrilus planipennis*, is setting a new standard for destruction by this insect family. The native oaks of Southern California are now also at risk to the recently introduced goldspotted oak borer (*Agrilus coxalis*).

(a)

(b)

FIGURE 14-31

(a) Larva of the bronze birch borer, *Agrilus anxius*, a flatheaded borer that develops under the bark of birch trees. (b) Larva of the flatheaded appletree borer, *Chrysobothris femorata*, a common wood borer of many hardwood trees. Photographs by Whitney Cranshaw/Colorado State University.

The brilliant colors of some buprestids attract the attention of humans for their use in ornaments and jewelry. The most commonly collected species for this purpose is *Sternocera aequisignata*, the green jewel beetle of Thailand. The coloration of the insects also is used for mate recognition, as was all too clearly made evident with the experience involving the Western Australian species, *Julodimorpha bakewelli*.

This species was threatened by a unique human threat, discarded beer bottles. Certain dimpled brownish "stubbies" (375 ml beer bottles) apparently reflected light in very much the same manner as the adult females, so the males would concentrate their attention on these bottles, attempting vainly to mate with what they perceived to be giant females. Slight alteration of beer bottle design resolved the problem.

● Emerald Ash Borer—An Ecological Train Wreck in Progress

The emerald ash borer, *Agrilus planipennis*, is a brilliant green, attractive insect, but one that has enormous potential for harm to ash trees (*Fraxinus* spp.) that grow in North America. Native to China, where it causes little injury to the ash species native to Asia, it was recently introduced accidentally into North America, likely in wooden packing material.

The insect was first detected in the Detroit area in the summer of 2002 when numerous ash trees were observed to be dying from a previously unknown cause. It soon became clear that this was a serious new threat and it was quickly evident that all North American species of ash trees were extremely susceptible to this new insect. A massive effort at containment was put into place, sometimes including the cutting and removal of all ash trees within a quarter mile of a new occurrence. Despite heroic efforts to eradicate this insect, it was too late and the emerald ash borer has spread so that now (2012) it is known to occur in over a dozen states east of the Mississippi. Tens of millions of ash trees have already been killed by this insect in the brief period since its establishment was first discovered.

It is too early to know the ultimate effects resulting from the unfortunate introduction of this exotic insect,

FIGURE 14-33
Damage done by the flatheaded borer larval stage of the emerald ash borer. Photograph courtesy of Eric Day/VPI/Bugwood.org.

FIGURE 14-32
Adult of the emerald ash borer, *Agrilus planipennis*. Photograph courtesy of David Cappaert/Michigan State University/ Bugwood.org.

but likely they will be grim. Ash is an important component of forests in much of North America, and as the emerald ash borer continues its march through the continent, populations of ash trees will be decimated. Forest ecosystems will take yet another blow, such as they sustained when chestnut blight (caused by a fungus introduced prior to 1904) essentially wiped out American chestnut, and Dutch elm disease (caused by a fungus spread by a bark beetle, introduced sometime before 1930) largely eliminated American elm.

FAMILY ELATERIDAE—CLICK BEETLES

Click beetles are ovoid to narrowly elongate beetles that are usually dullish gray or brown; however, some species are brightly colored and patterned. They show a wide range in size, between 1 and 60 mm, with one of the largest being the eyed elater (*Alaus oculatus*) that is found associated with rotting logs in eastern North America.

FIGURE 14-35
A click beetle. Photograph courtesy of Jim Kalisch/University of Nebraska.

FIGURE 14-34
The eyed elater, a large click beetle that develops in rotting wood. Photograph courtesy of Gerald Lenhard/Louisiana State University/Bugwood.org.

FIGURE 14-36
A click beetle larva (wireworm) feeding on potato piece. Photograph courtesy of Ken Gray/Oregon State University.

The most defining feature of many common click beetles is the ability of adults to flip through the air. Unlike most beetles, their prothorax is hinged, rather than fixed, to the rest of the body and a spine on the underside can snap into a notch on the underside of the next segment of the thorax (mesosternum). Tension is placed on the spine and when released, the beetle may jump up to 15 cm, often with an accompanying audible clicking noise. This behavior apparently has developed as a means to avoid predators and it also allows them to right themselves if they are placed on their back.

Larvae of click beetles are elongate-bodied with an exoskeleton more heavily sclerotized than most insect larvae. Their feeding habits vary with many predators of other small arthropods, particularly those larvae that are found in rotting wood. Soil-dwelling species often feed on roots and belowground parts of plants. Among these are important crop pests that damage germinating seeds, tubers, and roots; these are known as wireworms.

Several tropical species are bioluminescent, an impressive feature that they share with the fireflies of the beetle family Lampyridae and a few other insects. Typically, the luminescent click beetles have paired light organs on the prothorax that produce a greenish light. In addition, there is a spot on the upper side of the abdomen that glows yellow orange but is only exposed when the wing covers open for flight. Larvae may also glow, and a Brazilian species that lives on termite mounds produces light that attracts insect prey.

FAMILY LAMPYRIDAE—FIREFLIES OR LIGHTNINGBUGS

The fireflies (or lightningbugs) are the best known of the light-producing insects with adults of many species capable of producing bright "taillight" flashes. Despite their common names, they are beetles, and fireflies are usually found in marshy areas or near bodies of water. About 125 known species occur in North America with the most conspicuous species concentrated in the eastern and southern states.

Adult fireflies are elongate-bodied insects, somewhat flattened, with elytra that are unusually

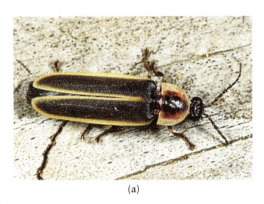
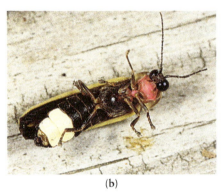

(a)   (b)

FIGURE 14-37
(a) A *Photinus* species of firefly. (b) The light-producing organ is visible on the tip of the abdomen. Photographs courtesy of Tom Murray.

quite flexible for beetles. The prothorax is broad, extends to cover the head when viewed from above, and is usually of a different color (yellow, red, orange) from the wings. Adult females of some species lack wings and largely resemble the larvae (larviform). These are commonly known as "glowworms," although that term is sometimes applied to other insects that produce light such as female phengodid beetles and some fungus gnat larvae that are insect predators.

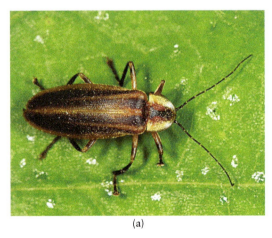

(a)   (b)

FIGURE 14-38
(a) A *Photuris* species of firefly. (b) The light-producing organ is visible on the underside of the beetle. Photograph courtesy of Tom Murray.

Firefly larvae are primarily predators of soft-bodied invertebrates, favoring snails, slugs, and earthworms. Others have broader food habits and may eat various soft-bodied insects, scavenge upon dead insects, and occasionally feed on soft, ripe fruit. Larvae of all fireflies are thought to be capable of producing light.

In the adult fireflies, that most spectacular light-producing organ is located at the tip of the

FIGURE 14-39
Larva of a *Pyractomena* species of firefly. Photograph courtesy of
Tom Murray.

FIGURE 14-40
Larva of a *Photuris* species of firefly. Photograph courtesy of
Tom Murray.

abdomen and involves thousands of specialized cells,
known as photocytes. These are infiltrated with
tracheoles that can deliver substantial quantities of
oxygen. Within each cell are compartmentalized two
key compounds—**luciferin** and **luciferase**. Light
production occurs from a two-stage process that first
causes luciferin to be catalyzed by the luciferase
enzyme in the presence of ATP. Introduction of
oxygen moves it to the next stage, where light is
produced with almost 100% efficiency producing
very little heat. The length and timing of the reaction
is largely controlled by the flow of oxygen, which is

adjusted by the airflow through the tracheoles. Flash
color can vary. The genus *Photinus* usually produces
yellow flashes, *Photuris* usually produces green
flashes, and many *Pyractomena* species produce a
flash that is orange yellow or amber.

FIGURE 14-41
A *Pyropyga* sp. of firefly in flight. *Pyropyga* spp. are among the
fireflies that are non-luminescent, do not produce flash patterns,
and fly during the day. Photograph courtesy of Jon Yuschock/
Bugwood.org.

Timing of the light flash is critical for fireflies as it
is used to attract and detect an appropriate mate. Each
species of firefly produces its own unique flashing
pattern. For example, males of *Photinus pyralis* make
undulating flights near the ground and flash at 1/2 s
intervals, producing a distinctive "J-pattern" flash that
is repeated at 6 s intervals. Receptive females respond
with a 1/2 s flash 2 s later; the male only responds if
the female flash is produced at the proper interval.
Many of the North American fireflies cannot be
distinguished by physical appearance, but their
flashing patterns are unique and defining.

The use of firefly light to attract mates has long
been recognized and was elegantly expressed in an
1860 poem by James Montgomery:

> *When evening closes Nature's eye,*
> *The glowworm lights her little spark*
> *To captivate her favorite fly,*
> *And tempt the rover through the dark.*
> *Conducted by a sweeter star*

Than all that deck the fields above,
He fondly hastens from afar,
To sooth her solitude with love.

Fireflies were the subject of "Das Glühwürmchen" ("The Glow-Worm") sung in the 1902 operetta *Lysistrata*:

Shine, little glow-worm, glimmer, glimmer
Shine, little glow-worm, glimmer, glimmer!
Lead us lest too far we wander. Love's sweet voice is
*   calling yonder!*
Shine, little glow-worm, glimmer, glimmer
Shine, little glow-worm, glimmer, glimmer
Light the path below, above. And lead us on to love.

These lyrics, in a substantially jazzier treatment by the Mills Brothers, were used fifty years later in the #1 Pop Chart Hit "Glow Worm."

Not all fireflies produce flashes. In western North America, bright lighting species are quite uncommon and males do not produce light. Females are capable of glowing but most are wingless and remain near ground burrows. Mate finding by these species primarily is performed through the use of sex pheromones produced by the female.

On the other hand, flashing is intensified by some southeast Asian/East Indies species that synchronize their flash. This is done by the foldedwinged fireflies (*Pteroptyx* spp.) that usually form small groups of several individuals clustered on a single plant. In this species, such groupings are composed of one beetle per leaf and all within an area flash in unison. When these insects are abundant, they may cover trees, producing "firefly trees" and their brilliant flashing can be seen a kilometer away. The purpose of this synchronized flashing is still unclear but is likely useful for advertising the presence of males in the dense jungle vegetation where these species occur. Synchronized flashing is also known from some fireflies that develop with the region of the Great Smoky Mountains of the eastern United States.

Not all firefly flashing is for mate recognition; some use it for attracting prey, including other fireflies. Females of the genus *Photuris* are notorious in this regard, and some have the ability to appropriately mimic the signals of other female firefly species. This allows them to lure in unsuspecting males, who are then eaten.

The remarkable light production of fireflies has also been exploited for many scientific purposes. Genes for production of luciferase have long been included in genetic transfer technology. Subsequent exposure to luciferin will cause the organism to glow if the gene transfer was successful. Luciferin and luciferase, extracted from collected beetles, are also useful for detecting the presence of ATP.

● Bioluminescence

The ability of living things to generate light is a curiously widespread phenomenon known as **bioluminescence**. It has been most commonly recorded among various marine forms of life but also is found in land-dwelling organisms, including bacteria, fungi, mollusks, earthworms, millipedes, insects, and springtails. Many of these have independently evolved bioluminescent abilities, and at least eight different systems of light production have been documented among land-dwelling species alone.

Over 2,300 hexapods have been documented to produce light. These produce colors that range from a blue light produced by larvae of certain fungus gnats (Diptera: Keroplatidae, Mycetophilidae) to the reddish "headlights" produced by the railroad worms (Coleoptera: Phengodidae). Most light-producing hexapods are beetles, but light production is also known among springtails, a few kinds of flies, and a South American cockroach.

Bioluminescence appears to be able to serve many different functions. The best known of the light-generating insects are beetles of the family Lampyridae, the fireflies or lightning bugs. These primarily use light to detect mates, but some may also use light to attract prey or for warning. The eyespots of luminescent click beetles (Coleoptera: Elateridae) also seem to be used as warning signals or to startle predators.

Attraction of prey is the primary purpose of the predatory fungus gnats that produce light. This is most dramatically illustrated by the fungus gnat *Arachnocampa luminosa* (Diptera: Keroplatidae) that colonizes moist banks and cave ceilings in New Zealand, attracting flying insects to sticky threads using a blue-green light. A distantly related North American species (*Orfelia fultoni*) (Diptera: Mycetophilidae) lives along Appalachian river banks and attracts insects to ensnaring webs with a bluish light. Light produced by the *Phrixothrix* railroad worms (Coleoptera: Phengodidae), flightless beetles that roam the floor of tropical forests, is particularly bright and may be used to actually find their millipede prey.

(continued)

FIGURE 14-42

(a) The sticky strands produced by the New Zealand glowworm, *Arachnocampa luminosa*. This is a larva of a type of fly that uses bioluminescence to lure insects to the hanging strands. (b) In North America, predatory fungus gnats of the genus *Orfelia* also luminesce and attract prey to sticky threads. Photograph of the glowworm courtesy of Marie McDonald/Lincoln University. Photograph of the adult *Orfelia* courtesy of Tom Murray.

FAMILY DERMESTIDAE—DERMESTID OR SKIN BEETLES

Among the many insects that function as macrodecomposers that recycle dead plant and insect matter, dermestid beetles are often the last of the "cleanup crew" to leave the job. Although they are general scavengers, many species have particular fondness for animal materials, and they are some of the very few insects (along with clothes moths) that can digest the protein keratin, a major component of skin and hair.

Adults have a compact body form that is usually oval but may be nearly round; the antennae are short and end in a club. On close inspection, the body is covered with fine scales. The larger species (*Dermestes* spp., *Trogoderma* spp.) tend to be dull colored, but some of the smaller "carpet beetles" (*Anthrenus* spp.) have colored scales and interesting patterning. The larvae are distinctively very hairy, and most have long spines, often concentrated on the rear of the insect. Encountering old larval skins, which largely resemble larvae, is often the best indication that developing dermestid beetles have been present.

FIGURE 14-43

The larder beetle, *Dermestes lardarius*. Photograph courtesy of Jim Kalisch/University of Nebraska.

FIGURE 14-44
Varied carpet beetles, *Anthrenus verbasci*.
Photograph courtesy of Jim Kalisch/University
of Nebraska.

Outdoors, dermestid beetles can be commonly found in abandoned bird nests, animal dens, old bee and wasp nests, where collections of dead insects occur, and among the dried remains of carrion. They frequently enter buildings and normally feed and reside on collections of lint, insect debris around lights or behind walls, or around old spider webs. They sometimes also damage various stored products. In pantries, they may feed on spilled grains but most commonly attack high protein materials, such as dried pet foods and meat-based products. As they also can feed on wool, they may also damage woolen fabrics, and in drier areas of the country, they are more damaging to wool than are clothing moths. Their ability to severely damage woolen rugs and carpeting has led to the common name "carpet beetle" being given to some species.

FIGURE 14-45
Larva of the varied carpet beetle, *Anthrenus verbasci*.
Photograph courtesy of Jim Kalisch/University of Nebraska.

FIGURE 14-46
Larva of the warehouse beetle, *Trogoderma variabile*.
Photograph courtesy of Jim Kalisch/University of Nebraska.

There are about 700 described species of dermestid beetles, and they are likely to be about the best described of the beetle families since they are often found feeding on insect collections, animal skins, and other materials common to zoology departments of museums. Insect collections must constantly be protected from their damage, either through insect-proof containers, fumigant insecticides, temperature treatment, or by a combination of methods. On the other hand, certain hide beetles, such as *Dermestes maculatus*, are frequently used by museums and taxidermists to clean skeletons of hair, skin, and other dried tissues prior to specimen preparation.

FAMILY COCCINELLIDAE—LADY BEETLES

Lady beetles ("ladybugs," "ladybirds") are, along with butterflies, the most recognizable and widely popular insects. In a great many cultures, they are considered symbols of good luck or blessings, and their common name is due to their being considered "Ladies of the Virgin Mary" (Of course, half of all lady beetles are male.). There are nearly 6,000 species known worldwide, about 475 of which occur in the United States and Canada. Six states have named one species of lady beetle or another as their State Insect (Appendix I).

With very few exceptions, lady beetles develop as predators of other insects. The larvae, which are usually dark colored with prominent thoracic legs, are active hunters and consume tremendous numbers of aphids, mealybugs, scale insects, spider

(a)

(b)

FIGURE 14-47
(a) The sevenspotted lady beetle, *Coccinella septempunctata*, feeding on aphids. Photograph courtesy of Jim Kalisch/ University of Nebraska. (b) A twicestabbed lady beetle, *Chilocorus stigma*. Photograph courtesy of Tom Murray.

mites, small insect larvae, and insect eggs. Adults may also feed on insects, although many shift their diet to include pollen, nectar, and honeydew. A significant species that deviates from this habit is the Mexican bean beetle (*Epilachna varivestis*), the "bad apple" of the lady beetle family that feeds on leaves and pods of beans and is a major pest of that crop.

(b)

(a)

FIGURE 14-48
(a) A mass of convergent lady beetles, *Hippodamia convergens*. Photograph by Scott Bauer/USDA ARS/ Bugwood.org. (b) A pinkspotted lady beetle, *Coleomegilla maculata*. Photograph by Whitney Cranshaw/Colorado State University.

Adult lady beetles often have a round body form, although some are elongate oval. Although many of the smaller species are dark colored, the most commonly seen lady beetles have bright colors and patterns. These are **aposematic** (warning) colors that advertise to potential predators that they are distasteful. When disturbed, lady beetle can secrete a yellow defensive fluid (isopropyl methoxy pyrazine) from their joints that is bitter, has an unpleasant odor, and can stain.

(a)   (b)   (c)

FIGURE 14-49
(a) Larva of the multicolored Asian lady beetle, *Harmonia axyridis*. (b) Larva of the convergent lady beetle, *Hippodamia convergens*. (c) Larvae of the Mexican bean beetle, *Epilachna varivestis*. Photographs by Whitney Cranshaw/Colorado State University.

During periods when prey insects become scarce, the adults of many common species move to cover and enter dormancy (diapause). With some species migration occurs during hot, dry summer weather; other lady beetles go dormant in late summer and early autumn. Lady beetles that earlier developed in yards and gardens may sometimes be found during fall in large aggregations under piles of leaves, behind loose bark, or among wood piles.

One of the most common species, the convergent lady beetle (*Hippodamia convergens*), may make long distance migrations, often ultimately to high elevation areas where they seek out cool, humid sites to survive and mate during the "off-season." In particularly favorable sites, tremendous numbers of convergent lady beetles may be present in aggregations that cover tree trunks, boulders, and the forest floor. Several such aggregation sites are best known in the Sierra Nevada mountains of California and collectors have annually been visiting these sites for almost a century, scooping up bags full of the insects. They can later sell these for the popular, but dubiously effective, practice of release by gardeners for insect control. Few of these field-collected beetles stay long in the area upon release, as their normal behavior is to disperse long distances, as they would make during the return to the distant Central Valley.

In recent decades, one species of lady beetle has emerged as a household pest in parts of North America, the multicolored Asian lady beetle

FIGURE 14-50
Different life stages of the convergent lady beetle. At the bottom is a full-grown larva that has settled prior to pupation (prepupa). In the center is a pupa shortly after the molt from the larval stage and at the top is an older pupa, which has colored and hardened. Photograph by Whitney Cranshaw/Colorado State University.

FIGURE 14-51
Massed convergent lady beetles in an aggregation site in the Sierra Nevada Mountains. Photograph courtesy of James Solomon/USDA Forest Service/Bugwood.org.

(*Harmonia axyridis*). Its name reflects its variable coloration and patterning, which can be yellow orange to red and have up to 19 black spots, or have no spots at all. The name also reflects its exotic origin. (There is still some debate on how it became established in North America, whether by intentional introductions or accident.). Regardless, it has emerged as a major new lady beetle species following its first detection in 1988 and now is among the most abundant lady beetles over broad areas of the United States. It is particularly common as a predator of aphids on trees and shrubs and has become a very important insect in managing the soybean aphid (*Aphis glycines*), which itself is another recently established insect. Unfortunately, the multicolored Asian lady beetle also has the habit of entering homes in fall, where it may occur by the hundreds, and has become a source of unusual bad publicity for these popular insects. In addition, it has provided a new challenge to some wine producers, as the beetles sometimes get included during grape harvest, adding an off-flavored "ladybug taint."

FIGURE 14-52
A small mass of multicolored Asian lady beetles found by a window sill. This is the only lady beetle that regularly enters buildings. Photograph by Whitney Cranshaw/Colorado State University.

The Vedalia Beetle, the Cottony Cushion Scale, and the Birth of Classic Biological Control

The recognition that some insects can be valuable in the control of insects is now widely appreciated. It was the sensational success of a lady beetle, the vedalia beetle, which first firmly established the use of **biological control** as a method to control insect pests.

A serious agricultural problem developed when the cottony cushion scale (*Icerya purchasi*) (Hemiptera: Margarodidae) was accidentally introduced into southern California, probably around 1868. Within two decades, it had spread throughout the citrus-producing region of the state with devastating effect, forcing many farmers to abandon their orchards.

The insect was thought to have originated from Australia or New Zealand, and a plan was developed to try and explore the region for the parasites and predators that attacked cottony cushion scale in its native regions. The idea was untested and novel, and there were difficulties in getting congressional approval for the project. Difficulties were sidestepped by sending the entomologist Albert Koebele under the guise of a State Department representative to the International Exposition in Melbourne, Australia.

While there, Koebele scoured the area for insects that attacked cottony cushion scale and found several. He was particularly interested in a small parasitic fly (*Cryptochaetum iceryae*), and about 12,000 of these were sent back to California. He also encountered a small lady beetle (*Rodolia cardinalis*) known as the vedalia beetle. A small number (129 total) of these beetles were also included in the shipments to California

(*continued*)

FIGURE 14-53
The cottony cushion scale, *Icerya purchasi*. This insect is native to Australia but has been accidentally introduced into many regions, where it often seriously damages citrus crops. Photograph by Jack Kelly Clark and provided courtesy of the University of California IPM Program.

FIGURE 14-54
The vedalia beetle, *Rodolia cardinalis*, feeding on cottony cushion scale. Photograph by Jack Kelly Clark and provided courtesy of the University of California IPM Program.

FIGURE 14-55
The small parasitic fly, *Cryptochaetum iceryae*, was also introduced into California with the vedalia beetle and helps to control cottony cushion scale. Photograph by Jack Kelly Clark and provided courtesy of the University of California IPM Program.

for release and sent back to the United States during late fall and winter of 1888.

The vedalia beetle adapted spectacularly to the southern Californian environment. Released onto an infested orange tree in Los Angeles, it almost completely destroyed the cottony cushion scales on the tree within 4 months. From there, it spread, originally aided by humans but soon dispersing on its own. In orchard after orchard, wherever this new beetle became established, it rapidly controlled the cottony cushion scale. Orchard production rebounded immediately and the number of train carloads of oranges sent out of the area went from 700 to 2,000 within a single year after the initial introduction of the vedalia beetle. The total cost of the project was $1,500.

(continued)

The success of the vedalia beetle introduction was hailed as a miracle of entomology and clearly established the potential of insect natural enemies for the control of plants pests. The vedalia beetle continues to control the cottony cushion scale in California to this day, and it has also been successfully used to control the insect in many other countries similarly infested. Since that time, there have been thousands of additional programs worldwide to establish insect natural enemies to control previously introduced pest insects—a method known as "classical biological control."

FAMILY TENEBRIONIDAE—DARKLING BEETLES

The darkling beetles are a large family of beetles (about 1,000 known North American species) and can be common insects; however, their generally dull appearance and quiet life histories rarely draw much attention. Most are black or brown but other features are difficult to generalize. Some can fly, but many are wingless having fused elytra that make them particularly well armored. The body shape is often elongated, but often, they are of oval shape. Their wings may be smooth, or very rough, and the overall body size ranges from slightly more than 1 mm up to 80 mm.

FIGURE 14-56
Larva, pupa, and adult of the yellow mealworm, *Tenebrio molitor*. Photograph courtesy of Jim Kalisch/University of Nebraska.

Adults are scavengers of plant matter that they usually find on the soil surface. The larvae also feed on plant matter and possess elongated bodies that are slightly armored. Species that develop within the soil are sometimes called "false wireworms" because of their superficial resemblance to the larvae of click beetles. Larvae of other species develop in rotting wood.

The most widely recognizable member of this family is the common mealworm (*Tenebrio molitor*) that is sold as a pet food staple. It is also about the simplest insect for anyone to rear. (A box of oatmeal periodically supplemented with an occasional fruit slice or potato piece is usually sufficient to keep a population of mealworms happy.) More recently, another larger member of this family has appeared in pet stores as fish bait, *Zophobas morio*. This insect is sometimes marketed as the "giant mealworm" or "superworm" (These need a bit of dried meat to successfully develop.).

The common mealworm and its close relative the dark mealworm (*Tenebrio obscurus*) are occasional pests of stored foods; however, much more common denizens of kitchen pantries are the flour beetles. Both the confused flower beetle (*Tribolium confusum*) and the red flour beetle (*T. castaneum*) are among the top four or five insects that occur within foods stored in homes. They are tiny beetles, only about 3.5 mm, which feed on flour. Because of their small size and ease of rearing, they have often been used in research involving insect behavior and ecology.

FIGURE 14-57
A confused flour beetle, *Tribolium castaneum*. Photograph courtesy of Peggy Greb/USDA ARS/Bugwood.org.

Outdoors, darkling beetles are most conspicuous in arid areas. In desert and semidesert areas of the southwestern United States, large black beetles of the genus *Eleodes* commonly can be seen walking about the soil surface. When disturbed, they will put their head down, point the tip of the abdomen upward, and can discharge a smelly, slightly caustic fluid. Locally, they go by various names, including "skunk beetle," "stink beetle," "circus beetle," or "pinacate." Another interesting darkling beetle of the region is the ghost beetle, *Asbolus verrucosus*, which can cover its body with wax, allowing it to blend better with its surroundings and conserve moisture.

Extreme adaptation to desert life occurs with darkling beetles that occur among the sand dunes of the Namib Desert along the southwestern coast of Africa. This region is one of the driest on earth, receiving an average of only 10 mm of rainfall annually; however, the area does often receive fogs coming off the cold Atlantic coast. The darkling beetles known as the Namib fog baskers (*Onymacris rugatipennis*, *O. unguicularis*) take advantage of the fog by perching on dune crests at

FIGURE 14-58

An *Eleodes* sp. darkling beetle in defensive posture. Photograph by Whitney Cranshaw/Colorado State University.

dawn and allowing the fog to collect on their bodies. They then acquire sufficient water by drinking the dew that trickles down their bodies to their mouth. These beetles also have large bodies and long legs that allow them to run very quickly as the desert sands heat.

● Living Jewelry—The Makech

Although wing covers of brightly colored beetles and butterfly wings have often been incorporated into artworks and ornamentations, only one insect can be considered a "living jewel"—the makech. This is a flightless black beetle (*Zopherus chilensis*) with very hard wing covers that are typical of the beetle family Zopheridae, the "ironclad beetles." The makech is native to the Mayan areas of the Yucatan where the larvae develop on rotting wood.

Its transformation to jewelry involves gluing rhinestones and other colorful objects to its back. A small gold chain with a pin at the end allows it to be used as a slowly walking broach. It is long lived and if occasionally fed a bit of apple slice can live over a year.

Its history is lost but is related to a Mayan legend. In this story, a princess fell in love with a man she was forbidden to marry. She became so deeply distressed that a shaman, seeing her predicament, changed her into a beautiful beetle. In this form, she could be kept next to the heart of her loved one.

FIGURE 14-59

A makech, a bejeweled ironclad beetle sold in the markets of Merida, Mexico. Photograph courtesy of Karen and Wayne Passmore.

FAMILY MELOIDAE—BLISTER BEETLES

Blister beetles are best known for their production of a powerful toxin, known as **cantharidin**. This chemical circulates in their blood and can be exuded through leg joints in a behavior known as reflexive bleeding. Cantharidin is very effective at repelling ants and most other predators. Species that contain the highest levels of cantharidin can also cause skin blisters if they are handled, although most North American species possess more modest amounts of the chemical.

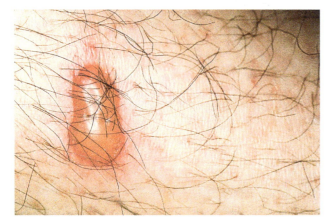

FIGURE 14-60
Blister produced from contact with cantharidin, produced by a blister beetle. Photograph courtesy of Jim Kalisch/University of Nebraska.

FIGURE 14-61
The black blister beetle, *Epicauta pennsylvanica*, is one of the most common blister beetles in North America. It produces only low levels of cantharidin. Photograph by Whitney Cranshaw/Colorado State University.

Although chemically defended, blister beetles lack the hard exoskeleton common to most other beetles. Their abdomen is particularly soft, and their elytra are better described as leathery. In general appearance, they are usually moderate sized beetles, with prominent legs adapted for walking. Their head is broader than their prothorax. Many are drab colored but some have bright patterns, and those of the genus *Lytta* are typically shiny, attractive insects.

Blister beetle larvae prey on either grasshopper egg pods or ground nesting bees in developing immature stages. Newly hatched larvae are active insects called **triungulins** with prominent legs. Those that develop on grasshopper egg pods can dig into soil; predators of ground nesting bees often crawl onto flowers and attach to visiting adult bees that then carrying them back to the nest (a behavior known as **phoresy**).

FIGURE 14-62
The active first instar larva of a blister beetle, known as a triungulin. Photograph courtesy of Ken Gray/Oregon State University.

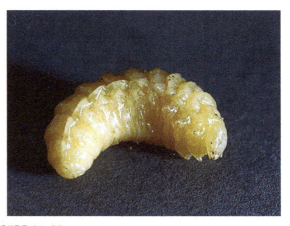

FIGURE 14-63
A sedentary, late-stage larva of a blister beetle. Photograph courtesy of Ken Gray/Oregon State University.

Developing blister beetle larvae undergo considerable changes in body form (**hypermetamorphosis**) as they develop. After the triungulin stage (Instar I) has located the ultimate food source, they molt through a series of largely legless grub-like forms when they consume their prey. The final larval stage, produced the following spring, again has legs, and pupation follows. The entire development process occurs within the soil.

Adult blister beetles feed on leaves and flowers of various plants, sometimes appearing in large aggregations. They are sometimes minor plant pests but perhaps more importantly, particularly when found in alfalfa, are a concern for the potential of cantharidin to cause livestock poisoning. Occasionally large numbers of blister beetles may accidentally be crushed during alfalfa harvest and, if subsequently ingested by horses, can produce painful illness and even death.

FIGURE 14-64
The striped blister beetle, *Epicauta occidentalis*, is the species most commonly associated with accidental poisoning of livestock when incorporated into alfalfa hay. Photograph courtesy of Jim Kalisch/University of Nebraska.

The irritating properties of cantharidin, its ability to produce "heat" and blisters, were long considered valuable in medicine. In the Mediterranean region, the most common source was *Lytta vesicatoria*, an emerald green species known as the 'Spanish fly' or 'cantharides', a species that contains very high concentrations of cantharidin. Its European use goes back at least 2,500 years, and Hippocrates (460–377 BC) mentioned it for treatment of stroke (dropsy). At one time or another, it was recommended for everything from snakebite, rabies, rheumatism, leprosy, and gout to earache, warts, and "lethargies." China concurrently developed its own uses of blister beetles in traditional medicine, using local *Mylabris* species.

FIGURE 14-65
The "Spanish fly," *Lytta vesicatoria*, a European blister beetle with high levels of cantharidin that was formerly used for medical purposes and as an aphrodisiac. Photograph courtesy of Gyorgy Csoka/Hungary Forest Research Institute/Bugwood.org.

Spanish fly also had more notorious applications. Its ability to produce irritation of the urinary tract often had the observed side effect of producing painful, but prolonged, erections. This led to a reputation as an aphrodisiac, and it was sometimes used at Roman orgies. (It was also used by the notorious Marquis de Sade, who was imprisoned for incorporating it into anise candies fed to unsuspecting women, almost killing some.) Cantharidin is extremely toxic to humans, being able to cause death in doses of only 0.5 mg/kg of body weight. Sublethal doses can cause kidney failure, abortion, internal hemorrhaging, abdominal cramping, and convulsions. Because of its extreme danger and lack of proven medical effectiveness, all medical uses are currently banned in the United States, with the exception of topical treatments to warts.

FAMILY CERAMBYCIDAE—LONGHORNED BEETLES

Longhorned beetles include some of the largest and most striking of all insects. Their namesake feature is the prominent antennae that extend well beyond the end of the body in many species. Most longhorned beetles possess an elongate or cylindrical body form and the day-active species may be brightly colored.

(a)   (b)

FIGURE 14-66

(a) The locust borer, *Megacyllene robiniae*, and (b) the whitespotted pine sawyers, *Monochamus scutellaris scutellaris*, are representative longhorned beetles. Note that the antennae of the male whitespotted sawyer is much longer than that of the slightly larger and heavier female. Such sexual differences are common with members of this family. Photographs by Whitney Cranshaw/Colorado State University.

Larvae develop as borers of plants and are known as roundheaded borers. They have an elongate body that is generally round in the cross section, a small head comprised largely of the massive mandibles, and thoracic legs that are highly reduced or entirely absent. They are very well adapted to chewing hard wood and the biomechanics of their jaws provided the inspiration for the first chain saw.

FIGURE 14-67

Larva of the locust borer, within a black locust trunk. Photograph courtesy of David Leatherman.

A few longhorned beetles will develop in nonwoody (herbaceous) plants, such as the red-colored milkweed longhorns (*Tetraopes* spp.) found most everywhere milkweed grows. (One of these graces the cover of this book.) Most all of the largest and most conspicuous species are associated with trees, particularly trees that are in decline or recently felled. A few species aggressively attack living trees, and one such species of current concern is the Asian longhorned beetle (*Anoplophora glabripennis*), which threatens maples and some other hardwoods. Asian longhorned beetles are native to China but were accidentally introduced in commerce and their populations became established in parts of New Jersey, New York, and Massachusetts; a population found in Chicago was eradicated following sustained efforts. In California, two other species of longhorned beetles (*Phoracantha semipunctata* and *Phoracantha recurva*) were accidentally introduced that can seriously damage eucalyptus, a tree that is presently widespread and very common in California since its introductions from Australia."

Adult longhorned beetles also feed on plants but do little injury. Pollen and flowers are preferred by some of the most colorful species, such as the bright yellow and black-striped locust borers (*Megacyllene robiniae*) common on goldenrod in the late summer. Others do a bit of minor chewing of leaves, needles, or bark of twigs. In one case, this latter behavior can cause some problems. Longhorned beetles known as pine sawyers

FIGURE 14-68
The Asian longhorned beetle, *Anoplophora glabripennis*, an accidentally introduced species that is a threat to eastern hardwood forests. Photograph courtesy of Kenneth R. Law/USDA APHIS PPQ/Bugwood.org.

(*Monochamus* spp.) feed on twigs as larvae. The damage results in wounds into which nematodes (*Bursaphelenchus xylophilus*) may enter and produce pine wilt disease. Pine wilt disease is a disease native to the United States and, although native North American pines are resistant, it can rapidly kill many of the susceptible pines that are now widely planted in North America (e.g., Austrian pine, Scots pine). Pine wilt also has spread to other areas of the world and is a serious threat to the native pines in Japan and Korea.

The largest North American species include various dark brown longhorned beetles of the genus *Prionus* that can reach 60 mm. These individuals, particularly males, possess unique and unusual serrated antennae. Larvae develop at the base of various trees and are sometimes known as "giant root borers." A slightly larger species, *Trichocnemis spiculatus*, known as the ponderous borer, is associated with Douglas-fir and ponderosa pine in western states. Even these species are dwarfed by some longhorned beetles that occur in the American tropics. *Titanus giganteus*, found in rainforests of French Guiana and Brazil, is a potential title holder for largest insect, with a body length of 167 mm and some impressive jaws. A prized beetle among collectors and admirers of insects is the harlequin

beetle (*Acrocinus longimanus*) that occurs in forests from southern Mexico to northern Argentina. Its name refers to the interesting red, black, and gray patterning (reminiscent of the gaily colored clowns of the European Middle Ages known as harlequins) on its elytra, but its most unusual feature is perhaps the extremely long forelegs of the males. Despite its appearances and its prominent use to scare Indiana Jones in his *Temple of Doom* movie, it is harmless.

FIGURE 14-69
Male and female of the ponderous borer, *Trichoneumus spiculatus mexicanus*, the largest longhorned beetle in North America. Photograph by Whitney Cranshaw/Colorado State University.

FIGURE 14-70
Larva of the ponderous borer. Photograph courtesy of David Leatherman.

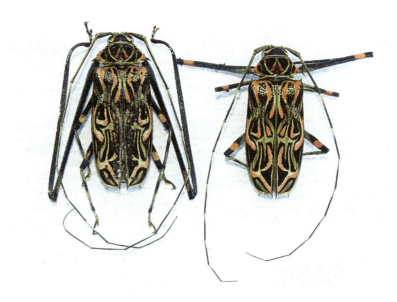

FIGURE 14-71
Male (left) and female harlequin beetle, *Acrocinus longimanus*. This longhorned beetle is native to forests from southern Mexico to Brazil. The forelegs are extremely long and can extend fully to the front of the insect as it walks among trees. Photograph by Whitney Cranshaw/Colorado State University.

FAMILY CHRYSOMELIDAE—LEAF BEETLES

The family Chrysomelidae is one of the largest (c.35,000 species) and all members develop by feeding on plants. Most feed on leaves as the common name suggests, but representative "leaf beetles" also may be found affecting roots and seeds. Adults are beetles of small to medium size, and most have an elongate-oval body form. Their short antennae often help distinguish them from longhorned beetles, and many have colorful, patterned wing covers. Larvae of leaf beetles that chew leaves are often dark colored and may have an elongate body similar to lady beetle larvae. Root-feeding species have pale-colored wormlike larvae, and those that develop within leaves, as **leaf miners**, are pale-colored and flattened.

Most crop-damaging leaf beetles primarily injure plants by chewing leaves. Perhaps the best example of this is the Colorado potato beetle (*Leptinotarsa decemlineata*), the world's worst insect pest of potato. Both the chunky-bodied red-orange larvae and the colorfully striped adults chew leaves,

(a)

(b)

FIGURE 14-72
(a) Elm leaf beetle, *Xanthogaleruca luteola*, larva. Photograph by Whitney Cranshaw/Colorado State University. (b) Cottonwood leaf beetle adult and larvae. Photograph courtesy of Jim Kalisch/University of Nebraska.

sometimes eating plants into the ground. It is a native insect of certain species of wild nightshade such as buffalo bur (*Solanum rostratum*) that grow on the plains west of the Mississippi and into Mexico. When potato was introduced into the area, this was found to be acceptable as well and Colorado potato beetle thrived on the new food source and expanded its range. Ultimately, it not only colonized much of the United States but also accidentally made its way across the ocean where it has become especially damaging through Eastern Europe into Russia. A significant contributing problem with Colorado potato beetle is that they have had an unusual ability to rapidly evolve resistance to insecticides, making control of this pest all the more difficult.

Among the smaller leaf beetles are the flea beetles (subfamily Alticinae) that have large hindlegs allowing them to jump. Adults are active insects that readily find new plantings of susceptible crops. Some of the more important species of flea beetles are associated with either plants in the cabbage family

FIGURE 14-73
Colorado potato beetles, *Leptinotarsa decemlineata*, on buffalo bur. Photograph by Whitney Cranshaw/Colorado State University.

MARY F. BENSON

FIGURE 14-74
Diagram of the various life stages of the Colorado potato beetle. Illustration by Mary Benson/USDA Smithsonian Department of Entomology.

(e.g., cabbage, broccoli, and canola) or plants in the nightshade family (e.g., potato, tomato, and eggplant). Somewhat unusual is that adult flea beetles are often the most damaging stage as they chew small pits ("shot holes") into leaves that stunt growth and may kill seedlings. Larvae are root feeders that produce more negligible injuries. Larger flea beetles occur on trees and shrubs and most of these have larvae that also chew the leaves.

The most costly species to US agricultural production have been the various "corn rootworms," notably the western corn rootworm (*Diabrotica virgifera virgifera*) and northern corn rootworm (*Diabrotica barberi*). Most of the damage by these insects is produced by the larval stages that chew corn roots, and hundreds of millions of dollars are spent annually for their control. Crop rotation, such as following corn with soybean, has been used successfully to manage this insect as eggs are laid in midsummer at the base of corn plants and only corn roots can support the young that hatch the following year. Unfortunately, in some areas where this practice has been intensively followed for decades, strains of corn rootworms that resist the practice of rotational cropping have evolved, either by producing eggs that hatch in the second year after being laid or by laying eggs in soybean fields.

FIGURE 14-75
Crucifer flea beetles, *Phyllotreta cruciferae*, damaging a broccoli plant. Photograph by Whitney Cranshaw/Colorado State University.

FIGURE 14-76
Western corn rootworm, *Diabrotica virgifera*. Photograph courtesy of Jim Kalisch/University of Nebraska.

FIGURE 14-77
Damage to corn roots done by larvae of the western corn rootworm. Photograph courtesy of Jim Kalisch/University of Nebraska.

Most leaf beetles are quite specific on the type of plants they will accept as food. Because of this, several species have been used to control introduced plants that have become serious weeds. Among the more prominent leaf beetles currently being used to manage weed problems are *Galerucella calmariensis* and *G. pusilla* for the control of purple loosestrife in the eastern United States and *Diorhabda elongata deserticola* to control tamarisk in some western states.

FIGURE 14-78
The black margined loosestrife leaf beetle, *Galerucella calmariensis*, a species purposefully introduced into North America for the control of the invasive weed purple loosestrife. Photograph courtesy of David Cappaert/Michigan State University/Bugwood.org.

● Tortoise Beetles as Color Magicians and Peddlers

Insects use various mechanisms to produce color including permanent pigments and microsculpturing of their body surface. The latter is a source of iridescent hues that may shift depending on the angle of view. Tortoise beetles use a different approach that allows them not only to produce brilliant color but also to change it rapidly.

Tortoise beetles occur within the leaf beetle family, but their rounded body form often causes them to be mistaken for lady beetles. Two brightly colored species most often attract attention in the United States—the golden tortoise beetle (*Charidotella sexpunctata bicolor*) and the mottled tortoise beetle (*Deloyala guttata*), both of which may be common on leaves of morning glory, sweet potato, and field bindweed. The golden tortoise beetle is usually brilliant metallic gold when first seen on a plant but fades quickly to a duller reddish brown beetle with spots if captured and killed. This color shift is achieved by introducing, or withdrawing, moisture to the surface of the exoskeleton. When hydrated, a perfect reflecting surface is produced, but in the absence of the moisture the underlying colors become visible.

FIGURE 14-79
A mottled tortoise beetle, *Deloyala guttata*. Photograph courtesy of Tom Murray.

Tortoise beetle larvae also chew on leaves. They are rarely observed and have certain features that discourage closer inspection. These larvae are flattened, spiny insects with an elongated moveable

(continued)

fork at the end of the body. They also have an eversible anus that they use to deposit their excrement on the back of the body, often mixed with old larval skins. These "peddlers" then carry with them a sort of moveable parasol that covers the body and helps deter potential enemies.

FIGURE 14-80

A golden tortoise beetle, *Charidotella sexpunctata bicolor*. Photograph courtesy of Tom Murray.

FIGURE 14-81

Larva of the mottled tortoise beetle, carrying old larval skins and feces on its back. Photograph by Whitney Cranshaw/ Colorado State University.

FAMILY CURCULIONIDAE—WEEVILS

The weevils, with over 60,000 described species, are an enormous animal family that alone outnumbers all the vertebrates (e.g., mammals, birds, fish, reptiles, and amphibians) combined. Also called the "snout beetles," the most distinctive feature of most weevils is an elongated projection of the head (rostrum) that is tipped with chewing mouthparts. This adaptation allows many weevils the ability to chew deeply into seeds, plant stems, and other plant parts. All weevils develop as plant feeders, and most are specialists that utilize a narrow range of food plants; often, a single host plant species serves as the sole source of food.

Weevil larvae are legless and those that develop within plants are pale-colored, somewhat resembling a piece of puffed rice with a dark head. Many of these are found boring into seeds or fruit, such as the pecan weevil (*Curculio caryae*) and plum curculio (*Conotrachelus nenuphar*). A small number have a more exposed habit, such as the alfalfa weevil, which tunnels stems and buds of alfalfa when young then moves to chew on the emerged leaves.

One weevil has had a particularly serious impact on American agriculture, the boll weevil,

FIGURE 14-82

Head of a maize weevil, showing the prominent extension of a weevil head, tipped with chewing mouthparts. Photograph courtesy of Gary Alpert/Bugwood.org.

Anthonomus grandis. A native of central Mexico, where it fed on native cotton species, it began to move northward and was found in southern Texas in 1892. From there, it found the wide-open

FIGURE 14-83
A nut-feeding weevil, *Curculio* sp. Photograph courtesy of Susan Ellis/USDA APHIS PPQ/Bugwood.org.

FIGURE 14-85
The clover root curculio, *Sitona hispidulus*, feeding along the edge of a leaf. Photograph courtesy of Ken Gray/Oregon State University.

FIGURE 14-84
Larva of a billbug, *Sphenophorus venatus*, feeding within a grass stem. Photograph courtesy of Ken Gray/Oregon State University.

FIGURE 14-86
Boll weevil on a cotton boll. Photograph courtesy of USDA ARS/Bugwood.org.

cotton fields of the Cotton Belt states to provide ideal conditions. Within 30 years, it had reached Roanoke, Virginia, and left devastation in its wake.

The boll weevil threatened the entire economy of the South, requiring huge expenditures in control measures, yet high crop losses continued. When insecticides that could have some effect were discovered, they were applied in massive amounts. To date, no single insect in US history has ever had more insecticide aimed at it. For decades, over 40% of all insecticide applications made in the United States were directed solely at the boll weevil. Incidental to this insecticide use were untold environmental problems that ultimately helped expose some of the limitations associated with insecticides. Recognition of these problems with insecticide use later led to changes in their use and availability.

Over time, with intensive research efforts, many alternate methods of boll weevil control were developed, including the use of pheromones to monitor its presence, prescribed dates to plant and plow down crop debris that prevented successful reproduction of the insect, and insecticide use that targeted critical life stages. By 1971, control and management methods began to be coordinated in area-wide efforts to eradicate boll weevil. Ultimately, these programs have had such success that the boll weevil has been eliminated from large areas of the southeastern and southwestern states where it used to dominate cotton production concerns.

The economic and social effects of boll weevil depredations profoundly changed the southern United States. Employment opportunities vanished as farms were bankrupted by losses, providing a key trigger to the massive migration of African Americans to northern cities in the years surrounding World War I and the following decade. The boll weevil percolated into the identity of the south, featured commonly in songs, such as *The Boll Weevil Song* and *Boll Weevil Rag*. The term "boll weevil" was also used as a political term for much of the late 1900s to describe conservative southern Democrats.

A few can be said to have benefited from the boll weevil. A pioneering aviation company, Huff Daland Dusters, thrived, developing the new technology of aerial application of insecticidal dust for boll weevil control; later, expanding into different services, they became the parent of today's Delta Airlines. Brook Benton took *The Boll Weevil Song* to a #2 hit on the Pop Charts in 1961. But, perhaps most important, the boll weevil exposed the region's overdependence on the cotton crop. Farmers were forced to diversify, often into peanuts and vegetables, a change that ultimately benefited local economies. Testament to this is the 1919 Boll Weevil Statue in Enterprise, Alabama, dedicated "In Profound Appreciation of the Boll Weevil and What It Has Done as a Herald of Prosperity."

Bark beetles. The bark beetles are a subfamily (Scolytinae) of the weevils that include the most important forest insects of North America. Particularly damaging are various *Dendroctonus* species, including the southern pine beetle (*D. frontalis*) in southern pine plantations, mountain pine beetle (*D. ponderosae*) in the Rocky Mountains and Black Hills, and spruce beetle (*D. rufipennis*) occurring over a wide swath extending from northern Arizona to Alaska. These insects periodically produce massive outbreaks that kill millions of trees.

FIGURE 14-87
The boll weevil statue erected in the town center of Enterprise, Alabama. Photograph courtesy of USDA ARS/Bugwood.org.

FIGURE 14-88
A mountain pine beetle attempting to tunnel into the trunk of a pine. The tree is producing a large amount of pitch in response that may ultimately "pitch out" the beetle and kill it. Photograph by Whitney Cranshaw/Colorado State University.

Adult females bore into trees and attempt to cut a tunnel under the bark that serves as the egg gallery, along which the eggs are deposited. The emerging larvae then develop by tunneling the living wood, producing girdling wounds roughly perpendicular to the egg gallery. The larval tunneling cuts off the flow of nutrients and water throughout the tree, causing the tree to rapidly "fade" as the dying tree changes to reddish brown.

Vigorously growing trees can usually defend themselves from bark beetle tunneling by drowning invading insects with a flow of thick resinous sap (pitch) or other chemical defenses. Trees become most susceptible to bark beetle attacks when under stress from drought or overcrowding, or when attacked by a massive numbers of beetles. One key strategy used by bark beetles to improve their chances of success in colonizing a tree is to produce **aggregation pheromones** that "call" other beetles to a single tree. Through their combined efforts of tunneling, pheromone-coordinated mass attacks can often overcome tree defenses.

Essentially all bark beetles are also associated with fungi. Adult beetles carry the fungi with them when

FIGURE 14-90
Blue staining of pine logs produced by blue stain fungi. Photograph courtesy of Ronald F. Billings/Texas Forest Service/Bugwood.org.

they emerge from the wood and subsequently inoculate new trees with the fungi as they tunnel into the trunks. Those beetles that develop within conifer trees typically introduce various fungi collectively known as "blue stain fungi" due to the blue-gray discoloration they produce on the wood. Fungi and the consequent blue stain can extend deep into the plant.

Those bark beetles known as ambrosia beetles cut only shallow tunnels into the trunk of their host. These tunnels are then colonized by special "ambrosia fungi" carried in special pouches on the body of the adult beetle. Larvae of the ambrosia beetles do not tunnel but live within chambers previously constructed by the adults and live by consuming this fungus. Such relationships show strong mutualistic benefits as both the fungus and beetle are wholly dependent on each other. Currently, one very serious introduced destructive ambrosia beetle is the redbay ambrosia beetle (*Xyleborus glabratus*), which transmits a deadly fungus to plants in the Laurel family (Lauraceae), including redbay, sassafras, *Litsea* spp., among others.

A much more casual relationship between a bark beetle and fungus occurs with the smaller European

FIGURE 14-89
Tunnels produced by mountain pine beetle under the bark. The large vertical central egg gallery was produced by the adult; larval tunnels radiate from the egg gallery. The dark coloration was produced by blue stain fungi. Photograph courtesy of William Ciesla/Bugwood.org.

FIGURE 14-91
A Colorado lodgepole pine forest showing effects of mountain pine beetle infestation. Successfully attacked trees "fade" and turn to dark reddish-brown as they die. Photograph by Whitney Cranshaw/Colorado State University.

FIGURE 14-92
Tunnels produced by an ambrosia beetle. The side chambers are "cradles" in which the young develop. Photograph courtesy of John Moser/USDA Forest Service/Bugwood.org.

FIGURE 14-93
The smaller European elm bark beetle, *Scolytus multistriatus*. This bark beetle is the primary vector of Dutch elm disease. Photograph courtesy of Pests and Diseases Image Library (Australia)/Bugwood.org.

elm bark beetle (*Scolytus multistriatus*) and *Ophiostoma novo-ulmi*, the agent that can produce Dutch elm disease. The bark beetles get little benefit from this fungus but help greatly in its spread by introducing its spores into feeding wounds they make in small twigs. Dutch elm disease has largely eliminated American elm, once the premier street tree in much of the United States, following the accidental introduction of fungus-contaminated elm bark beetles near Cincinnati in the 1930s.

Order Neuroptera— Alderflies, Dobsonflies, Fishflies, Snakeflies, Lacewings, Antlions, and Owlflies

Entomology etymology. The name is generally interpreted as "nerve" (neuros) "wing" (ptera), a reference to the prominent venation. It is also commonly argued that at the time the order was originally described, the word "neuros" was more likely used as "sinew," that is, the veins of the wings were strengthened as by sinews. Collectively, insects of this order are sometimes referred to as neuropterans.

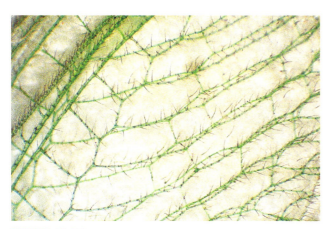

FIGURE 14-94
Close-up of the wing of a green lacewing, a commonly encountered representative of the order Neuroptera. Photograph courtesy of Jim Kalisch/the University of Nebraska.

Neuroptera adults are soft-bodied insects with an elongated body form and long antennae. Their wings are membranous, and among the more common families, there is an extremely large number of veins, rivaling the Odonata in this regard. Most neuropteran adults are weak fliers, although there are exceptions (e.g., owlflies).

Essentially, all Neuroptera are predators of insects or other arthropods. (One family feeds on

freshwater sponges.) The larvae have prominent, forward-projecting jaws. The fearsome appearance of these larvae has made them particularly favored models for science fiction movies.

Neuroptera is a rather small order, with only about 6,000 described living species, but it is ancient, having extensive fossil records that include many extinct families. The surviving species present a considerable range in physical and behavioral features that make them particularly challenging to taxonomists. Many arrangements accept three suborders within the order Neuroptera: Megaloptera (alderflies, dobsonflies, and fishflies), Raphidioptera (snakeflies), and Planipennia (lacewings, antlions, and relatives). Other commonly accepted taxonomic arrangements consider them all to be separate orders. Regardless, substantial differences occur between these three groups, particularly involving the larvae.

SUBORDER MEGALOPTERA— DOBSONFLIES, FISHFLIES, AND ALDERFLIES

The suborder Megaloptera includes most of the aquatic species among the Neuroptera. The larvae have an elongated fleshy body with legs well designed for their habit of crawling about to search for prey. (Larger species known as "hellgrammites" are highly prized as bait by fisherman.) Along the sides of the abdomen are seven to eight tapering abdominal projections that help absorb oxygen from the water; they may also possess gills. Their curved jaws are relatively short but powerful, and these insects can be among the top insect predators in many lakes and riverways. When fully grown, the larvae move to shore, often in a synchronized emergence, and disperse from 2 to 10 m away from the waterway. There, they will excavate a shallow cavity in the soil and pupate.

Adults emerge a few weeks later. They have long membranous wings and the hindwings are much wider, but pleated and folded underneath the forewings when not in flight. The adults are not predators but will sustain themselves on pollen if they feed, though many do not. Females lay eggs over the water, often on plants but sometimes in conspicuous masses on bridges.

The alderflies (Sialidae family) and fishflies (Corydalidae family) are moderate sized insects ranging from 15 to 30 mm. The family Corydalidae also

(a)

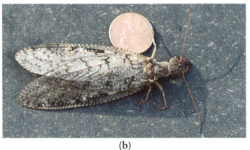

(b)

FIGURE 14-95

(a) Dobsonfly, *Corydalus cornutus*, adult male. (b) Dobsonfly, *Corydalus cornutus*, adult female. Photographs by Whitney Cranshaw/Colorado State University.

FIGURE 14-96

Egg mass of a dobsonfly. Photograph by Whitney Cranshaw/ Colorado State University.

FIGURE 14-97

A hellgrammite, the aquatic larval stage of a dobsonfly. Photograph courtesy of Tom Murray.

includes the dobsonflies, which are among the largest and most striking of all aquatic insects; some dobsonflies have a wing span exceeding 130 mm. The jaws of the female are impressively large and capable of a sharp pinch, but the jaws of males are often bizarrely elongated and spear-like. Despite this fearsome appearance, the male dobsonflies cannot bite and use the mandibles solely to help grasp the female during mating.

SUBORDER RHAPHIDIOPTERA— SNAKEFLIES

Snakeflies are curious-looking insects found in wooded areas of western North America. They have a very long prothorax that is flexible and allows the head to rise, somewhat in a manner of a striking snake. Although they are predators, they have rather short mandibles and can only catch small insects and mites.

Females have a very long ovipositor, used to lay eggs in cracks in bark, making the sexes easy to

FIGURE 14-98

A female snakefly. Photograph courtesy of Tom Murray.

distinguish. The larvae are active hunters and feed on a wide range of insects. Worldwide, there are 220 described species, 17 in North America.

SUBORDER PLANIPENNIA—ANTLIONS, LACEWINGS, OWLFLIES, MANTIDFLIES, AND RELATIVES

The overwhelming majority of the Neuroptera are found in the suborder Planipennia. Adults have delicate wings of similar size and do not possess ocelli on the head, as do Megaloptera.

Greater differences are found with the larvae. Planipennia possess long, thin sickle-shaped mouthparts with an internal "blood groove" that allows them to drain fluids of their prey into the mouth; little solid material is ingested. Curiously, the larvae do not have their midgut and hindgut connected and cannot excrete solid feces until after these portions of the gut are joined during pupation. (Liquid nitrogenous wastes can be excreted by larvae.) The Malpighian tubules and hindgut are modified in these insects to allow the production of silk, which is used to produce a cocoon to surround the pupal stage.

The insects known as lacewings are found in two families: green lacewings (Chrysopidae) and brown lacewings (Hemerobiidae). The former are particularly abundant (84 known North American species, about 1,500 described species worldwide) and are among the most widely recognized of the insect predators that help manage garden pests throughout North America. They are also quite important as general predators used for the biological control of other plant-feeding pest insects in a variety of cropping situations. Their larvae, often called "aphid lions," hunt among plants and feed on a wide range of soft-bodied insects and eggs. In some species, their hunting abilities are enhanced through the use of camouflaging debris, including carcasses of their prey that they pile on their back and secure with long hairs.

Green lacewings also are notable in the unique manner in which they deposit their eggs. Each egg is deposited atop a tall silken stalk that may rise 1 cm or more off the plant surface. The base of the stalk is treated by the adult with a defensive fluid to deter ants, their primary natural enemy, and cannibalism.

With over 2,000 known species, the antlions are the largest family (Myrmeleontidae) of this order. Those found in North America (92 known species)

FIGURE 14-99
A green lacewing adult. Photograph by Whitney Cranshaw/ Colorado State University.

FIGURE 14-100
Eggs produced by a green lacewing. Photograph courtesy of Brian Valentine.

FIGURE 14-101
A green lacewing larva. Photograph courtesy of Brian Valentine.

tend to be most abundant in the drier areas of the country in the south and west. The adults have a very slender abdomen with long wings and are frequently mistaken for damselflies of the order Odonata. Unlike damselflies, antlions possess prominent clubbed antennae and can fold their wings over their back. Antlions are weak fliers that make fluttering flights and are usually active after dusk.

FIGURE 14-102
An antlion adult. Photograph by Whitney Cranshaw/Colorado State University.

FIGURE 14-103
An antlion larva. Photograph courtesy of Joseph Berger/Bugwood.org.

The larvae are ambush hunters that impale passing prey with their prominent jaws. Many species lie in wait, hidden under plant debris or other types of cover, and may make chase. Better known are those that construct soil pits to help trap passing insects.

Sometimes called "doodlebugs," antlion larvae that construct soil pits often leave meandering trails in loose soil as they search for a suitable place to build their pit. Pits are usually constructed under some overhang to provide protection from rainfall. During the construction, the antlion larva first encircles the boundary of the pit, which they then excavate by flicking away the soil with their head. Digging progresses as a spiral toward the center and the pit is completed when it reaches its angle of repose, the point where the soil particles lining the sides of the pit barely remain stable within the steep conical shape. The antlion larva then burrows into the center of the pit and waits with jaws ready to snap on an insect (often an ant) that falls within reach.

The mantidflies (Mantispidae family) are a fairly small family (14 North American species) and uncommonly encountered. Mantidflies possess notably prominent grasping front legs and an elongated prothorax, a feature shared with the mantids (order Mantodea). In addition, one of the most common species, known as the wasp mantidfly (*Climaciella brunnea*) closely mimics paper wasps in size and coloration. The adults are predators of other insects and are most commonly seen when visiting lights.

Mantidfly larvae develop as parasites within egg sacs of spiders, or less commonly, on ground nesting bees and wasps. The first larval stage is an elongate-bodied, highly active, triungulin form that seeks out its ultimate hosts. Once safely established with its future food, it molts to a much less active grub form. Such a dramatic change during larval development is known as hypermetamorphosis and is also seen with blister beetles.

The owlflies (Ascalaphidae family) are the fastest flying of Neuroptera but active only during brief periods around dusk, during which time they hunt for flying insects. Owlflies have long wings similar to antlions, but their heads bear conspicuously long antennae that terminate in a club. During daytime, they rest on branches, projecting their thin abdomen outward such that they resemble a twig. Only six species are known to occur in North America (about 430 known species worldwide), and they are largely restricted to the southern and southwestern states.

Adult females lay eggs in small groups on twigs. Immediately below the egg mass, the female will encircle the twig with a physical barrier or wall,

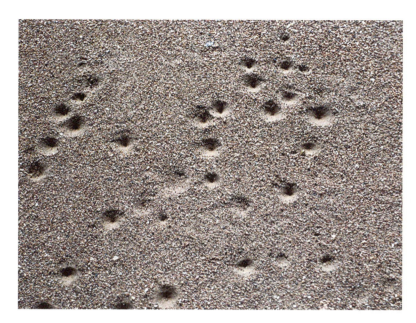

FIGURE 14-104
Soil pits produced by antlion larvae. Photograph courtesy of Howard Ensign Evans.

FIGURE 14-105
A wasp mantidfly, *Climaciella brunnea*, feeding on a blow fly. Photograph by Whitney Cranshaw/Colorado State University.

FIGURE 14-106
An owlfly. Photograph courtesy of Howard Ensign Evans.

called a **repagula**. The repagula is composed of dozens of abortive eggs that form raised projections on the surface of the twig. Additionally, these aborted eggs are covered with a shiny defensive fluid that is highly repellent to ants. Upon hatching, the larvae remain together through the first molt. These first instars position themselves such that a defensive ring is formed with their huge jaws directed outward. After developing to the second instar, larvae lower themselves to the soil surface on silken threads where they develop further as solitary ambush hunters. Once on the soil surface, they conceal themselves under stones, among loose sand, or on tree trunks and lie in wait for prey. Their huge sickle-like jaws quickly snap up a passing insect. Owlfly larvae also inject venom that can further help subdue the prey.

15

• City Builders That Rule

Ants, Bees, Wasps, and Sawflies

Social behaviors are well developed in the order Hymenoptera and present in many of the insects that humans most regularly encounter: insects, ants, bees, and wasps. Many members of this order are also well recognized as beneficial to human interests, functioning as important pollinators or as natural enemies of pest insects. Less favorably considered, the Hymenoptera contains all of the insects that sting. It is also one of the largest orders, with about 150,000 species described and an unusually high percentage remaining undescribed.

Entomology etymology. The name of the order is derived from the words "Hymen" (a Greek god of marriage) along with "pteron" (wings). This refers to the mechanism by which the hindwings are hooked to the forewings for flight. When in flight, the two pairs of wings function as a single wing. Insects within this order may be described as hymenopterans.

The features of the wing are the primary means that distinguish the Hymenoptera from other orders of insects. Both wings are membranous, but the hindwing is considerably smaller. A row of hooks (called **hamuli**) along the leading edge of the hindwing allow it to fasten to the front wing; this allows the two wings to beat together in flight. Most hymenopterans also possess a spur on the tibia of the prolegs, which is used to clean antennae, and females possess a prominent ovipositor.

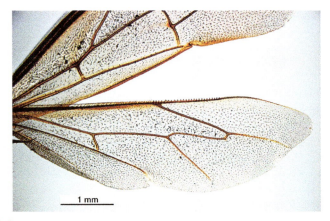

1 mm

FIGURE 15-1
Close-up of the wing of a honey bee showing the hamuli along the leading edge of the hindwing.
Photograph courtesy of Pest and Diseases Images Library (Australia)/Bugwood.org.

Because of their diversity, using various subdivisions of the order helps in describing and discussing insects in the order Hymenoptera. The members of Hymenoptera that exhibit the most ancient features include insects known as sawflies; one of the families present today (Xyledidae) has a truly impressive fossil record extending back into the Triassic period (230 mya). Along with the more recently evolved sawflies, as well as insects known as horntails, these insects are placed in the suborder Symphyta. The larvae of all Symphyta develop on plants, either feeding on foliage or developing as borers in wood or plant stems. All members of Symphyta also have a broad area where the thorax attaches to the abdomen, typically giving the adults a cylindrical body form. The ovipositor of females is quite pronounced in some, particularly wood wasps, and is used to insert eggs into plant tissues (leaves, stems, wood). These three characteristics (phytophagous feeding habit, broadly joined abdomen, and distinct functional ovipositor) collectively define Symphyta within the order Hymenoptera.

The other important body modification found in Hymenoptera was the evolution of the "wasp waist." This is a sharply narrowed constriction of the first two segments of the abdomen, and has the decided advantage of allowing the tip of the abdomen, having the ovipositor, to be twisted in many directions. This feature was first used to great effect by parasitoid wasps that use the ovipositor to pump an egg into or near some other insect that serves as a host for their developing larvae. The "wasp waist" feature is now widespread among the most commonly encountered members of the order (e.g., bees, wasps, ants).

FIGURE 15-2
A xyledid sawfly, the balsam shootboring sawfly (*Pleroneura brunneicornis*). Photograph courtesy of Ronald S. Kelley/Vermont Department of Forests, Parks, and Recreation/Bugwood.org.

FIGURE 15-4
Parasitic wasp ovipositing in an aphid. Photograph courtesy of Alex Wild.

FIGURE 15-3
An adult tenthredinid (common) sawfly. Photograph courtesy of Brian Valentine.

FIGURE 15-5
Side view of a cornfield ant, *Lasius* sp., showing the constriction at the base of the abdomen. Photograph courtesy of Brian Valentine.

Another adaptation in insects, the **stinger**, evolved still later. Insects capable of stinging (known as the aculeate Hymenoptera) have an ovipositor that does not function for depositing eggs but instead serves to inject venom. Eggs are deposited not through an ovipositor but via an opening at the base of the stinger. The stinger has many functions and probably originally served to subdue prey, a function that continues to be used by many ants and wasps. Stingers are also used for defense, which is the sole function they serve for the pollen- and nectar-feeding bees.

Within the Hymenoptera, larval forms are rarely observed due to their habits. Larval forms are easily observed only in leaf-feeding sawflies and might commonly be seen in a garden or forest setting; a few sawflies can be important defoliators of trees, shrubs, and flowers. Larvae of other Hymenoptera develop out of sight, either within plant parts, in the

FIGURE 15-6

A baldfaced hornet, *Dolichovespula maculata*, in position to sting. Photograph courtesy of William Hantsbarger.

bodies of other insects, or hidden within colonies. Silk (from modified labial glands in the larvae) is produced by most hymenopterans and is used to construct a cocoon around the pupa.

● The Paradox of Altruistic Behavior in Social Insects

The conspicuous existence of social insects long provided a conundrum for biologists, because it appeared to involve **altruistic behaviors**. In these colony-producing insects, the great majority of individuals are the workers who are infertile females. The queens are the only few females that reproduce. The rest of them sacrifice reproduction and instead dedicate their activity to assisting the functions of the colony. Because so many do not reproduce themselves, the obvious related question is how such a trait could possibly evolve. This puzzle is longstanding and was even addressed by Charles Darwin in the *Origin of Species*; Darwin could not resolve this issue and had to note that it seemed to be a special case. A century later the British biologist W. D. Hamilton came up with an elegant explanation with the theory of **kin selection**.

Understanding the implications that result from the unusual means by which sex is determined among the social insects within the Hymenoptera (ants, bees, wasps) is fundamental to this theory. Females are produced through normal sexual reproduction, which is known as **haplodiploid sex determination**, arising from a fertilized egg with two sets of chromosomes (diploid). Males, however, are produced asexually and possess only a single set of chromosomes (derived from their mother); they are haploid. Males arise from unfertilized eggs.

These insects produce colonies that are essentially all related females (sisters). The mother of the colony (known as a queen) is diploid and thus contributes half of her genes to all colony members via a meiotically

FIGURE 15-7

Honey bees on the comb of a hive. The workers surround the queen, which is seen in the center of the photograph. Although all are adult females, only the queen is capable of reproduction. Photograph courtesy of Carl Dennis/Auburn University/Bugwood.org.

produced egg. The sperm of the common haploid father is produced by mitosis so that each sperm contains identical genes. The result of this type of reproduction is a colony of full sisters sharing not the normal 50% but 75% of their genes (all of their father's, and half of their mother's, genes are in common). What this means is that the daughters (female workers) are more genetically related to each other than they would be to any

(continued)

daughters they themselves might produce (daughters only share 50% of their genes with their mothers).

Because they are so genetically similar, non-reproducing workers have their genes especially well represented by their reproducing kin, sisters who become queens and establish new colonies. Essentially, workers reproduce by proxy, and their ability to dedicate activity to other colony-wide functions (e.g., foraging, defense, rearing the young) favors the well-being and continuation of the colony, resulting in yet more of their sisters being able to ultimately reproduce. Under these conditions, their apparent sacrifice does not appear altruistic but instead provides best means to have their genes represented in the next generation.

SUPERFAMILY TENTHREDINOIDEA—SAWFLIES

Sawflies are stout-bodied wasps whose larvae develop on plant foliage. Female sawflies have a bladed ovipositor designed to insert eggs into leaves, needles, or soft stems. Adults of the conifer sawflies (Diprionidae family) possess prominent serrated antennae whereas the cimbicid sawflies (Cimbicidae family) possess antennae that end in a rounded club. Such distinguishing features are less consistent with the largest family of sawflies, the common sawflies (Tenthredinidae family). The majority of common sawflies are dark colored with long antennae.

(a)   (b)

FIGURE 15-8

(a) A diprionid sawfly, the blackheaded pine sawfly (*Neodiprion excitans*), (a) adult and (b) larvae. Photographs courtesy of Gerald Lenhard/Louisiana State University/Bugwood.org.

Sawfly larvae look very much like the larvae of Lepidoptera; they even have similar fleshy prolegs on the abdomen as do butterfly and moth larvae. (Sawfly larvae can be distinguished by the presence of six to eight pairs of prolegs, each lacking the crochets that tip the two to five pairs of prolegs found in the larvae of Lepidoptera.) While feeding, sawfly larvae typically grasp the leaf or needle with their thoracic legs and curl the abdomen into a J- or loose S-shaped form. Many species are gregarious and feed as groups, which often may collectively twitch when disturbed as a defensive behavior. When fully developed, sawfly larvae drop to the upper soil and spin a cocoon within which they later pupate.

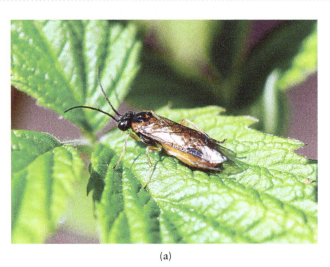

(a)   (b)

FIGURE 15-9

A tenthredinid (common) sawfly, the currant sawfly (*Nematus ribesii*), (a) adult and (b) larvae. Photographs by Whitney Cranshaw/ Colorado State University.

The most damaging sawflies that occur in North America feed on conifers and include pests of forests as well as landscape plants. Important pest species include the European pine sawfly (*Neodiprion sertifer*), read-headed pine sawfly (*Neodiprion lecontei*), and larch sawfly (*Pristiphora erichsonii*). Other sawflies feed on the leaves of broad-leaved trees, shrubs, and flowers. Sawflies are also among the most common insects that develop as **leafminers,** living between the leaf surfaces of plants such as birch, elm, and alder. Many of the galls that occur on willow trees are also produced by sawflies.

FAMILY SIRICIDAE—HORNTAILS OR WOOD WASPS

The horntails are large wasps, typically at least 25 mm long, with a cylindrically shaped body. Females possess a long thin ovipositor that is supported by a stout spike-like projection at the tip of the abdomen, producing its "horn tail." The ovipositor is used to drill through the tough bark of host trees and eggs are deposited within the wood. Larvae develop as wood borers, extensively riddling the heartwood.

FIGURE 15-10

The pigeon tremex, *Tremex columba*, a horntail that develops in deciduous trees. Photograph by Whitney Cranshaw/Colorado State University.

Most species develop in conifers, with the notable North American exception of the pigeon tremex (*Tremex columba*). The pigeon tremex is the most common species of horntail found in eastern North America and is associated with hardwoods such as maple, elm, beech, or ash. All horntails typically deposit their eggs in trees of poor health, which they can determine by its moisture content.

Horntails also have mutualistic association with various species of white rot fungi. When females drill into trees with their ovipositor, either to lay eggs or during exploratory probes, they introduce the fungus into the wood. The growing fungus works to break down the cellulose, lignins, and other compounds within the wood, rendering it more digestible to the developing horntail larvae. Prior to pupation, female larvae incorporate some of the fungus into special pouches on their abdomen where it will be available when laying eggs in the future.

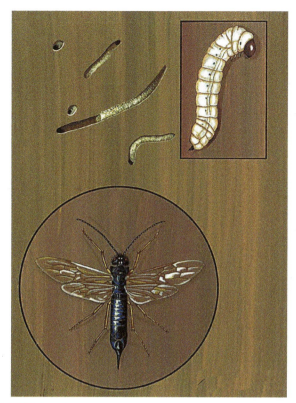

FIGURE 15-12
Life cycle of *Sirex noctilio*, a wood wasp. Image provided by Robert Dzwonkowsi/Bugwood.org.

FIGURE 15-11
A wood wasp, *Urocerus gigas gigas*, ovipositing in a log. Photograph courtesy of Vicky Klasmer/Instituto Nacional de Tecnologia Agropecuaria/Bugwood.org.

The life cycle of horntails often takes 2 years to complete and may be extended if wood conditions are less favorable. They are not uncommonly encountered in new homes constructed with lumber that was infested by horntails before milling, with adults that developed in the lumber sometimes emerging several years following construction of the building.

PARASITIC HYMENOPTERA

The most common life style found among the Hymenoptera is parasitism of other insects. In these species, the larvae live either on, or most commonly within, the body of another **host** insect, consuming the tissues and fluids of the host. Ultimately, these wasps kill their host and thus are known as **parasitoids**, in contrast to non-lethal parasites such as lice. (Parasitoids kill their hosts; parasites often do not.) With parasitoid wasps, the host succumbs when the larva of the parasitoid wasp has nearly completed its development. Delaying the death of the host is a behavior that is highly efficient for the survival of the parasitoid wasps, which have perfected their parasitic life.

FIGURE 15-13
A parasitic wasp ovipositing in a gypsy moth caterpillar. Photograph courtesy of Scott Bauer/USDA ARS.

FIGURE 15-15
Aphids that have been killed by parasitic wasps are known as "aphid mummies." Once the wasp has to emerge it cuts an exit hole in the back of the host. Photograph courtesy of Ken Gray/Oregon State University.

developmental stage of a host insect. Most will develop within only a single species or genus of insect, and their life cycles are often well synchronized to that of their host.

FIGURE 15-14
Parasitic wasp larva exposed from within its aphid host. If undisturbed, the wasp would complete its development and emerge in its adult form, cutting an exit hole from the back of the aphid host. Photograph by Whitney Cranshaw/Colorado State University.

FIGURE 15-16
Trichogramma wasps lay their eggs in the eggs of caterpillars, which their larva will consume during development. They are among the tiniest of all insects, not much bigger than the period ending this sentence. Photograph by Jack Kelly Clark and provided courtesy of the University of California IPM Program.

A great many families of wasps are parasitoids with some of the most prominent being the ichneumon wasps (Ichneumonidae), braconid wasps (Braconidae), aphelinid wasps (Aphelinidae), eulophid wasps (Eulophidae), encyrtid wasps (Encyrtidae), and trichogramma wasps (Trichogrammatidae). Their remarkable diversity is due in large part to so many species being highly specific in their choice and

Life as a parasitoid can be quite complex as many species of wasps may, in turn, be parasitoids of parasitoids all within a single host insect. A parasitoid of a parasitoid is known as a **hyperparasite**. There are even examples of parasitoid wasps that attack these hyperparasites, making the former a parasite of a parasite's parasite.

Given their specialized habits, there is a considerable size range found among the various parasitic wasps. In general, species that attack the late stages of large insects (e.g., caterpillars, white grubs) are much larger than those that attack small insects (e.g., aphids, scale insects), or small life stages, such as an egg. In North America, the largest parasitic wasps are ichneumon wasps of the genus *Megarhyssa* that attack horntail larvae; they may have a body length of 40 mm with an ovipositor that is twice as long as the body. Conversely, the smallest of all insects occur within the parasitic Hymenoptera, tiny fairyflies (Mymaridae family). Larvae of fairyflies develop within the eggs of other insects and achieve a body size limited by the size of their host's egg.

Parasitic Hymenoptera are tremendously important in the biological control of insects, including most of all of the major agricultural pests. Worldwide, there are a great many parasitic wasp species that have been purposefully introduced to help manage pest species that originated elsewhere, a technique known as Classical Biological Control. Insect management practices also often are modified so that the activities of these beneficial insects are conserved, such as through use of more selective insecticides, or enhanced, by providing cropping systems that favor maintaining or increasing their population densities.

● Polydnaviruses and Their End Run Around Host Defenses

Many insects are involved in transmitting viruses. Several plant diseases are the result of viruses being transmitted from plant to plant by aphids, leafhoppers, thrips, beetles, and other insects. Mosquitoes are critical to the spread of disease viruses in humans and other vertebrates (e.g., yellow fever, dengue fever, and West Nile virus). In none of these relationships does the virus substantially affect the insect vector. The situation is different with some parasitic wasps that may apparently use viruses to subvert the defenses of their host.

Two families of parasitic wasps (Braconidae and Ichneumonidae) have a unique association with the viruses in the polydnavirus family. These viruses apparently have had a very long association with these insects such that virus genes are now found incorporated into those of the wasps. The virus only replicates within the wasps, specifically within the calyx cells of the ovary as the wasp matures its eggs. At this time the virus forms packages (nucleocapsids) containing its tightly coiled DNA. Recently, several researchers have now begun to question whether in fact this relationship between the wasp and the virus has not coevolved to the point where the virus has forever become part of the wasp and now functions as an organelle of the wasp.

The virus (or organelle) produces no adverse effects in the wasp where it originates. In contrast, its effects occur when it is introduced into a host insect along with the wasp egg during oviposition. At that time the polydnavirus is activated and affects many kinds of cells in the new insect host. Most significantly, these viruses suppress the ability of the host's protective blood cells, the phagocytes, to recognize and destroy the egg of the parasitic wasp. Normally the phagocytes respond quickly to the presence of a foreign object, clumping, and eventually, encapsulating it. The polydnavirus stops

FIGURE 15-17

Cocoons of a parasitic wasp on the body of its caterpillar host. Earlier, the larval stages of these wasps developed within the caterpillar, which ultimately dies from these injuries. Photograph by Jim Kalisch/University of Nebraska.

this process, allowing the wasp egg to survive and hatch. This ultimately results in the death of the host insect but allows the perpetuation of the wasp parasitoid, and the polydnavirus genes it carries.

FAMILY CYNIPIDAE—GALL WASPS

The ability to induce plants to produce unusual growths known as galls occurs among many groups of insects (gall midges, psyllids, phylloxerans, aphids, sawflies) and the eriophyid mites. But it is the gall wasp that produces the most diverse and complex gall structures. Gall wasps produce galls that are highly **determinant**—their appearance is radically different from the plant tissues from which they originate. For example, some galls wasps produce round balls ("oak apples" or "roly-poly galls") that may grow to the size of a small hen's egg. Spiny growths on rose, "horned galls" on oak, or densely hairy "mossy galls" are some of the other forms produced by gall wasps. Gall colors may be from brilliant red to creamy white, contrasting sharply with the green foliage on which they occur. In North America over 800 known species of gall wasps occur, far exceeding the number of species found in other gall-producing families.

Despite possessing a large number of species, gall wasp galls are rarely found on plants other than certain species of rose (*Rosa* spp.) or oak (*Quercus* spp.). On these plants more than one type of gall can be produced; many species of gall wasps have complex life cycles that can be alternately found on different plant parts. For example, the rough bulletgall wasp (*Disholcaspis quercusmamma*)

FIGURE 15-18
Gall produced on oak by the wool sower gall wasp, *Callirhytis seminator*. Photograph courtesy of Eric Day/VPI/Bugwood.org.

FIGURE 15-19
Gall produced on rose by the mossy rose gall wasp, *Diplolepis rosae*. This gall has been cut open to expose three developing gall wasp larvae. Photograph courtesy of Ken Gray/Oregon State University.

produces conspicuous rounded galls on twigs of bur oak and swamp white oak. In fall, around Halloween, an all-female generation emerges from the twig galls, and these individuals lay eggs in the dormant buds of their host plant. The following spring, feeding by the developing wasp larvae results in tiny galls that resemble buds. Ultimately, both male and female wasps emerge from these spring-produced galls a few weeks after leaves emerge. The females then insert eggs into the growing twigs, which ultimately results in the next cycle that involves the twig galls.

Although many galls have a bizarre appearance, few of these species significantly affect the health of their host plant. Some stunting of branches may occur with twig galls, but most galls merely cause minor changes in appearance that is of little consequence to the plant. A common gall that draws attention of a different nature is the jumping oak gall associated with the wasp, *Neuroterus saltatorius*. The galls themselves are small and seed-like, and ultimately drop from the leaf to the soil. The developing larvae within the galls twist and turn, causing the gall to curiously move about in a manner similar to a miniature Mexican jumping bean (page 365).

Some of the large "oak apple" galls have historically been a very important item of human commerce as they are rich in tannic acid. These galls serve as the base material for **iron-gall ink**, a permanent rich purple-black ink produced from

(a)   (b)

FIGURE 15-20

(a) Adult of the oak rough bulletgall wasp, *Disholcaspis quercusmamma*, leaving a twig gall on oak. Only females come out of these large knot-like twig galls, which then lay eggs in dormant buds in autumn. (b) The tiny spring bud galls produced by the oak rough bulletgall wasp. Both males and females emerge from these galls, which appear in spring. Photographs by Whitney Cranshaw/Colorado State University.

processing the crushed galls with ferrous sulfate. Until the twentieth century, these galls produced the highest quality inks used in the United States and Europe. The inks were well adapted for use on the vellum parchments used in important documents. Sketch books of Leonardo da Vinci and Rembrandt, as well as all the seminal documents of United States (Declaration of Independence, Constitution, Bill of Rights) were written using iron-gall ink.

FIGURE 15-21

Jumping oak galls produced by the gall wasp, *Neuroterus saltatorius*. The galls drop from the leaf and the insects inside cause the galls to twitch and move if they are located in an unfavorable location. Photograph courtesy of Ken Gray/Oregon State University.

FIGURE 15-22

A type of oak apple gall. Photograph courtesy of Steven Katovich/USDA Forest Service/Bugwood.org.

FIGURE 15-23
A type of oak apple gall that is cut away to expose the internal structure. The gall wasp larva lives in a small, round cell in the center of the gall. Photograph by Whitney Cranshaw/Colorado State University.

FAMILY MEGACHILIDAE—LEAFCUTTER BEES AND MASON BEES

Unlike the honey bees and bumble bees, the great majority of bee species do not produce colonies. Most bees have a solitary habit where individual females construct and provision a nest. Such solitary bees are found in many bee families, including the ground-nesting digger bees (Anthoprinae), andrenid bees (Andrenidae), sweat bees (Halictidae), and the wood boring carpenter bees (Xylocopinae). Often the most conspicuous solitary bees are the leafcutter bees and mason bees of the family Megachilidae.

Leafcutter bees construct their nests within existing hollow cavities. They can also excavate tunnels in rotten logs or in twigs of plants with soft pith such as roses, brambles, and caneberries. Once a nest site is established the female proceeds to produce a series of cells in which to rear her young. Each of these cells will then be constructed of leaf fragments that the mother collects from nearby plants. Leaf cutting is done rapidly and is rarely observed, but the distinctive semicircular cuts along leaf edges are commonly noted by gardeners on such favored plants as rose, lilac, ash, and Virginia creeper.

The ultimate nesting chamber consists of several cells aligned end to end. Each cell somewhat resembles a miniature cigar butt and is constructed from dozens of leaf pieces, each cut specifically to construct different parts of the chamber. At the

FIGURE 15-24
Life cycle of an alkali bee, a type of solitary bee that nests in soil. Drawing by Art Cushman/USDA Department of Entomology. Image courtesy of Bugwood.org.

FIGURE 15-25
A leafcutter bee, *Megachile* sp., visiting a sweet pea flower. Leafcutter bees, and all bees of the family Megachilidae, carry pollen on the underside of the abdomen. Photograph by Whitney Cranshaw/Colorado State University.

FIGURE 15-26
A leafcutter bee, *Megachile* sp., cutting a leaf piece from lilac. Cut pieces of leaves are then used to line their nest cells. Photograph by Whitney Cranshaw/Colorado State University.

bottom are three to four elongate, rectangular pieces that are crimped to form the base of the structure. Oval-shaped leaf fragments are then cut and cemented to produce the sides of the structure, the smooth side of the leaf face inwards. When the chamber is complete, near perfectly round cut leaf pieces form the top. To complete a single nest cell over 500 cm^2 of leaf material may be used.

Leafcutter bees provision each cell within a chamber with a mixture of nectar and pollen, the latter carried on the underside of the abdomen in specialized structures (scopa). The pollen is mixed with nectar to produce a damp paste, which they place in individual cells. When a cell is about two thirds full of this material, an egg is laid on top and the cell sealed.

Leafcutter bees are extremely important pollinators, and many native plants are dependent for pollination on their activities. These bees also have proven to be particularly good at pollinating some crops; one species (*Megachile rotundata*) is now routinely used in the production of alfalfa grown for seed as it is far more efficient than honey bees for this purpose.

Other leafcutter bees do not cut leaves but instead use plant resins to line and form nest cells. Among those with this habit is the world's largest bee, *Megachile pluto*, which occurs in parts of Indonesia. This bee, sometimes referred to as Wallace's giant bee, has a wingspan up to 63mm and forms its nests within active termite mounds. Leafcutter bees of the genus *Anthidium*, known as wool carder bees, are common residents around many gardens and use plant hairs to line their nests.

Mud is the primary construction material of the mason bees (*Osmia* spp.). Often an existing cavity with be used for the nest site but the individual cells are constructed of mud rather than leaf fragments. Some of the more common species of mason bees can be induced to produce nests if provided bundled drinking straws, pieces of dried bamboo, or similar small diameter tubes. There has been increased interest in the use of such nest sites as a means to increase mason bee activity as pollinators of fruit trees.

FIGURE 15-27
Leafcutter bee larvae developing in nest cells constructed in a wooden block of drilled wood. Photograph courtesy of Ken Gray/Oregon State University.

FIGURE 15-28
Blue orchard mason bees, *Osmia lignaria,* use mud instead of leaf fragments in nest construction. These bees are nesting in a "bee board" made by drilling wood and lining it with soda straw inserts. Photograph courtesy of Scott Famous/Bugwood.org.

FAMILY APIDAE—HONEY BEES, BUMBLE BEES, CARPENTER BEES, DIGGER BEES, STINGLESS BEES

The family Apidae includes most of the insects commonly recognized as bees. These insects are hairy, rear their young on condensed nectar (honey) and pollen, and include many species that are critical for plant pollination. The overwhelming majority of these species live a solitary life style with females individually constructing nests and assuming all chores for rearing the young. Some widely recognized social insects also are present in this family, notably the honey bees, bumble bees and stingless bees.

Honey bees. The familiar honey bee introduced into the Americas is *Apis mellifera*, known variously as the "western honey bee," "European honey bee," or, most correctly, just "honey bee." It is native to Africa, Europe and parts of western Asia and was first introduced into North America in 1622 by English colonists. The strain of the original honey bee that was brought to North America was a predominantly dark brown-bodied strain known as the "dark bee" (*Apis mellifera mellifera*) that is common to northern Europe. Subsequently, many other strains of honey bees have been introduced into North America, usually because they possessed desirable qualities such as improved disease resistance, increased honey production, or gentleness when handling. Currently, the most common strains of honey bees cultured in North America are the yellow-orange Italian strain (*Apis mellifera liguistica*) followed by the Carniolan strain (*Apis mellifera carnica*), an orange–brown honey bee notable for gentleness.

In the wild, honey bee colonies are established in large pre-existing cavities that are protected from wind. Hollows of trees and fallen logs are typical locations used by wild bees and considerable care is taken in selecting a home. For example, work by Thomas Seeley of Cornell University found that honey bees prefer to nest in sites well above ground (protection from predators), in cavities of between 15 and 75 L capacity, with a small diameter entrance hole (easy to defend, easy to control temperature) that is located at the bottom of the cavity (easy to regulate temperature). Although "feral bees" that have escaped from managed apiaries readily establish colonies in naturally available cavities that meet their needs, the overwhelming majority of honey bees in North America are housed in hives provided by beekeepers.

FIGURE 15-30
A honey bee hive constructed behind the wall of a shed. Photograph by Whitney Cranshaw/Colorado State University.

Honey bee colonies are perennial, with the colony persisting from one year to the next. During the spring peak in reproduction, a healthy honey bee colony can be expected to have about 50,000 individual bees. Lower numbers occur during fall as the colony temporarily ceases reproduction and survives the cold months of winter on honey stores.

FIGURE 15-29
Two honey bees foraging, showing difference in coloration. Photograph by Whitney Cranshaw/Colorado State University.

Honey bees construct their nest from wax that is produced by special glands of the abdomen and secreted as small flakes. These flakes are then chewed and formed into cells that are used to rear young and store food (honey and pollen). Honey bee cells are hexagonal, a shape that is optimally efficient as it allows the cells to be most densely packed using a minimum of wax. Sheets of these cells align vertically in the hive and are known as comb.

FIGURE 15-31
Honey bees make wax comb in which larvae are reared and food is stored. Photograph by Whitney Cranshaw/Colorado State University.

Like other bees, nectar and pollen collected from flowers provides their fundamental food source. Pollen primarily provides proteins, fats, and vitamins while nectar is the energy source composed mainly of sugars. As necessary, honey bees also will collect water to help air condition the hive or liquefy honey

FIGURE 15-32
Honey bees make wax in special glands in the abdomen. The wax is secreted as wax flakes that are then formed by the mandibles into comb. Photograph courtesy of Susan Ellis.

that has become crystallized. Plant resins, known as **propolis**, also are collected to function as "bee glue" use to seal small gaps and spaces within the hive.

Honey bees undergo complete metamorphosis with all developmental stages occurring within the hive. The queen bee will deposit eggs individually in the base of the wax cells. On hatching, the developing honey bee larvae are fed continuously by adult bees until they are fully grown. They then spin a silk lining (silk produced from modified labial glands) in which they pupate and subsequently emerge as a winged adult.

FIGURE 15-33
Honey bees on a frame of comb. The raised cappings on some of the cells are produced by pupae and indicate that these cells are being used for rearing young. Photograph by Whitney Cranshaw/Colorado State University.

Honey bee castes. A honey bee colony produces individuals of 3 different castes: workers, drones, and a queen. The **queen** is a fully developed, fertile female and only one queen is present in a normal colony. New queens are produced when needed, usually after the previous queen leaves in a swarming event. During larval development a queen is reared in a specially built and spacious cell and fed exclusively on **royal jelly**, a highly nutritious substance produced in specialized glands located in the heads of some workers. Because of this high quality diet, the time for development to the adult stage is shortened; only 15 days are required for the egg to develop to an adult of the queen caste.

FIGURE 15-35
A honey bee drone. Photograph courtesy of Susan Ellis.

FIGURE 15-34
A honey bee queen being tended by workers. Photograph courtesy of Jessica Lawrence/Eurofins Agrosciences Services/Bugwood.org.

About a week after emerging from the cell as an adult, the virgin queen will make one or more short mating flights outside the hive. While on a mating flight, the queen will release sex pheromones that attract adult males (**drones**). A queen will mate on average with 7–17 drones during these mating flights, storing about three to five million sperm in her spermatheca; the sperm will remain available to fertilize all of the eggs she subsequently produces. After mating, the queen will return to the colony. Unless the colony splits during a swarming event or abandons the old hive (absconding), the queen never again leaves the hive following the mating flight.

A few days after mating, the queen will begin laying eggs and will continue to do so for the remainder of her life. Oviposition is only interrupted for a few months beginning in late summer and then resumes in early winter. Up to 200,000 eggs may be produced in a year, but egg production declines with age and queens usually only live 2 or 3 years. As her fertility decreases with age, workers begin to produce new queen cells that may allow a replacement queen to be produced through a process known as **supersedure**.

The overwhelming majority of the colony is composed of **workers**, which are infertile females. Honey bee workers develop from fertilized eggs that may be identical to those that can produce a queen. However, during rearing the larvae of future workers are provided poorer diet than is a queen. After initially receiving some royal jelly, worker larvae are switched to a diet of less nutritious "bee bread," a mixture of nectar and pollen. This difference in diet prevents the development of the ovaries of workers and extends the time of larval and pupal development. Workers typically require about 6 more days to reach the adult stage than does the queen.

On emergence as adults, workers usually go through a sequence of different hive duties that change depending on hive needs and the age of the worker. During the first few days a newly emerged worker will likely help to clean and cap cells. A few days later, their behavior shifts, and they serve as nurse bees that tend to and feed the larvae. Workers then enter a period when they have more general in-hive duties, such as transferring pollen, nectar or other materials brought in by foragers and converting (ripening) the raw nectar to form

FIGURE 15-36
A honey bee worker. This one is collecting pollen, which is carried on special pollen sacs on the hindlegs. Photograph courtesy of David Cappaert/Michigan State University/Bugwood.org.

honey. At this time some workers may specialize to the task of wax production and new cell construction. Only after they are about 2 weeks old do workers shift to tasks requiring them to be out of the hive.

Ultimately, most honey bee workers spend the last days of their lives functioning as foragers, collecting materials needed by the colony; pollen, nectar, water, and propolis. Foragers dedicated to pollen collection are recognizable by the conspicuous pollen baskets on the hindlegs. Foraging is a wearing and dangerous business; few workers live more than a couple of weeks after they have begun foraging. Any of those workers surviving until the end of the season will remain within the hive and may survive through winter.

Swarming. As all individuals function only as a dependent member of the larger colony, individual honey bees cannot survive for long. Consequently, the honey bee colony needs to be considered as a "superorganism" that reproduces in a unique way, forming new colonies through a type of budding process known as swarming.

The swarming of a bee colony is a spectacular event to observe. A swarm of honey bees typically comprises about two thirds of the workers along with the queen. Emerging from the old hive in a flying frenzy, the swarm often settles on a nearby tree in a large clump composed of chains of individuals with their legs interlocked. The appearance of a buzzing swarm is often alarming, but during the period when honey bees are in a swarm phase they are quite non-aggressive as they

are in transition without a hive to defend. Swarms are often sought by beekeepers, who can shift them easily into a constructed hive box and move them to their bee yard (apiary).

FIGURE 15-37
A honey bee swarm that has come to rest on a small tree in an orchard. Photograph courtesy of Harold Larsen/Colorado State University.

If undisturbed, the primary activity occurring within the swarming bees is the location of a new, permanent home. To this end scout bees fly off searching for suitable locations. Returning scouts will communicate the location of potential sites to other bees by means of dances. Eventually consensus is reached among the members of the swarm and they leave *en masse* to their ultimate home.

Back at the old hive, the remaining bees tend to be the older foragers along with some developing young, including new queens that were being reared prior to the swarming event. When the first of these queens emerges, she scours the hive to locate the other queen cells and will attempt to destroy them by stinging potential rivals to death. A few days later, the sole remaining virgin queen will leave the hive to engage in mating flights, returning a short time later to initiate her egg production as the new queen of the colony.

● The Bee Dance "Language"

One of the most unique abilities of the honey bee is their ability to communicate complex information to one another; returning foragers are able to transmit precise information as to the location of food or other useful resources to other members of the colony. This is accomplished through performing special dances in the darkness of the hive and is known as the "dance language."

FIGURE 15-38
A waggle dance may be performed by honey bee foragers after they return to the hive having visited a site of desirable resources (e.g., nectar, pollen). The dance can communicate direction and distance and is conducted on the combs within the hive. Direction is determined by the angle of the center line of the dance from the vertical, which correlates to the angle of the resource in relation to the sun. Distance is determined by the relative intensity and speed of the dance, with the central waggle portion repeated more rapidly when the resource is close. Drawing by Matt Leatherman.

A "tail-wag dance" is used when communicating locations approximately 100 m or more from the hive. This dance involves a short central straight line dance during which the abdomen is shaken back and forth. At the end of the line, the bee turns sharply in a half circle back to the original point and again repeats the wagging straight line. Reaching the end, it veers tightly in the opposite direction again returning to the start. This process results in an elongated figure-of-eight pattern with the wagging occurring between the two loops of the "eight." The dance is repeated again and

again, attracting the attention of other workers and may stop only intermittently as the returned dancing bee shares a bit of collected nectar.

This tail-wag dance communicates two things: the direction from and the distance to a nectar (or pollen, water, propolis) source. Direction from the hive is indicated by the angle of the straight line portion of the dance. Its angle to the vertical surface of the comb corresponds to the same angle of the location of the resource relative to the direction of the sun as viewed from the hive entrance. For example, a honey bee dancing at 45° angle from vertical is communicating direction to something that is at a 45° angle from the sun when leaving the hive entrance.

The speed of dance performance provides information on the distance; the closer the resource is to the hive, the more rapidly dances are repeated. A honey bee communicating a location that was 100 m distant will repeat the straight line portion of the tail-wag dance

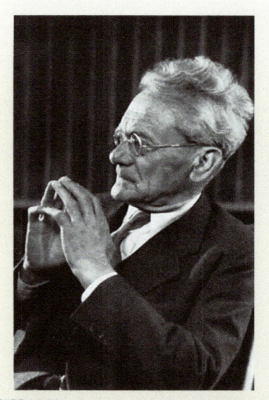

FIGURE 15-39
Karl von Frisch studied the behavior of honey bees and was the first to understand how information was communicated by use of in-hive dances.

(continued)

at a frantic rate of 9–10 times every 15 s. A location that is quite a distance, e.g., 1,000 m, will generate a more leisurely dance rate of four straight line runs in 15 s. Such a leisurely dance will also attract the attention of fewer workers and, thus, will recruit fewer foragers to the site of resources. Furthermore, returning scouts that have located a particularly rich food source likely will repeat the dance cycle more frequently and for longer duration.

Different dances are produced for resources located closer to the hive. Within about 10 m, a rapidly run "round dance" is performed, which stimulates other bees to search generally in the vicinity. As the distance from hive to resource gradually increases, the shape of the round dance progressively becomes more sickle-shaped and approaches the wag-tail dance form. Honey bees are further guided to resources by odor cues, and marking pheromones left by foragers efficiently directs them to newly discovered resource sites.

The cracking of the bee dance language was largely the result of research coordinated by Karl von Frisch (1886–1982) and brought to general public attention in his book *Dancing Bees: An Account of the Life and Senses of the Honey Bee*. For his work with communication in insects and comparative behavioral physiology, he received the 1973 Nobel Prize for Medicine or Physiology, along with fellow pioneers of animal behavior Nikolaas Tinbergen and Konrad Lorenz.

● Honey

Honey is the primary commercial product extracted from honey bees, and for most of human history, it has served as the most important and abundant sweet foodstuff available. Most honey bees (*Apis* spp.) produce substantial amounts of honey. Small amounts of honey are produced by many of the stingless bees (*Melipona* spp. and *Trigona* spp., primarily) and at least two species have long been cultured in Mayan Yucatan for honey and wax.

Most honey originates as plant nectar, although other sweet materials such as honeydew excreted by aphids and other hemipterans is sometimes collected as well. A foraging honey bee collects very large loads of nectar and stores it in a specialized crop known as the honey stomach. A full load may weigh up to 85% of the weight of the bee (70 mg nectar/82 mg honey bee) and typically results from visiting 50–100 flowers although this varies greatly depending on nectar content. The foraging bee then returns to the hive and regurgitates the nectar to waiting "house bees" that carry it away for processing and subsequent colony use. The forager may then make another foraging flight for a new load of nectar; most honey bee foragers average 7–13 such trips per day.

FIGURE 15-41
Over one million flower visits were needed to collect the nectar that is ultimately converted to the eight fluid ounces of honey held in a typical "honey bear" container. Photograph courtesy of Scott Bauer/USDA ARS.

FIGURE 15-40
A honey bee feeding on sugar water. Photograph courtesy of Brian Valentine.

(*continued*)

The transformation of nectar to honey begins as enzymes in the saliva of the bees mix with the nectar, converting sucrose into its simple sugar components, fructose and glucose. As nectar is about 60% water, steps are also taken to concentrate the material. Droplets of the nectar may be painted along the sides of cells allowing water to evaporate better with the exposed larger surface area. Bees will also accelerate the evaporative process by fanning their wings or repeatedly folding and unfolding droplets with their mouthparts. Ultimately, honey is considered ripened when the moisture content has dropped to 18% or less, a critical level that inhibits the growth of yeasts and other microorganisms. The resulting material is then placed in cells and capped with a thin wax layer.

Honey is primarily used as an energy source of simple carbohydrates. In addition to fructose and glucose, other sugars are present in lesser amounts, including maltose and sucrose. About 0.5% of honey consists of proteins, amino acids, various vitamins, and minerals. It is the end result of much hard work as the lifetime collections of a single worker average in the resulting production of less than 0.5 ml of honey. A typical "honey bear" container (8 fl oz.) purchased at a grocery store approximates the sum total effort of over 570 nectar-collecting bees, making some 14,400 foraging trips and visits to over a million individual flowers—plus substantial in-hive processing.

● Modern Beekeeping and the Moveable Frame Hive

Cave and rock paintings discovered from Europe to South Africa document an ancient history of robbing honey from wild honey bee hives. The practice continues in some areas and involves breaking into the hive to extract the honey-filled comb, often with the use of smoke to help subdue the angered bees. Robbed colonies are seriously weakened and many die off from the extensive destruction produced when honey is gathered in this manner.

Limited management of honey bees also has long been practiced. Beehives made of unbaked clay and straw were recently found in an archeological excavation of the Jordan River Valley and date to 900 BCE and there is some evidence of man-made hives in use in the fifth Dynasty of Egypt (ca 2400 BCE). Honey bees have also long been housed in more perishable materials such as hollow logs or woven baskets (**skeps**). All of these early bee hives were designed to provide a space for a swarm of bees to establish; the hive was later destroyed to extract the valuable honey and wax.

In the mid-1800s beekeeping was transformed through the effects of a simple discovery—the "bee space." It had been observed that there is a critical spacing within honey bee hives of about 5–8 mm that the bees maintain for movement. Wax comb is built in wider spaces and narrow spaces are usually plugged with propolis, the "bee glue" honey bees collect from plant resins. Rev. Lorenzo Langstroth (1810–1895) took this information and created a revolutionary new design for a bee hive, the **moveable frame hive**. Langstroth patented the idea in 1852 and further popularized it in the 1853 first edition of *The Hive and the Honey Bee*,

FIGURE 15-42

"The Man of Bicorp," an 8,000-year-old cave painting near Valencia, Spain that illustrates a human collecting honey from a wild hive of bees.

which provided practical information on these new methods of beekeeping.

By constructing hives using frames that maintain the proper spacing, honey bees can construct combs that can then be withdrawn, moved about, and replaced

(continued)

FIGURE 15-43
A woven straw skep used to house a swarm of bees. To extract the honey the colony is usually destroyed when the honey comb is removed. Photograph by Whitney Cranshaw/ Colorado State University.

without harming the bees. This allowed for an examination of the bees and their activities and provided a means of increasing the size of the hive as needed by stacking additional frame-bearing boxes on top of one another. Most importantly, it allowed a honey-filled comb to be removed, have the honey extracted, and then returned to the hive. Currently, essentially all beekeeping systems use some variation of the moveable frame hive. Because of his fundamental contributions to the development of the modern bee-keeping hive, Langstroth is widely recognized as the "Father of Modern Beekeeping."

FIGURE 15-44
Reverend Lorenzo Langstroth developed the idea of making hives with removable frames, an idea he developed after noting that honey bees naturally maintained a "bee space" of 5–8 mm between the combs in their hives. The moveable frame hive provided tremendous advantages in beekeeping and Rev. Langstroth is considered to be the "Father of Modern Beekeeping."

FIGURE 15-45
Movable frame hives can allow individual areas of a comb to be withdrawn and examined. Frames containing honey may be removed, the honey extracted, and then returned. Photograph courtesy of Scott Bauer/USDA ARS.

Honey Bees—Social Insects with Social Diseases

Honey bees live in a constant crowd. Even when new colonies are established it occurs as a mass event, in the form of a swarm. This is in contrast with other social insects, the ants, termites, bumble bees, and social wasps, which start new colonies with a single foundress queen. Unfortunately, the continuous crowding of individuals within a honey bee hive also provides ideal conditions for diseases to establish and spread.

(continued)

TABLE 15-1 Some of the more common diseases of honey bees (*Apis mellifera*) in North America.

DISEASE	CAUSAL ORGANISM	TYPE OF PATHOGEN	EFFECTS ON HONEY BEES
American foulbrood	*Bacillus larvae*	Bacterium	Lethal and highly infectious disease of larvae. Often the most important disease of honey bees.
European foulbrood	*Melissococcus plutonius*	Bacterium	Lethal disease of very young larvae (>48 h).
Nosema	*Nosema apis*	Protozoan	Disrupts digestive system of adult bees producing reduced vigor and activity.
Nosema	*Nosema ceranae*	Protozoan	Disrupts digestive system of adult bees producing reduced vigor and activity. A relatively recently identified disease that originated from the wild Asian honey bee (*Apis cerana*).
Chalkbrood	*Ascosphaera apis*	Fungus	Fungal disease of the gut of larvae. Often lethal but generally considered a minor disease.
Acute paralysis disease/Israeli acute paralysis disease		Virus	Two of the more widespread viral diseases of adult honey bees. Infections may lead to stress and reduced vigor that weaken colonies.
Deformed wing virus		Virus	Virus disease first described in the early 1980s that has increased sharply in incidence in recent years and spread widely, along with its primary vector, *Varroa destructor*. The virus prevents normal wing development.
Sacbrood		Virus	A disease of late stage larvae. Often lethal but generally considered a minor disease.

(continued)

DISEASE	CAUSAL ORGANISM	TYPE OF PATHOGEN	EFFECTS ON HONEY BEES
Varroa mite	*Varroa destructor*	Mite	A large, external parasite of larvae, pupae and adults. Debilitating growth irregularities result from infestation of developing larvae. Some viruses are vectored by this mite. A devastating and recent parasite of the honey bee. Varroa mite originated from the Asian honey bee (*Apis cerana*), on which it is a minor pest.
Tracheal mite	*Acarapis woodi*	Mite	A minute mite that infests the tracheae of adult bees, producing stresses. One result from heavy colony infestations is reduced ability to survive winter.
Small hive beetle	*Aethina tumida*	Beetle	Native to sub-Saharan Africa the small hive beetle was first found in the United States in 1996. Larvae feed on stored pollen, honey and comb.

Because of their importance as pollinators, and incidentally as honey producers, honey bees are also extensively moved and manipulated. Hives that are transported into new regions can be exposed to new pests and diseases, such as occurred a few decades ago (probably in the Philippines some time during the 1960s) when the varroa mite (*Varroa destructor*) first moved from the Asian honey bee (*Apis cerana*) into colonies of the honey bee (*Apis mellifera*). Human transport of honey bees or pathogen-containing materials can allow these diseases to rapidly circumvent natural barriers and subsequently spread across national boundaries. Furthermore, honey bees may be periodically concentrated together in dense aggregations, such as occurs when honey bees from around other parts of the United States move to California to provide pollination services for almonds and citrus. At these times, pathogens rapidly spread as bees drift between colonies or transmit pathogens to one another while visiting the same flowers.

FIGURE 15-46
Varroa mites, *Varroa destructor*. This mite is an external parasite of bees in the genus *Apis*. It was originally a parasite of the Asian honey bee (*Apis cerana*) but moved onto the honey bee (*Apis mellifera*) about 50 years ago and has since spread across the globe. In addition to being highly destructive as it feeds, the varroa mite can also spread viral diseases within honey bee colonies. Photograph courtesy of Scott Bauer/USDA ARS.

(continued)

FIGURE 15-47
Varroa mite on the body of a honey bee pupa. Photograph courtesy of Scott Bauer/USDA ARS.

Many diseases of honey bees have been recognized for centuries, such as the bacterial "foulbrood" diseases that destroy bee larvae. Other pathogens and pests have only recently attached themselves to honey bees, notably varroa mites and the viruses they vector (e.g., Israeli acute paralysis disease). Human-assisted spread has also helped move many existing pests to new areas; tracheal mite, small hive beetle, and *Nosema ceranae* only arrived in North America during the past few decades.

The effects of these pathogens and pests have greatly impacted the health of honey bees in many

FIGURE 15-49
The small hive beetle is one of the insects that has recently found its way into North American honey bee colonies and is potentially very destructive. Photograph courtesy of James D. Ellis/University of Florida/Bugwood.org.

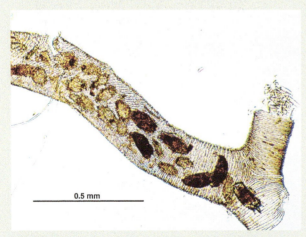

FIGURE 15-48
Tracheal mites within the trachea of the prothorax of a honey bee. Tracheal mite infestations can severely stress honey bees as they interfere with oxygen flow into the body. Photograph courtesy of Pests and Diseases Images Library (Australia)/Bugwood.org.

FIGURE 15-50
One method of controlling varroa mite and other in-hive pests is to use pesticides such as this Apistan strip. Photograph courtesy of Florida Division of Agriculture and Consumer Services/Bugwood.org.

(continued)

FIGURE 15-51
A honey bee worker showing effects of infection with deformed wing virus. Photograph courtesy of Susan Ellis.

areas of the world, including the United States. Major declines in honey bees in North America followed the near concurrent (accidental) introduction of both varroa mites and tracheal mites in the mid-1980s. More recently additional pathogens have been discovered (e.g., Israeli acute paralysis virus, *Nosema ceranae*) and a new insect pest, the small hive beetle (*Aethina tumida*) has become established in North America. The social behaviors of the honey bee that are so important to their success, and the high value of honey bees to humans that has caused them to be spread and nurtured throughout much of the planet, has clearly come at a cost.

● Africanized Honey Bees—A Tough Bee for Tough Places

The familiar species of honey bee (*Apis mellifera*) is a native of Europe, Africa, and western Asia. It was brought to the Americas with European settlement, first arriving within a couple years following the Pilgrim settlement in Massachusetts. Throughout its broad geographic range, various strains of the species occur, each adapted to local conditions. Furthermore, the culture of honey bees (**apiculture**) has a very long history during which beekeepers have selected strains of the insect that have desirable characteristics suitable to various regions. Presently there are about two dozen different recognized strains of honey bees.

One of these strains is native to southern and eastern Africa, *Apis mellifera scutellata*. Known as the African honey bee, it was introduced purposely to Brazil for breeding as it possesses many desirable traits that make it suitable for tropical areas. Unfortunately, it accidentally escaped in the late 1950s and soon became well established as a "wild" honey bee. Well-adapted to the climate, it spread rapidly, moving throughout South America and Central America. In 1990 it was first found in the United States, in southern Texas, and is now established in several southwestern states.

During its spread through the Americas, the introduced African honey bee strain bred with various wild European strains whose ancestors had been imported and spread earlier by beekeepers. The crossing of the African strain with the European strain produced a hybrid generally known as an "Africanized honey bee." Although genetically related to both strains, the original African genes have predominated in crosses between the strains, carrying with it a

FIGURE 15-52
An Italian strain honey bee (foreground) and an Africanized honey bee (rear). There are no obvious differences in size between Africanized honey bees and other strains long present in North America. Color differences also are not very useful, as many other honey bee strains are similar in coloration. Photograph courtesy of Scott Bauer/USDA ARS.

variety of African honey bee traits, some of which are desirable, some of which are not.

Among the tropically advantageous adaptations of Africanized honey bees are the abilities to rapidly establish new colonies and vigorously forage for nectar and pollen. This strain also readily produces swarms,

(continued)

and whole colonies may periodically move (**abscond**) and disperse. This strain is very beneficial in regions where available foods are irregular as it allows the entire colony to move to areas of improved resource availability. Africanized honey bees also display an increased inclination to defend the hive, a trait needed where numerous threats exist (including humans) to their desirable honey stores.

These latter behaviors have resulted in a few serious problems regarding honey bee–human interactions. The aggressive response of Africanized honey bees colonies to disturbance greatly increases the likelihood of stinging. This has led to numerous highly sensationalized accounts, most notoriously the description of these as "killer bees." Establishment of Africanized bees has also caused problems for beekeepers, as these bees are more difficult to work with and too readily swarm.

Nonetheless, in the half century since it arrived in the Americas, humans have adapted as well to this new strain of bee. Beekeepers have learned to manage it and, in some areas, honey production has improved as the hard-working Africanized honey bee collects nectar for honey stores at a faster rate than the European strains do. The Africanized bee has also greatly slowed its expansion in recent years as it reaches the geographic edges where its tropical adaptations fail it. Furthermore, "wild" Africanized colonies that are not maintained by beekeepers have suffered a heavy toll from varroa mites and other honey bee diseases that have more recently spread into North America.

Bumble bees. Bumble bees (*Bombus* spp.) are large, heavy-bodied, and very hairy bees that are common in the northern temperate areas of the world. Yellow and black are common colors, but there are species with orange, red and white markings. Fifty-seven species of bumble bees occur in the United States and Canada.

Bumble bees have truly social behaviors that involve a colony with overlapping generations providing mutual care to the offspring of a queen. However, unlike honey bees, bumble bees produce annual colonies that require re-establishment every season; the only bumble bee stage that survives through winter is a fertilized female queen. In spring, she initiates a new colony that is typically located in an insulated cavity. An abandoned rodent burrow or bird nest is a common site for bumble bees to establish new colonies.

The large overwintered queen is the foundress that must initially perform all colony functions. She forages for nectar and pollen and subsequently produces the wax used to fashion a roundish pot into which the first eggs are laid. The queen will cover these first eggs with her body, keeping them at a relatively stable temperature and allowing them to hatch more rapidly. She will then continue to collect pollen and nectar to feed the developing larvae. Due to the competing demands of the queen, these first bumble bees produced are usually poorly fed and relatively small in size. They assist the queen in colony maintenance and brood care and, as the colony grows in number, the queen soon only performs reproductive duties and remains within the colony. Her newly produced daughters will take over all tasks except egg production. Spherical-shaped pots of wax are continuously constructed and reconstructed to rear larvae and store pollen or nectar, as needed.

Through the season the colony population expands, and the size of the daughters produced later

FIGURE 15-53
A bumble bee (*Bombus huntii*) and a honey bee foraging at the same flower. Photograph by Whitney Cranshaw/Colorado State University.

in the following season. A few males are also produced in late summer, as a result of eggs deposited by these non-fertilized daughters. Males and their sisters, the future queens, leave the colony at season's end and mate repeatedly. The females then feed to produce fat reserves allowing them to survive the winter, which is done in some protected location such as a hole in the ground or hollow log. Within the original colony, all males, all workers, and the queen of the previous season perish at the end of the year.

FIGURE 15-54
A foraging bumble bee. Bumble bees carry pollen in special pollen baskets on the hindlegs, as do honey bees and other bees in the family Apidae. Photograph courtesy of Jim Kalisch/ University of Nebraska.

FIGURE 15-56
Bumble bees can show a considerable range in size, as is indicated by these *Bombus huntii* individuals. Smaller bees are produced when they are fed less during larval development and they tend to occur early in the season, as colonies become established. As the season progresses and colonies become larger, the bumble bees tend to be better nourished as larvae which produces larger adults. The largest bees produced late in the season can function as future queens and are the only stage that survives over winter. Photograph by Whitney Cranshaw/ Colorado State University.

FIGURE 15-55
Top view of a bumble bee colony. Photograph courtesy of Ken Gray/Oregon State University.

in the season tends to increase in response to their improved nutrition during larval development. By late summer, there are some full-sized, fertile females present that will ultimately become potential queens

Bumble bees are very important in the pollination of many species of plants. Bumble bees pollinate plants in a different manner than do other bees, known as **buzz pollination**. When visiting a flower, a bumble bee will grasp it with its mandibles and vibrate, causing the pollen to be shaken out onto its body. This pollination approach is critical for some plants, notably many in the nightshade family (Solanaceae), and bumble bees are used widely as pollinators of greenhouse-grown tomatoes and peppers. Many native plants require pollination by bumble bees and some bumble bees adapted to alpine conditions are critical to the reproduction of many flowering plants that occur at the highest elevations.

FIGURE 15-57
Bumble bees are particularly good at pollinating plants that require "buzz pollination." Greenhouse-grown tomatoes and peppers make extensive use of bumble bee hives in production. Photograph courtesy of Jim Kalisch/University of Nebraska.

FAMILIES SPHECIDAE AND POMPILIDAE— HUNTING WASPS AND SPIDER WASPS

Several kinds of wasps hunt for prey to feed their young. Some of these first subdue the prey (another insect or spider) with a paralyzing sting, then transport it whole to a nest site where the prey will serve as the food source for a developing wasp larva. Wasps that have this life history primarily occur within two families, the hunting wasps (Sphecidae) and the spider wasps (Pompilidae). Additionally, some species in the family Vespidae also hunt and cache paralyzed prey for rearing young.

The hunting wasps (Sphecidae) are by far the most common and behaviorally diverse with over 1,100 known species found in North America alone. Each species specializes in the hunting of some kind

FIGURE 15-58
A hunting wasp, *Tachysphex* sp., with grasshopper prey. These wasps nest in soil. Photograph courtesy of Howard Ensign Evans.

of insect. The largest hunting wasps are the cicada killers (*Sphecius* spp.). These are very large wasps that specialize in hunting dog-day cicadas; cicadas are cached in previously constructed holes and tunnels in the ground. Smaller species of hunting wasps (e.g., *Pemphredon* spp.) specialize in hunting small insects such as aphids and leafhoppers. In between these extremes are hundreds of species that vary in hunting their specific hosts, which may be a kind of katydid, weevil, fly, bug, or some other insect.

FIGURE 15-59
Eastern cicada killer, *Sphecius speciosus*, capturing a dog-day cicada (*Tibicen* sp.). The cicada will be taken to a previously prepared nest cell in a soil burrow and will nourish a developing cicada killer larva. Photograph courtesy of Howard Ensign Evans.

Each species of hunting wasp establishes a nest at some location before it begins hunting prey. Nests may be above ground or located in the soil. The latter is typical of the cicada killers that produce large mounds around their tunnels. Various *Ammophila* spp. that hunt large, naked caterpillars and sand nesting *Bembix* wasps that prey on flies also nest in soil. Other hunting wasps use tunnels that they locate or excavate from plants and then stuff their paralyzed prey into these previously excavated cells.

In North America, a conspicuous hunting wasp that exhibits relatively unique behaviors in regard to nesting is the yellow and black mud dauber, *Scleriphron caementarium*; this species constructs nests in the form of tubular cells made of mud. This wasp specializes in hunting spiders, most commonly crab spiders and orb weavers. A half dozen or more spiders may be collected to provision each cell of a mud dauber's nest.

Spiders are the only prey collected by the spider wasps of the family Pompilidae. These wasps attack

FIGURE 15-60
The black and yellow mud dauber, *Scleriphron caementarium*, constructs nest cells out of mud, which are often noticed under eaves of buildings. After a nest cell is completed they hunt for spiders to fill the cell. Photograph courtesy of Howard Ensign Evans.

FIGURE 15-61
A larva of the black and yellow mud dauber exposed with some remaining spider prey. The larvae develop quickly and usually consume all their food within about a week of hatching from the egg. Photograph courtesy of Jim Kalisch/University of Nebraska.

spiders that are large enough to provide food for their young. After being paralyzed with a well-placed sting to the nerve cord, the spider is often dragged into the nearest available crack or crevice whereupon the female deposits an egg. The site is then sealed off with soil and pebbles and the young wasp larva will consume the paralyzed spider after the egg hatches.

Among the 290 spider wasps that occur in North America, each specializes on hunting a separate group of spiders. Human attention is most often given to the largest spider wasps that hunt the largest spiders. These are the blue-black and orange winged wasps of the genera *Pepsis* and *Hemipepsis* that specialize in hunting tarantulas (and large wolf spiders), often referred to as "tarantula hawks." Other spider wasps specialize in other, smaller types of spiders and are consequently substantially smaller. As a group, the spider wasps are able to produce some of the most painful stings of any insect, but they are not aggressive and will sting only if handled or confined.

FIGURE 15-62
Tarantula hawks are large spider-hunting wasps. Females hunt and paralyze tarantulas or other large spiders for their young, which then feed on the immobilized spider prey. Photograph by Whitney Cranshaw/Colorado State University.

● Stings and Pain

One of the most defining features of bees, wasps, and many kinds of ants is their capacity to sting. The stinger is a special modification of the ovipositor so that it no longer functions for egg laying but does possess an associated venom gland. Because the ovipositor is a characteristic of females only, only female insects can sting. Although the original purpose of the stinger was probably to help hunting wasps subdue their prey, it is

(continued)

also now more widely used to defend the colonies of social insects. The purpose of the stinging event is to produce memorable pain, rather than cause mortality.

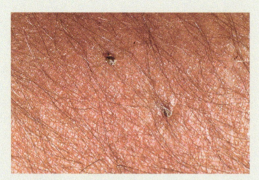

FIGURE 15-63

Stingers of a honey bee embedded in an arm. Honey bees have a barbed stinger and when pulling away after stinging leave the poison sac and stinger behind. Other stinging insects do not have a barbed stinger and do not leave a stinger. Photograph by Whitney Cranshaw/Colorado State University.

Pain perceived by humans from various stings can be affected by many things. The size of the stinger and how deeply it penetrates is important. For example, many harvester ants (*Pogonomyrmex* spp.) can potentially produce very painful stings but their stinger is rather blunt and usually cannot penetrate thicker areas of skin. The amount of venom injected is another factor influencing pain as little or no venom may be injected.

Honey bee workers have a unique system to maximize venom injection; they utilize a barbed stinger. Such a stinger remains embedded within the targeted flesh, and when the worker pulls away, the stinger, the associated venom glands, and the associated musculature remain behind, continuing to pump venom into the victim. (Alarm pheromones in the form of isopentyl acetate are also released, attracting other honey bees to sting near the same site.) All other stinging insects lack the barb and may sting repeatedly.

A wide range of different venoms are employed by insects that use stingers. These venoms are capable of producing a wide range of pain in humans. In an attempt to quantify this, Justin Schmidt of the USDA Carl Hayden Bee Research Lab developed a pain index that was first presented in 1984 and later refined. The Schmidt Sting Pain Index ranked the relative pain of stings produced by 78 insect species on a sliding scale of 0 (no pain) to 4 (traumatic pain). In addition he provided subjective comments that were unusually descriptive.

Among the least painful insect stings was that of a sweat bee (1.0 rating) described as "light, ephemeral, almost fruity. A tiny spark has singed a single hair on your arm." A bit higher ranking (1.2) was given to fire ants capable of producing pain that was "sharp, sudden, mildly alarming. Like walking across a shag carpet and reaching for the light switch."

Outside the areas where fire ants are common, yellowjackets (*Vespula* spp.) are usually the most common stinging insects in the United States and they received a pain index rating of 2.0. ("Hot and smoky, almost irreverent. Imagine W. C. Fields extinguishing a cigar on your tongue.") This was about the same as the honey bee ("Like a matchhead that flips off and burns on your skin.") The *Polistes* species of paper wasps, also a common cause of stings, were given a Schmidt Sting Pain ranking of 3.0 described as "Caustic and burning. Distinctly bitter aftertaste. Like spilling a beaker of hydrochloric acid on a paper cut."

Several stinging insects rounded out the top with 4.0 rankings. The large spider wasps of the genus *Pepsis*, commonly known as "tarantula hawks," have a sting described as "Blinding, fierce, shockingly electric. A running hair drier has been dropped into your bubble bath (if you get stung by one you might as well lie down and scream)." The giant tropical bullet ant (*Paraponera clavata*) found in parts of Central and South America produces "pure, intense, brilliant pain. Like walking over flaming charcoal with a 3 inch rusty nail in your heel." Finally, the Florida harvester ant *Pogonomyrmex badius* that occurs in the southeastern United States produced pain likened to "someone turning a screw into the flesh" or "ripping muscles and tendons."

FAMILY VESPIDAE—PAPER WASPS, YELLOWJACKETS, HORNETS, MASON WASPS, POTTER WASPS

The family Vespidae includes most of the insects the general public recognizes as "wasps," and many that are unfortunately mistaken for "bees." Most commonly encountered are the paper wasps and yellowjackets. These are social species that produce colonies maintained within nests constructed of paper. All can produce a painful sting and, with the notable exception of areas where red imported fire ants are common, are the most frequent culprits involved in stinging humans.

The yellowjackets (*Vespula* spp. and *Dolichovespula* spp.) include the most pestiferous of the social wasps. They construct nests from paper, usually produced by chewing weathered wood and mixing it with saliva. Its young are reared in hexagonal cells that are produced in horizontal

FIGURE 15-64

The western yellowjacket, *Vespula pensylvanica*, is a notorious scavenger that disrupts outdoor dining. Meats and sweet materials, such as the syrup of pancakes, are favored foods. Photograph by Whitney Cranshaw/Colorado State University.

FIGURE 15-66

A close-up of the surface of a yellowjacket nest. Nests of yellowjackets are constructed from chewed wood fibers and the patterning varies depending on the source. Photograph courtesy of Jim Kalisch/University of Nebraska.

comb. The entire nest is covered with a gray-colored paper envelope and is continuously being reconstructed as the colony expands.

FIGURE 15-65

A German yellowjacket, *Vespula germanica*, with a small nest produced early in the season. A couple of eggs are present in the paper nest cells. Photograph courtesy of Gary Alpert/Bugwood.org.

Nesting location varies with the two most common genera of yellowjackets. *Vespula* species almost always nest below ground or in protected cavities. *Dolichovespula* species typically produce exposed above ground nests that may be attached to tree branches or the sides of buildings.

Yellowjackets produce nests that are established anew each spring. The only stage that survives winter is a large fertilized female that is a potential foundress queen. On warm days in spring, these over-wintered females emerge from protected sites used for shelter during the cold winter months and begin searching for locations in which to establish a colony. Initially working alone, the founding queen does all the necessary colony chores of nest construction, foraging, and rearing young. About a month after starting the colony, adults begin to emerge that the queen has reared. These are infertile female workers, smaller than the queen, and the workers ultimately take over all colony functions, except egg laying. During the summer the size of the colony expands dramatically and comes to include hundreds and even thousands of workers by late summer. As the season progresses increasing numbers of males and reproductively viable females (future queens) are produced. Mating occurs among the latter, and the newly fertilized females disperse to find winter sheltering sites. At the end of the year all males, all workers and the original queen die, and the nest is permanently abandoned.

Two main feeding behaviors occur among the various species of yellowjackets. Some, including all the aerial nesting species, are predators that feed almost entirely on living insects such as caterpillars and flies. These can provide substantial control of pest insects. Predatory yellowjackets cause relatively few nuisance and stinging problems.

Serious problems with nuisance and stinging do occur with several species of yellowjackets that are primarily scavengers. These kinds of yellowjackets

FIGURE 15-67
An underground nest of the prairie yellowjacket, *Vespula atropilosa*, exposed by a digging skunk. Most yellowjackets (*Vespula* spp.) nest below ground. Photograph courtesy of Karen Renneker.

feed on carrion, dead insects, earthworms, as well as most any type of meat- or fish-based meal they can find. Scavenging species also forage frequently for sugars that they acquire from nectar, insect honeydew, soft drinks, or hummingbird feeders. Human encounters are all too frequent during the "yellowjacket season" of late summer, at which time attempts at outdoor dining may be temporarily suspended due to these nuisance wasps.

Stings also occur when nests are accidentally disturbed, a common occurrence as the nests are usually well hidden and have only a small entrance. Among the most important of the pestiferous yellowjackets are the western yellowjacket (*Vespula pensylvanica*), the eastern yellowjacket (*Vespula maculifrons*), the common yellowjacket (*Vespula vulgaris*), the southern yellowjacket (*Vespula squamosa*), and an introduced species of expanding range in North America, the German yellowjacket (*Vespula germanica*).

Almost all yellowjackets are bright yellow with black markings. Very often they are mistaken for bees and are very frequently the actual cause of alleged "bee stings." They can probably be most easily distinguished from bees by having a significantly less hairy body and they do not collect pollen for rearing young, although adults will occasionally feed on

nectar and pollen. Also, yellowjackets lack a barbed stinger allowing them to repeatedly sting and do not leave behind a stinger, as does the honey bee.

To add a bit to the confusion with these social wasps, there is one common species of yellowjacket that lacks yellow and instead is black and white, the baldfaced hornet (*Dolichovespula maculata*). These make large semi-spherical nests in trees and shrubs. The paper envelope has a scalloped surface, often of multiple colors that reflects different sources of wood used for construction. Baldfaced hornets are

FIGURE 15-68
A baldfaced hornet, *Dolichovespula maculata*, chewing on weathered wood. This material is mixed with saliva and used in nest construction. Photograph courtesy of Jim Kalisch/University of Nebraska.

predators of insects, particularly of caterpillars, and do not cause problems as nuisance scavengers.

Nest of a baldfaced hornet cut open to show the layers of paper cells in which the young are reared. Photograph courtesy of Ken Gray/Oregon State University.

Despite the common name, the baldfaced hornet is technically not a hornet, a designation properly given only to social wasps of the genus *Vespa*. One "true hornet" now occurs in parts of North America, an introduced species known as the European hornet (*Vespa crabro*) that now ranges through much of the northeastern quadrant of the United States. It produces large, brown paper nests that are rather brittle because of their use of rotten wood for construction material. Like the baldfaced "hornet," the European hornet is a predator that feeds on other

A European hornet (top), *Vespa crabro*, compared to a yellowjacket. Photograph courtesy of Jim Baker/North Carolina State University/Bugwood.org.

insects and rarely causes nuisance. Workers will fly at night and sometimes are found at lights.

The paper wasps (*Polistes* spp., *Mischocyttarus flavitarsis*) also produce paper combs for nesting, but these species produce only a single-layered comb and do not envelop it in paper. Paper wasp colonies tend to be small, rarely more than a total of 200 individuals through the season, often considerably fewer. Because of the smaller colony size, paper wasps can nest in small cavities such as playground equipment, outdoor electrical boxes, an outdoor grill, or the various small nooks provided by many buildings. These hidden locations often may result in the nest being accidentally disturbed, and the wasps will sting in defense.

A European paper wasp, *Polistes dominula*, tending a nest. Larvae are present in some of the cells. Photograph courtesy of Joseph Berger/Bugwood.org.

Paper wasps are predators of other insects and can be extremely important in control of garden pests. Most are readily distinguishable from other social wasps by the slender shape of their body, long trailing hindlegs, and general brownish coloration. However, a recently established species, known as the European paper wasp (*Polistes dominula*), has yellow and black markings that make it more difficult to distinguish from yellowjackets.

The paper wasps have a slightly less rigid social structure than yellowjackets and hornets. New colonies are established in spring by overwintering fertile females but, on occasion, more than one female wasp will work together to create a colony. In these cases, one usually becomes the dominant queen

FIGURE 15-72
Polistes fuscatus tending a nest. Photograph courtesy of Jim Kalisch/University of Nebraska.

while the other reverts to a status more aligned with workers. Both females may produce eggs; however, the queen produces the large majority of them. As with other social wasps, males and new potential queens are produced at the end of season, but there is very little difference in size between the workers and queens at any time of the year.

FAMILY FORMICIDAE—ANTS

Ants are often the most commonly observed of all insects and very few areas of Earth fail to support some representative of the 9,500 known species of ants. A few minutes' examination of a piece of ground almost always will allow one to see ants as they go about their activities of scavenging dead insects, collecting seeds or honeydew, preying on other insects, or engaging in occasional wars between colonies.

All ants are social insects that establish persistent colonies with identifiable specialized castes. Wingless sterile workers comprise essentially the entire population of most colonies. Workers are all females, and in many ant species, the workers are all roughly the same size (**monomorphic**). Other ant species are **polymorphic** and produce workers that may have a wide range of size. Amongst the polymorphic species, the tiniest forms usually specialize in nest chores within the colony, while the largest individuals are primarily involved in nest defense. In addition to size, workers may also vary in the shape of their head or jaws to allow some to more effectively bite, crack seeds, or perform heavy lifting.

FIGURE 15-73
A cast made of the nest of an ant, *Aphaenogaster cockerelli*. Ant colonies typically have numerous chambers that are used for tasks such as larval rearing, food storage, and preparation. Photograph courtesy of Gary Alpert/Bugwood.org.

FIGURE 15-74
Texas leafcutting ant showing a range of different sizes of workers and reproductive forms. This species is unusually polymorphic and shows a wide range of worker size. Photograph courtesy of R. Scott Cameron/Advanced Forest Protection, Inc./Bugwood.org.

Regardless of size, all ants with legs are adults. Larvae only occur within the confines of the colony and develop through the usual stages of complete metamorphosis: egg, several larval stages, pupa, and adult. Ant larvae are pale colored, legless and totally dependent on adult worker ants for their care.

Ant pupae occur within a silken cocoon, giving them a rather cylindrical shape. Silk is produced from specialized labial gland within the mouth of the larva. Pupae are sometimes seen when exposing an ant colony by lifting a rock or log.

Although all worker ants are female, their ovaries fail to develop during development. The task of reproduction falls to the caste of fertile females known as queens. Often there is only a single queen in each colony, but multiple queens are common with some species of ants. Queens are substantially larger than the workers and are long lived, typically surviving for a few years to, sometimes, decades. When an old queen dies, depending on the species, she is usually replaced with a new reproductive female. In species that have a mechanism for queen replacement, ant colonies can last for decades, and potentially forever.

FIGURE 15-75
Worker ants with pupae. Photograph courtesy of David Cappaert/Michigan State University/Bugwood.org.

FIGURE 15-76
Male (top) and female (bottom) winged reproductive stages of a carpenter ant, *Camponotus* sp. The female is a potential queen that will attempt to establish a colony after mating. The males die shortly after they emerge from the colony on their mating swarm. Photograph by Whitney Cranshaw/Colorado State University.

Actively ovipositing queens within the colony are wingless; however, they once possessed wings. Although individual ants come from eggs produced by the queen, an ant colony reproduces by the production of special forms capable of establishing new colonies. These consist of reproductive forms, the winged females (potential future queens) and winged males. These winged reproductive ants remain within the colony until forced out during periodic swarming events. On these occasions, hundreds of winged ants may suddenly appear around the nest entrance and fly away. Such ant swarms are often synchronized in an area and typically occur during sunny, calm days that shortly follow a heavy rain.

During mating flights winged ants frequently aggregate at certain sites, often near some prominent landscape feature such as a large tree, isolated building, or hilltop. Such behavior is known as **hilltopping**. There the sexes come together, and the males attempt to mate with the females. This sometimes produces rolling balls of males surrounding a single female. After mating, the female departs the aggregation; the males remain behind and soon die.

The newly mated queen attempts to establish a new colony by herself. After finding a suitable site, she breaks off her wings; the large flight muscles are then used for nourishment. Initially working alone, the queen is later assisted after the first larvae mature into workers. Ultimately, a new viable colony may become established; although, the odds of success are heavily against establishment. Among the tremendous numbers of winged females that emerge during a mating swarm only a tiny fraction, often less than 1 in a 1,000, will be successful in creating a new colony whereas the others perish.

Ants fundamentally affect the ecology of natural systems in myriad ways, and many species have important impacts on human activities as well. It is impossible to do justice to this family in brief treatment, and many excellent books are dedicated solely to the ant family or even specific types of ants. The following provides a brief sampling of species that show some of the range of behaviors and life styles found within the ant family.

Carpenter ants (*Camponotus* spp.). Carpenter ants can be found worldwide, and many species may be common in the forested areas. This genus includes the largest ants that occur in the North America, with some workers reaching a size of 13 mm in length. Most species are black and/or reddish in color

FIGURE 15-77
Prominent points in a landscape serve as "hilltopping sites" where winged ants and other insects aggregate to find mates. The site pictured is frequently used by harvester ants during periods of mating swarms. Photograph by Whitney Cranshaw/Colorado State University.

FIGURE 15-78
Mating balls of harvester ants found around the base of a tree used as a hilltopping marker. Photograph by Whitney Cranshaw/ Colorado State University.

FIGURE 15-79
A carpenter ant queen shortly after dropping her wings following the mating flight. She will use the muscles formerly used for flight to help nourish her as she attempts to initiate a colony. Photograph courtesy of Jim Kalisch/University of Nebraska.

and can be distinguished by having the back of the thorax smoothly curved in profile.

Most carpenter ants nest in tunnels excavated within wood, piling the stringy sawdust near the nest entrance. Fallen logs, hollow trees, or dead branches are typical nest sites but carpenter ants occasionally nests in buildings, particularly where wood has been softened by previous decay from water damage.

Although they may forage over 150 m from the colony, a lot of carpenter ant activity may go undetected as they are active at night and may travel most of the time along underground channels. Carpenter ants primarily feed on sweet materials, but during the peak periods when larvae are being reared they will shift their diet to include more dead insects and protein-rich materials.

Unusual behaviors have developed among some carpenter ant species. The Australian species *Camponotus inflatus* is one of the "honeypot ants" that collect sweet materials, notably honeydew excreted from local scale insects and psyllids. As honeydew is only sporadically available, when it is present the ants store the excess in the bodies of

FIGURE 15-80
Carpenter ants tending eggs. Photograph courtesy of Jim Kalisch/University of Nebraska.

FIGURE 15-81
Life stages and injury produced by carpenter ants. Graphic produced by Arthur Cushman and provided courtesy of USDA/Bugwood.org.

certain workers that function as honeydew storage vessels ("**honeypots**"). These ants have an abdomen that can distend greatly to the size of a grape when their crop is full. They dangle motionless from the ceiling of the colony, sometimes for many months, until called upon by their nestmates to regurgitate

the contents of their crop. Other species of ants similarly store sweet fluids in specialized "honeypot" workers including various *Myrmecocystus* spp. that occur in the desert areas of the southwestern United States. Both in North America and Australia these "honeypot ants" have long been used as a sweet delicacy by native peoples.

A Malaysian species of carpenter ant (*Camponotus saundersi*), popularly known as the "exploding ant" or "kamikaze ant," has an unusual defense. These ants possess huge glands running from the base of the mandibles to the end of their abdomen that are filled with a mixture of sticky, corrosive fluids. When attacked the ants have muscles that cause their body wall to burst open, allowing the fluids to spray over and immobilized attackers.

Field ants, wood ants, thatcher ants (*Formica* spp.). Ants of the genus *Formica* tend to be rather large ants and may be black, reddish brown, or a combination of these colors. Although they lack a stinger, they are aggressive ants that defend themselves by spraying formic acid from the tip of the abdomen. Many are important as predators of other insects, but they also readily collect honeydew from colonies of aphids and other sap-feeding Hemiptera. They do not nest within homes but may nest near building foundations and enter homes periodically in search of food or water.

FIGURE 15-82
Formica sp. ant tending aphids to collect honeydew. Photograph courtesy of David Cappaert/Michigan State University/Bugwood.org.

The many common names associated with ants of this genus—wood ants, field ants, thatcher ants to name a few—are suggestive of their widespread occurrence and various habits. They can be common

FIGURE 15-83
Western thatching ants, *Formica obscuripes*, feeding on a dead insect. Photograph courtesy of Howard Ensign Evans.

FIGURE 15-84
Large mound produced by western thatching ant, *Formica obscuripes*. Photograph courtesy of Howard Ensign Evans.

in gardens and forested areas where they may build conspicuous mounded nests. The species that occur in cooler climates may make especially large mounds constructed of small twigs, needles, and fallen leaves, reminiscent of roof thatching, hence the name thatcher ants. These materials help to conserve moisture and heat, but the primary reason for the mound is to capture heat from the sun, helping to warm the colony. The Allegheny mound ant (*Formica exsectoides*) of eastern North America can produce mounds nearly 1.0 m in height and even larger mounds are produced by some forest-dwelling species of northern Europe.

Multiple queens may be present in colonies of some field ants as new queens may return to a colony following the mating flight. These fertilized queens may later establish new nests of their own by departing their original colony and taking with them a group of workers, a behavior known as colony budding. As the individual ants remain related within such colonies, new colonies behave as extensions of the original, and the workers continue to interact freely. In this manner, "super colonies" are produced extending over extensive areas. One such colony mapped on the island of Hokkaido in northern Japan extended over 270 ha. It was further estimated to contain 306 million workers and 1 million queens, dispersed among the 45,000 interconnected nests.

Slave making (**dulosis**) occurs among some *Formica* species, which raid the colonies of other *Formica* species for their pupae. The raids involve the use of chemicals that the slave-maker produces in a large abdominal gland; these chemicals mimic the alarm pheromones of the species being attacked. Sprayed as the attack commences, they produce immediate confusion within the colony being attacked. This allows the raiding species to quickly slip in and out with minimal fighting. The adults that develop from the enslaved pupae appear to behave as normal workers alongside non-slaves and perform tasks as they would in their colony of origin.

Harvester ants (*Pogonomyrmex* spp.). The most commonly observed ants in the arid shortgrass prairie and desert areas of the United States are usually harvester ants. These are moderately large (5–7 mm) and may be reddish-orange, brown, black, or both red and black in color. Their nests

FIGURE 15-85
Nest mound of the western harvester ant, *Pogonomyrmex occidentalis*. Photograph by Whitney Cranshaw/Colorado State University.

FIGURE 15-86
Workers of the western harvester ant at the nest entrance. Photograph by Whitney Cranshaw/Colorado State University.

are particularly easy to spot as they typically appear as broad mounds, surrounded by an area extensively cleared of vegetation. The mounds often have incorporated within their surfaces small pebbles that the ants have removed from their extensive network of below ground tunnels. The mound, its surface covering materials, and the surrounding cleared area prevent shading and help warm the nest.

Mating swarms of the winged reproductive stages occur in summer, usually a few days following a heavy rain. Some species orient to prominent features of the landscape (hilltopping) and masses of winged ants sometimes have been observed atop tall buildings, over chimneys, and even farm equipment. The newly mated female attempts to establish a colony by herself and, if successful, can live for a decade or more.

Harvester ants primarily feed on collected seeds although they do collect dead insects, including winged termites following their mating swarms. If nests are directly disturbed, harvester ants vigorously defend the mound and are capable of very painful stings. Perhaps because of the stings, along with the reddish color of many species, they are often called "fire ants," an inaccurate name that causes confusion with the "true" fire ants of the genus *Solenopsis*.

Indeed, there are concerns that competition by red imported fire ant has greatly reduced populations of harvester ants in west Texas. This, in turn, imperils the Texas horned lizard ("horned toad") that feeds almost exclusively on harvester ants. Native plant communities may also be altered where harvester ants are eliminated, as many plants are dependent on harvester ants for dispersal of seeds.

Red imported fire ants (*Solenopsis invicta*). Since its establishment along the Gulf States in the 1930s, the red imported fire ant has greatly extended its distribution and developed into the most economically damaging ant species in the United States. It now occurs throughout most all of the southern tier states and is present along the southern Pacific Coast.

The primary concerns involving red imported fire ant are related to their ability to sting coupled with their very aggressive defensive behavior when disturbed. Although a relatively small ant, workers range from 2.4 to 6 mm, the red imported fire ant produces a very painful sting that quickly results in a characteristic pus-filled blister at the site of the sting. Human deaths from stings have occurred, particularly as a result of allergic reactions. Livestock deaths have also been reported, and populations of ground nesting birds and rodents

FIGURE 15-87
Red imported fire ant, *Solenopsis invicta*, crawling on a pencil. Photograph courtesy of USDA APHIS PPQ/Bugwood.org.

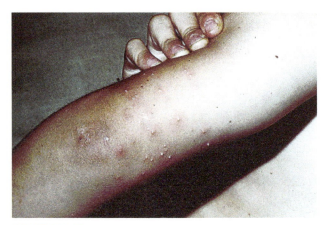

FIGURE 15-88
Pus-filled blisters are characteristic reactions to stings by the red imported fire ant. Photograph courtesy of USDA APHIS PPQ/Bugwood.org.

often are devastated where the red imported fire ant becomes abundant.

Red imported fire ant can cause many other kinds of damage. Tender tissues of certain vegetable and field crops may be eaten. The ants may produce hundreds of large mounds (over 40 cm across) per hectare that interfere with farm equipment and remove large areas of productive pasture. In addition, the red imported fire ant has proven to be a common cause of electrical equipment failures, their bodies and stings produce short circuits in air conditioners, street lights, and electrical switches.

The red imported fire ant is a highly invasive species that has proved to be a much more important

pest in the United States than in its native South America. The potential of the species as a pest was recognized fairly early, and during the 1960s and 1970s, there was a massive attempt made to eradicate it over much of the southern United States by aerial application of the persistent insecticide Mirex. These efforts ultimately failed to contain the spread of fire ants and also proved to be extremely controversial due to the detrimental effects of these treatments on **non-target organisms** (e.g., birds, pollinating insects). Today, red imported fire ants remain well established across a broad area of the southern United States and in Texas alone are annually estimated to result in $1.2 billion in damage and control costs.

Argentine ant (*Linepithema humile*). The Argentine ant is native to South America and was first introduced into the southern United States over a century ago. It adapts well to life around human residences and may be common under boards, stones, or within mulch or piles of plant debris. It will occasionally nest in homes. A common ant in many southern states, the Argentine ant is also the most commonly encountered ant in urbanized areas of

FIGURE 15-89
Argentine ants tending scale insects for honeydew. Photograph by Whitney Cranshaw/Colorado State University.

southern California. It feeds almost exclusively on sweet materials, visiting honeydew-producing insects and any sweet materials left open in homes. It is a fairly small ant (2.4–2.8 mm), and is light to dark brown in appearance.

The Argentine ant has also proved adaptable to many natural areas and has been spectacularly successful in colonizing southern California where it regularly outcompetes and eliminates native ant species. This appears to be related to a difference in the behavior of the ants in California from those in their native South America. Whereas Argentine ants normally defend individual nests in their area of origin, those found in California often work together, resulting in the formation of massive, cooperative "super colonies." Although the cause of this altered behavior remains a subject of discussion, the loss of native ants that are important in providing normal ecosystem services (e.g., seed dispersal, decomposition) has had serious negative effects on many plant species.

Weaver ants (*Oecophylla* spp.). The weaver ants are tree-nesting species that occur over wide areas of Africa, Australia, and Asia. They are large ants (ca. 8 mm), aggressive, and predatory, feeding on all manner of insects and vigorously defending their territory from other nearby weaver ant colonies. Their potential to control insect pests was recognized at least 700 years ago in China where their nests were sometimes transferred to orchards equipped with bamboo runways specifically constructed to allow ants to move freely from tree to tree. Husbandry of weaver ants to control crop pests continues to the present in parts of southeast Asia and Africa.

FIGURE 15-90
Weaver ants constructing nest from leaves tied together with silk. Photograph courtesy of Howard Ensign Evans.

Weaver ants construct their nests out of living leaves fashioned into a carton nest. These nests are produced by the ants collectively working to pull together two adjacent leaves, sometimes creating chains of living ants to bridge distances too wide for individual ants to reach. The leaves are then permanently fastened together in a truly unique manner; they are stitched together using silk. The source of the silk is from full-grown larvae ready to spin their cocoon. These are taken by workers to the construction site, and the larvae function as living shuttles being moved back and forth between the two leaves, the silk binding them together.

During this process, the larvae expend all of their silk and are unable to create a cocoon. However, they are returned back to the interior where they are given special care such that they usually can successfully pupate and continue development to the adult stage.

Army ants. Army ants can be described as "any species of ant that goes out in search of food in companies," rather than using scouts that pre-mark trails leading to discovered resources. Such massive foraging behaviors occur most prominently among the **legionary ants** (*Eciton* spp.) of tropical America and the **driver ants** (*Dorylus* spp.) of Africa.

Eciton burchellii is an example of one of the legionary ants of South America. This species does not establish permanent colonies but spends the night in temporary **bivouacs** consisting of chains of worker ants, linked together by their legs. The colony takes on the form of a large mass of ants that may contain 150,000–700,000 workers. Within the interior of the bivouac is a single queen and thousands of larvae and developing pupae. Workers range greatly in size (3–12 mm) with the largest forming a specialized soldier caste equipped with long, sickle-like mandibles.

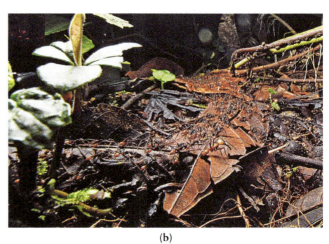

(a)

(b)

FIGURE 15-91

(a and b) Legionary ants (*Eciton burchellii*) creating a chain of their bodies. Legionary ants periodically conduct swarm raids that collect insects and other foods for return to their bivouac. Photographs courtesy of Alex Wild.

At daylight, the colony dissolves and becomes a raiding swarm. Some legionary ants are column raiders that move along various narrow trails where travel is easiest. *Eciton burchellii* is a "swarm raider" producing a broad raiding front that may be 15 m across. These ants will attempt to feed on anything that they encounter. As the legionary ants lack cutting mouthparts, larger prey may survive; however, they occasionally kill small birds and rodents through asphyxiation. The approaching swarms are easy to notice. First there is a noticeable and auditory rustling produced by the myriad insects attempting to escape the advancing swarm. This gives way to the advancing swarm often accompanied by the chirping of antbirds that feed on the insects being driven in front of the moving swarm.

(a)

(b)

FIGURE 15-92

Driver ants (*Dorylus* sp.) are polymorphic (a) with workers of various sizes that conduct different colony functions. The winged reproductive forms (b) that periodically leave the nest on mating flights are sometimes referred to as "sausage flies." Photographs courtesy of Alex Wild.

There are two main phases of activity that legionary ants periodically exhibit. While in the **nomadic phase** the ants make a new bivouac every night, carrying along the queen and developing larvae and pupae. New eggs are not laid during this time and a nomadic phase may last about 3 weeks. The nomadic phase is followed by a **stationary phase** when the bivouac remains in place. During this phase, reproduction occurs and the queen rapidly matures massive numbers of eggs. These are deposited within the bivouac over a period of a few days and are tended by the workers. When the eggs hatch, the behavior of the colony shifts once again to the nomadic phase. Except for a cycle during the dry season, only non-reproducing worker and soldier forms are produced during each stationary phase. Some winged males and potential queens may be produced during the dry season.

The driver ants of sub-Saharan Africa produce massive colonies that may contain over 20 million workers. They possess sharp mandibles with a shearing ability that can cut apart almost any prey. Meats and oils are their primary foods, and they have been known to kill humans who could not escape from their path. The common name driver ant derives from the effect of these ants to drive out anything that can move out of its oncoming path. Other names include "safari ant," "saifu," or "ensanfu."

A queen driver ant is the largest of all ants, 39–50 mm in length. It can produce up to one to two million eggs per month and lays them continuously, without breaks in the reproductive cycle such as occur with the legionary ants. To sustain this tremendous egg production, driver ant queens mate repeatedly throughout their lives. This is achieved by periodically bringing in new males. Male driver ants are also especially large and are known as "sausage flies." The males are winged and become attracted to the odors of an active colony and land nearby. If found by workers, their wings are removed, and they are then picked up and carried back to the colony. New colonies are produced by fission in a behavioral manner somewhat resembling honey bee swarming, with a new queen departing along with many workers.

Driver ants also produce bivouacs, but these are substantially more stable than those created by legionary ants, typically persisting for a couple of months. Bivouacs are also dug into the soil, producing large mounds. When weather permits,

daily raids issue forth from the colony in the form of numerous dark streams of rushing ants that coalesce across a broad front, advancing at about a rate of 20 m/h.

Leafcutting ants (*Atta* spp.). The leafcutter ants are the premier agriculturists of the insect world, culturing their food within their underground nests. They produce huge colonies that likely involve the most complex social systems of any ant, if not of all insects. Fifteen species of leafcutting ants occur from southern Louisiana to Argentina; the Texas leafcutting ant, *Atta texana*, is the only species of leafcutting ant found in the United States.

FIGURE 15-93
Leafcutting ants cutting leaves. Photograph courtesy of Susan Ellis.

FIGURE 15-94
Leafcutting ant carrying leaf fragment back to nest. Photograph courtesy of Scott Bauer/USDA ARS.

As their name indicates, the most visible behavior of these ants is cutting leaves. Their sharp jaws act as scissors to cut large pieces of leaves from foliage. Workers then return to the colony carrying the pieces of leaves over their backs. Once returned to the colony, other workers receive the leaf pieces from the foragers and cut them into very small fragments. The leaf fragments are, in turn, crushed into a damp paste that is used as the base material on which fungi are grown.

The sole food of leafcutting ants is fungi that grow on this prepared plant paste. Each leafcutting ant has its own associated fungi and these occur nowhere but within the chambers of the fungus gardens cultivated by these species of ants. The task of growing the fungus requires careful tending, with tiny workers acting as gardeners, spreading tufts of the desired fungus onto new plant paste. Other undesirable species of fungi are weeded out during the ants' constant patrols through the crop of fungi. As needed, the fungus is harvested and carried off to feed the developing young, the queen, and the workers.

The proper functioning of the leafcutting ant colony requires many different types of tasks and, as a result, different worker castes are produced for each. The largest are the soldiers (also known as majors) with huge heads and a body size that may be 300× that of the tiniest workers (minims) whose function is to tend the fungus gardens. Ants of different sizes patrol the nest area or cut leaves. One of the smaller

castes, the minors, may travel out of the nest with the workers involved in leaf cutting and return with them, riding on the cut leaf to ward off parasitic flies.

Colonies of these species can be tremendous, including millions of workers, and the founding queen may live for over a decade. Foraging columns seeking the leaves needed by the colony may travel 100 m or more. To maintain the colony, workers must continuously bring in new leaf fragments; consequently, these ants can have a tremendous impact on local plant life. A large colony may remove as much leaf matter in a day as would a cow, and when leafcutting ants invade agricultural areas they can be serious pests.

FIGURE 15-96
An excavation of a Texas leafcutting ant colony with the primary chambers preserved. Photograph courtesy of John Moser/USDA Forest Service/Bugwood.org.

Although in some situations their impacts are destructive to human activities, the leafcutting ants are species of critical ecosystem importance in most of their distribution. Their extensive tunneling and incorporation of organic matter has large beneficial effects with regard to soil creation, and they help greatly in accelerating the recycling of nutrients.

FIGURE 15-95
Fungus garden being tended by Texas leafcutting ants, *Atta texana*. Photograph courtesy of John Moser/USDA Forest Service/Bugwood.org.

Scale-Winged Beauties and Custom Homebuilders

Moths, Butterflies, and Caddisflies

The orders Lepidoptera (moths and butterflies) and Trichoptera (caddisflies) share many physical features indicating that they are closely related orders of insects. Larval stages of both orders produce silk from labial glands associated with the mouthparts and larval mouthparts are of chewing type. Wing structures are generally similar and the wings in both the orders are covered with setae, although these are hairlike in caddisflies while they are flattened into scales on the moths and butterflies. Greater differences develop when examining life histories. Caddisfly larvae live a life in water, where they may be scavengers or predators. Moths and butterflies are almost exclusively found on land, where their larvae (caterpillars) feed on plants with few exceptions.

Order Lepidoptera—Butterflies, Moths, and Skippers

Entomology etymology. The name Lepidoptera is derived from the Greek meaning "scale wing," referring to the scales that cover the wings of the adults. Collectively, members of this order are sometimes referred to as lepidopterans.

The order Lepidoptera is one of the most familiar and commonly encountered of all the insect orders, the moths and butterflies. It is also one of the largest orders of insects, with about 150,000 species currently described. Perhaps the most defining physical feature of the order is the presence of scales that cover the wings, yielding the patterning and colors that distinguish so many butterflies and moths. The scales dislodge easily when brushed, helping these insects slip from webs of spiders and aid escape from birds, lizards, or other predators.

FIGURE 16-1
Close-up of a moth densely clothed in scales.
Photograph courtesy of Brian Valentine.

FIGURE 16-2
The bright coloration of butterflies, such as this twotailed swallowtail, is due to the scales that cover the body. Photograph by Whitney Cranshaw/Colorado State University.

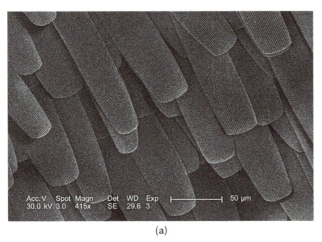

(a)

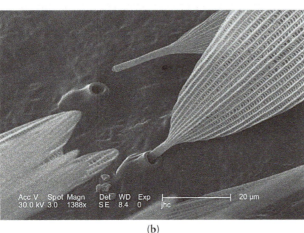

(b)

FIGURE 16-3
Electron micrographs of scales covering the body of a moth. Photographs courtesy of the Centers for Disease Control and Prevention.

Lepidoptera undergo complete metamorphosis. The larvae are known as caterpillars and they are distinguishable from other insect larvae by the presence of fleshy **prolegs** that protrude from the abdomen and are used in crawling. Five pairs of prolegs are typical, although the inchworms possess only two. At the tip of each proleg are a series of tiny hooks, known as **crochets**. With the exception of the plant-feeding wasp larvae known as sawflies, the larvae of Lepidoptera are the only plant-feeding insects with prolegs. Pupae are infrequently seen as they are usually well camouflaged or occur belowground, and many of the moths that pupate above ground cover their pupa with silk, producing a protective cocoon. All caterpillars possess labial glands capable of silk production, and silk may be used for other purposes (e.g., production of feeding shelters, resting mats during pupation) even among those that do not construct cocoons.

Caterpillars have chewing mouthparts, and almost all Lepidoptera develop by feeding on plants. (A small number are scavengers, and there are even a

(a)

(b)

FIGURE 16-4
(a) Full grown larva of an achemon sphinx, *Eumorpha achemon*, a species with five pairs of prolegs. Photograph by Whitney Cranshaw/Colorado State University. (b) A onespotted variant looper, *Hypagyrtis unipunctata*, a caterpillar with only two pairs of prolegs. Photograph courtesy of Tom Murray.

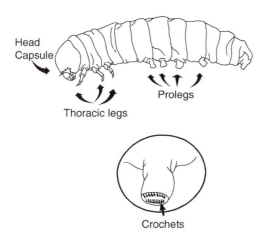

FIGURE 16-5
Illustration of abdominal prolegs on a caterpillar and the location of the crochets. Drawing by Matt Leatherman.

FIGURE 16-6
Ventral side of a Zimmerman pine moth caterpillar, *Dioryctria zimmermani*. The crochets on the abdominal prolegs are in circular pattern with members of this family (Pyralidae). Photograph courtesy of Jim Kalisch/University of Nebraska.

couple of species that are predators.) Included within this order are some species that are among the most important pests of gardens, forests, and crops such as the various armyworms, cutworms, budworms, hornworms, carpenterworms, and clearwing borers.

The habits of the adult butterfly or moth differ greatly from those of the juvenile caterpillar. Adult mouthparts have evolved into a simple tube, derived from the maxillae. Mouthparts are designed to suck fluids such as nectar from flowers. Many moths do not feed at all, instead living off the stored fat they acquired during their caterpillar stage.

Butterflies and colorful moths are arguably the most enchanting and widely appreciated of all insects. However, some people do not share such enjoyment of these insects. Fear of flying moths (and less commonly butterflies) is a common

phobia (mottephobia), though it is less widespread than fear of spiders (arachnophobia).

BUTTERFLY OR MOTH?

In popular discussion the order Lepidoptera is typically neatly divided into the butterflies and the moths, with the skippers sometimes added on as an afterthought. As more has been learned about this order and the features of its individual species, such simple divisions have become more difficult to defend. Basically, all should just be considered as lepidopterans, with the butterflies merely being considered to be "fancy moths." Nonetheless, in the general public perception of the Lepidoptera, the distinctions do continue and likely always will. Table 16-1 summarizes the features that are associated with insects commonly known as butterflies, skippers, and moths.

(a)

(b)

FIGURE 16-7
Life stages of the alfalfa caterpillar, *Colias eurytheme*: (a) adult; (b) larva; (c) pupa. Photographs of the larva and pupa courtesy of Ken Gray/Oregon State University. Photograph of the adult by Whitney Cranshaw/Colorado State University.

FIGURE 16-7
(*continued*)

(c)

(a)   (b)   (c)

FIGURE 16-8
Life stages of the tobacco budworm, *Heliothis virescens*: (a) adult; (b) larva; (c) pupa. Photographs by Whitney Cranshaw/Colorado State University.

(a)   (b)   (c)

FIGURE 16-9
Life stages of the silverspotted skipper, *Epargyreus clarus*: (a) adult; (b) larva; (c) pupa. Photographs courtesy of Jim Kalisch/University of Nebraska, David Cappaert/Michigan State University/Bugwood.org, and Sturgis McKeever/Georgia Southern University/Bugwood. org, respectively.

TABLE 16-1 Generalized differences between the Lepidoptera commonly known as butterflies, skippers, or moths.

	BUTTERFLIES	SKIPPERS	MOTHS
Families	Papilionidae, Pieridae, Lycaenidae, Nymphalidae	Hesperiidae	All remaining families of Lepidoptera
Coloration	Brightly colored and often conspicuously marked. The upper wing surface sometimes is more brightly colored than is the lower surface.	Usually brightly colored	Forewings are usually dull-colored and the overall appearance of most is dull and inconspicuous. Hindwings may be brightly colored. A few of the day-flying species may have bright coloration and/or mimic stinging insects.
Wing folding	When at rest the wings project from the back and do not fold over the body.	When at rest the wings project from the back and do not fold over the body.	Wings fold over the abdomen when the insect is resting.
Activity period	Day active adults	Day active adults	Most moths are nocturnal, with peak activity at dusk and early evening. A few moths fly during the day and these usually are mimics of wasps or bees.
Antennae	The antennae are elongated with a thickened tip described as clubbed.	The antennae are elongated with a thickened, clubbed tip ending with a hook.	Variable. Some have long, thin antennae, others have clubbed or plume-like antennae.
Larvae	Many are covered with short, very fine hairs giving them a somewhat felty appearance. A few have fleshy projections and others are naked without hairs.	Smooth-bodied with an unusual enlarged round head	Variable. Some are smooth bodied but many have prominent hairs and some are completely covered with hairs.
Pupae	Naked and never covered with a cocoon. The pupa may suspend downward from the tip of the abdomen or be held upright with a narrow swath of girdling silk. A butterfly pupa is known as a **chrysalis** or **chrysalid**.	Produced within a cocoon that is often among leaves and woven together with silk	Species that pupate above ground surround the pupa within a silken cocoon. Those that pupate belowground often do not produce a cocoon.

● Butterfly Etymology

The current English term "butterfly" derives from the Old English *buttorfleoge* that subsequently morphed in Middle English to *butterflie*. But where this all originated is subject to a bit of debate. One suggestion is that the original term refers to the cream-colored excrement (known as the **meconium**) excreted by some butterflies as they emerge from the pupal stage. German folk legends associate butterflies with milk-stealing witches.

A more prosaic/pleasant explanation involves the white-colored butterflies known as the cabbage whites. Two species are common in England, the large cabbage white (*Pieris brassicae*) and the small cabbage white (*Pieris rapae*), the latter now a commonly seen introduced species in North American gardens. These butterflies emerge in early spring, coincident with the time that cows began to produce milk again. (Formerly, milk cows ceased production during the winter months when food was scarce, resuming in the spring with calving.) The butterflies, white or cream colored, were an annual sign of the resumption of fresh milk and butter.

Butterflies have always attracted attention and interest so almost all languages likely have a special word for these. Some of these are summarized in table 16-2.

(a)

(b)

FIGURE 16-10
The cabbage butterfly (a), *Pieris rapae*, is one of the first butterflies seen in spring. One story of the origin of the name "butterfly" attributes the name to the coincidence of this cream-colored butterfly with the time when cows resumed milk production. The larval stage (b), is a common insect found feeding on cabbage family plants in gardens and is known in North America as the imported cabbageworm. Photograph of the adult courtesy of Brian Valentine. Photograph of the larvae by Whitney Cranshaw/Colorado State University.

TABLE 16-2 Terms used to describe butterflies from around the world.

LANGUAGE	TERM FOR BUTTERFLY	NOTES
Ancient Greek	psyche	Meaning "soul" or "breath." Now associated with "mind."
Modern Greek	petaloudia	Related to words "petal," "leaf," and "spreading out"
Chinese	hu die	
French	papillon	
German	schmetterling	Derived from the Czech word smetana for cream. This refers to the observation of butterflies hovering about milk pails and butter churns. Folk legend had them associated with witches that steal milk.

(continued)

LANGUAGE	TERM FOR BUTTERFLY	NOTES
Italian	farfalla	Also means "bow tie"
Japanese	chou chou	
Portuguese	borboleta	
Russian	babochka	Diminutive of baba or babaka meaning woman or grandmother
Spanish	mariposa	From "la santa maria posa" referring to the Virgin Mary alighting or resting
Norwegian	summerfugle	Summer fly
Yiddish	zomerfeygele	Summer bird

FAMILY NYMPHALIDAE—BRUSHFOOTED BUTTERFLIES

The brushfooted butterflies are moderately large butterflies with wings that are usually brightly colored and patterned. They derive their common name (brushfooted) from the fact that their front legs are greatly reduced in size, held by the head, and not used for walking. Brushfooted butterflies are common visitors to flowers though many have broader tastes that run to fermented fruit, ooze from plants, and even fresh animal dung. It is a large family (over 6,000 known species worldwide and over 200 in North America) including butterflies known as checkerspots, crescents, fritillaries, milkweed butterflies, ladies, admirals, and emperors.

The painted lady (*Vanessa cardui*) is one species that many school children have encountered, as rearing one of these is now almost a rite of passage in elementary school classes learning the concepts of metamorphosis. The painted lady is one of the most common butterflies of the world and the most widely distributed, found on every continent except Australia and Antarctica. The adults are brightly colored, with orange, black, and white markings that sometimes cause them to be mistaken for the monarch butterfly. The caterpillars, often known as "thistle caterpillars," feed on a wide variety of plants: thistles, mallows, some legumes, and sunflower.

(Artificial diets are available and used for most school projects.)

Painted lady butterflies are almost constantly in motion and migrate extensively. During winter, they vacate most of the United States, remaining active only in areas of Mexico and some bordering areas of southwestern states. In late spring, they move northward as host plants emerge and become available in spring. The size of these migrations varies tremendously from year to year and is most dependent on the occurrence of spring rains in their overwintering areas. When this moisture occurs, painted lady populations can explode, and it may become the most common butterfly over extensive areas of North America in the following summer.

The milkweed butterflies are also members of the Nymphalidae and are best represented in the United States by the queen (*Danaus gilippus*) and the monarch (*Danaus plexippus*). The latter is undoubtedly the best known of all US butterflies with flashy colors and an unusual life history. Found in all 50 states (it has been present in Hawaii since about 1850) the monarch has been promoted by the Entomological Society of America as the candidate for a national insect.

Monarch caterpillars develop almost exclusively on the leaves of milkweed plants (*Asclepias* spp.). This plant genus is poisonous to

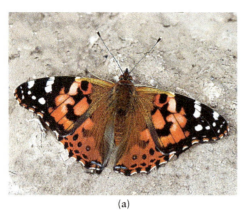

(a)   (b)   (c)

FIGURE 16-11
Life stages of the painted lady, *Vanessa cardui*: (a) adult; (b) larva; (c) pupa. Photographs by Whitney Cranshaw/Colorado State University.

FIGURE 16-12
Monarch butterfly, *Danaus plexippus*. Photograph courtesy of David Cappaert/Michigan State University/Bugwood.org.

FIGURE 16-13
Caterpillar of the monarch butterfly. The larval stage of this insect feeds on milkweeds, a plant that is toxic to a great many species. Toxins in the plant are stored in the body of the monarch and these render them distasteful to many potential predators, such as birds. As a result both the larvae and adults are brightly colored, a form of aposematic coloration. This advertises they are chemically defended. Photograph courtesy of Jim Kalisch/University of Nebraska.

almost all other animals, containing powerful chemicals known as cardiac glycosides that readily induce vomiting, among other effects. Not only does the monarch caterpillar feed and thrive on a diet of milkweed, it also converts the plant's own defenses to its own advantage. The cardiac glycosides are sequestered within the body of the caterpillar, making it exceptionally unpalatable, a feature it advertises with bright striping and warning coloration. The adult monarch is also colored brightly, orange and black, and it too contains high levels of toxic cardiac glycosides acquired as a juvenile.

Such bright warning colors, known as **aposematic coloration**, are found also in other insects that similarly can defend themselves through stings, poisonous hairs, toxins, or other means. Some color patterns in particular are used repeatedly, notably yellow/black and orange/black. When a wide range of insect species all use these same warning

signal colors/patterns this is known as **Mullerian mimicry**. Inexperienced predators soon associate such colors and patterns with bad outcomes and usually learn to avoid all insects with similar warning signals.

In some areas a butterfly of extremely close resemblance to the monarch, the **viceroy** (*Limenitis archippus*), may also be present. Unlike the monarch, caterpillars of the viceroy have a very different appearance, resembling a bird dropping. These larvae feed on willow, poplars, and cottonwood, plants that lack the toxins found in milkweeds. It has long been proposed that the viceroy, lacking its own defenses, uses its close resemblance to the monarch as a means to similarly avoid predation. Such mimicry of a

FIGURE 16-14
Comparison of (top) a viceroy butterfly (*Limenitis archippus*)
and (bottom) a monarch butterfly (*Danaus plexippus*).
The viceroy is a close mimic of the monarch that is usually
more abundant and well defended chemically, so few predators
attack it. Photograph by Whitney Cranshaw/Colorado
State University.

FIGURE 16-15
Larva of the viceroy. This species feeds on the foliage of a variety
of trees, notably willows, and thus does not contain the toxins
found in milkweeds. The larvae have an appearance somewhat
similar to a bird dropping, which allows camouflage.
The chrysalid of the viceroy also resembles a bird dropping.
Photograph courtesy of Sturgis McKeever/Georgia Southern
University/Bugwood.org.

defended/poisonous species by one that lacks such
defenses is known as **Batesian mimicry** and such mimicry
is widespread in many insect groups, notably among
some common groups of flies. (It has subsequently
been determined that some viceroy butterflies do
possess chemical defenses of their own, which can
further reinforce deterrence by potential predators.)

(a)

(b)

FIGURE 16-16
Batesian mimicry is widespread in the insect world, where a
harmless insect closely mimics a species that is well defended.
For example, the top picture (a) is of a flower fly (Syrphidae
family) that can neither sting, bite, nor has any chemical
defenses. It is a mimic of a yellowjacket (b), which readily stings.
Photograph of the flower fly courtesy of Tom Murray.
Photograph of the western yellowjacket by Whitney Cranshaw/
Colorado State University.

● The Amazing Migration of the Monarch

One of the most spectacular of all animal migrations is annually produced by the monarch butterfly (*Danaus plexippus*). During the summer, this insect is a common resident throughout much of the United States and southern Canada, present where there are growing milkweeds (the larval host plant). Found in all 50 states, the monarch has even been nominated by the Entomological Society of America as a candidate for a national insect.

During late summer and early fall, the monarch flees the northern and eastern areas of the United States and Canada for warmer southern areas. A tiny area in central Mexico, a forested area of fir trees at an altitude of above 3000 m, is the ultimate goal for the monarch populations that develop east of the Rocky Mountains. (The overwintering area is located along the border of the states of Michoacan and Mexico.) Monarch populations west of the Rocky Mountains move southward from the Pacific Northwest to coastal areas of southern and central California and northwest Mexico. About 25 sites are annually used by monarchs for winter refuge along the Pacific coast, with those around Pacific Grove, California being the best known.

The adults remain in their southern overwintering sites throughout winter, packed densely on trees in numbers sometimes approaching four million per acre. They occasionally drink and feed on nectar if available; however, they remain in a semi-dormant state. Only in late winter do they begin to resume normal activity, mating and then dispersing back northward to catch the first milkweed plants as they resume growth. Developing through several generations in spring and summer, northward migrations continue through the summer and may reach sites many thousands of miles from their winter quarters.

The approaching autumn produces changes that include aging milkweeds (declining host plant quality), cooler temperatures, and shorter day length. These cues trigger behavioral change, and it may be the great grandchildren of the spring migrants that then begin making the return migration to the southern winter roosts, typically arriving during early November.

(a)

(b)

FIGURE 16-17
(a and b) Images from the winter roosting area of the monarch in Mexico. Photographs courtesy of Bonnie Head.

FAMILY PAPILIONIDAE—SWALLOWTAILS AND PARNASSIANS

The swallowtails are large, often beautifully colored butterflies represented by over 600 known species worldwide. They are more abundant in tropical areas, and of the two dozen species reported from the United States, several are only occasional migrants originating from Mexico and Central America. All North American species of swallowtails have one or two projections off the hindwings, the "swallow tails" (resembling tails of the swallow birds); this feature is not present in many tropical species. Among the tropical species are the largest of all butterflies, the giant birdwings of Australia and southeast Asia. One rare species, Queen Alexandra's birdwing (*Ornithoptera alexandrae*), found in a small region of Papua New Guinea, has a wingspan that can exceed 25 cm. The popularity of swallowtail butterflies in North America is reflected in that three states (AZ, OR, VA) have proclaimed some species of swallowtail as the State Insect, and seven others have established a swallowtail as a State Butterfly (Appendix I).

(a)

(b)

FIGURE 16-19
An early stage (a) and late stage (b) larva of the twotailed swallowtail. The small young larva have an appearance of a bird dropping which helps them to camouflage among leaves of the trees on which they feed. The large late-stage larvae have different markings, including prominent eyespots behind the head that may be used in startle displays to deter predators.

FIGURE 16-18
A twotailed swallowtail butterfly, *Papilio multicaudatus*. Photograph by Whitney Cranshaw.

Swallowtails use a wide array of defenses against predators. Young caterpillars on leaves often closely resemble bird droppings, an excellent camouflage. Older caterpillars, too large to convincingly use this subterfuge, employ warning coloration or have large eyespots on the thorax. Their protection is further amplified by an eversible Y-shaped scent gland (**osmeterium**) that they can extend, releasing a noxious odor when disturbed. The overall effect of the bright markings, the suddenly protruded osmeterium, and its associated odors apparently give pause to potential predators.

Swallowtail pupae, in the form of a chrysalis, are usually attached at the base to a solid surface by a Velcro-like structure known as the **cremaster**. A thin silken girdle around the middle of the body holds the pupa upright. Pupae are capable of color change and

often blend very effectively against the background of bark or rocks where they occur.

Caterpillars and adults of species that develop on plants in the family Aristolochiaceae often are distasteful to birds and other predators because of chemicals (notably aristolochic acid) acquired from their larval food plants. For example, the pipevine swallowtail (*Battus philenor*) of eastern North America is a toxic species; it incorporates aristolochic acid from the various species of *Aristolochia* on which it feeds. Many other species of swallowtails do not develop on toxic plants and thus are palatable. The eastern tiger swallowtail (*Papilio glaucus*) is a palatable species, and a dark form of the females (but not the males) closely mimic the pipevine swallowtail butterflies where the two species overlap in range. This would be a form of Batesian mimicry.

(a)   (b)

FIGURE 16-20

Larva of (a) an eastern tiger swallowtail, *Papilio glaucus*, and (b) a black swallowtail, *Papilio polyxenes asterius*, shown everting the osmeterium. This is a Y-shaped gland that can be everted from an area behind the head. A strong scent is emitted when the gland is displayed. Photographs courtesy of Howard Ensign Evans and Susan Ellis, respectively.

FIGURE 16-21

Pupal stage (chrysalis) of the twotailed swallowtail. The coloration of the pupa blends with its background. Photograph by Whitney Cranshaw/Colorado State University.

Among other butterflies that range over a wide geographic area, mimicry patterns may change so that individuals resemble locally abundant toxic species. This is well demonstrated with the African swallowtail *Papilio dardanus* that occurs widely throughout the southern half of Africa, on Madagascar, and some other nearby islands. Females (but not males) of *Papilio dardanus* mimic various species of butterflies (species of *Danaus* and *Amauris*) that develop on toxic milkweeds. Throughout the range of *P. dardanus* dozens of different color pattern morphs occur, each mimicking some local species of toxic *Danaus* or *Amauris* butterfly.

(a)

(b)

FIGURE 16-22

Pipevine swallowtail (a), *Papilio troilus*, and the dark form of the eastern tiger swallowtail (b), *Papilio glaucus*. The pipevine swallowtail is chemically defended because the larvae feed on toxic plants of the genus *Aristolochia*. The eastern tiger swallowtail is not chemically defended and mimics the pipevine swallowtail where the two species overlap in range. Photographs courtesy of Sturgis McKeever/Georgia Southern University/Bugwood.org.

FAMILY NOCTUIDAE—CUTWORMS, ARMYWORMS, AND LOOPERS

The family Noctuidae is the largest within the order Lepidoptera, containing over 2,900 described species in North America alone. Adults of most are heavy-bodied moths of moderate size, usually marked inconspicuously with various shades of brown or gray.

Noctuid moths fly during dusk and early evening. As is with some other night-active moths, they possess tympana located on the metathorax that can detect ultrasonic sound frequencies. This ability allows them to help detect and evade bats, which use these frequencies as echo-locating sonar to detect their flying prey. In response to these signals, the moths will make erratic, evasive flights to avoid bat predators. (A similar bat-avoidance behavior is seen in the mantids.) Jingling keys or coins can usually produce ultrasonic frequencies to which noctuid moths also will respond.

The caterpillars are smooth bodied, and most are dull colored. They are usually active at night and hide about plants during the day. A few, including some climbing cutworms and loopers, remain on their host plants as they develop. Included among the noctuids are the most damaging species of caterpillars affecting US agriculture.

TABLE 16-3 Some noctuid caterpillars that are particularly damaging to crops in North America.

COMMON NAME	SCIENTIFIC NAME	INJURIES
Armyworm	*Mythimna unipuncta*	Primarily associated with grain crops, the caterpillars feed on leaves. During outbreaks caterpillars may migrate en masse in search of new food.

(a)

(b)

(c)

FIGURE 16-23

(a–c) Larvae of the black cutworm, *Agrotis ipsilon*, fall armyworm, *Spodoptera frugiperda*, and corn earworm, *Helicoverpa zea*. These are noctuid larvae among the most important caterpillar pests of crops in North America. Photographs courtesy of Clemson University, Jim Kalisch, and Whitney Cranshaw, respectively.

COMMON NAME	SCIENTIFIC NAME	INJURIES
Army cutworm	*Euxoa auxiliaris*	The most common cutworm in the high plains/Rocky Mountain region and particularly damaging to alfalfa, winter wheat, and garden seedlings in spring. Adults make annual migrations and locally are known as "miller moths."
Beet armyworm	*Spodoptera exigua*	A southern species that lays eggs as masses and feeds on many different vegetable crops. It has often proved to be one of the more difficult species to control because of repeated development of resistance to various insecticides.

(continued)

TABLE 16-3

COMMON NAME	SCIENTIFIC NAME	INJURIES

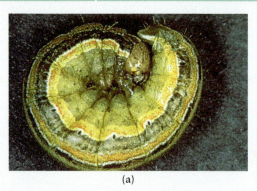

FIGURE 16-24
(a and b) Larva of the armyworm, *Mythimna unipuncta*, and damage it produced in a corn crop. The armyworm is one of several cutworm larvae that may migrate in bands during outbreaks. Photographs courtesy of Frank Peairs and R.L. Croissant/Colorado State University, respectively.

Cabbage looper	*Trichoplusia ni*	The most common looper in gardens, it feeds on vegetable and field crops as diverse as potato, cotton, peas—as well as cabbage family plants.
Corn earworm	*Helicoverpa zea*	Found throughout the United States, this climbing cutworm feeds on a wide variety of vegetable and field crop plants. It primarily damages fruiting structures and is known by three formal common names—corn earworm, tomato fruitworm, and (cotton) bollworm.

FIGURE 16-25
(a) Adults of an armyworm, *Mythimna unipuncta*, and (b) fall armyworm, *Spodoptera frugiperda*. Photographs courtesy of Whitney Cranshaw and Jim Kalisch, respectively.

Fall armyworm	*Spodoptera frugiperda*	A common caterpillar of corn, turfgrass, and grains in the southern states.

(*continued*)

TABLE 16-3

COMMON NAME	SCIENTIFIC NAME	INJURIES
Tobacco budworm	*Heliothis virescens*	A climbing cutworm that primarily feeds on flower buds and young fruit it is most damaging to tobacco, cotton, and a wide variety of flowering ornamental plants in the nightshade family.
Western bean cutworm	*Striacosta albicosta*	Primarily damaging to corn ears. Recently expanding range into eastern North America.

● Fatal Attraction—Moths to a Flame

Almost anyone who has watched as a moth spirals toward a light bulb, or even an open flame, wondered what is the reason for this obviously destructive behavior. Perhaps one's feelings on this may be summarized as stated in a Japanese haiku (author unknown):

> Frail and lonely moth,
> Seeking warmth in candle's flame.
> Goddamn idiot!

The problem arises because of humans' short-circuiting a navigation system that moths have likely used for tens of millions of years. Flying at night, they use the light of the moon, and perhaps the stars, as well as the difference in light intensity between the dark ground and the relatively brighter night sky to guide them. By keeping a constant angle to these light sources they are able to maintain a direct flight at night.

In very recent history, close light sources have become abundant in the form of street lights, campfires, candles, and other nighttime illumination. Fixing on these, as they would the moon or stars, the moths find that their angle of flight in relation to the light is constantly and rapidly changing. Correcting for this, the flight path is also constantly changed, resulting in a spiraling inward flight that ultimately results in contact with the flame.

Certain wavelengths of light are detected by night-flying moths, while other wavelengths are not, and this phenomenon can be exploited. Deep ultraviolet light is often particularly attractive to these animals, and entomologists use such "black lights" to improve the capture of night-flying insects that they are trying to sample. Alternately, red and yellow lights have little attraction to most night-flying insects. Sometimes incorrectly sold as "insect repellent" lights, they instead are simply not insect attractive.

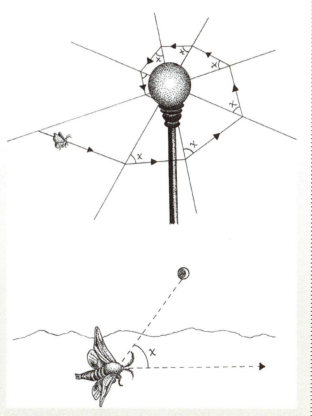

FIGURE 16-26

Night-flying moths, and other insects, are drawn to nighttime lights, often with fatal results. The reason for this behavior is that close points of lights, such as a candle's flame, are poorly handled by these insects that normally rely for navigation the distant points of light provided by stars and the moon. The distant celestial lights allow the insects to keep a constant angle to the light, but a terrestrial light requires the insect to constant correct direction cause it to spiral toward the light source. Drawing by Matt Leatherman.

● Moths on Vacation—Migrating Moths

When large numbers of butterflies migrate it is often an event of note that attracts wonder and appreciation. Many moths also migrate though their activities frequently go undetected. In part, this is due to the night-flying behavior of most moths and their (usually) dull colors. But moth migrations can be equally spectacular.

Some moths migrate in search of new foods or when favorable weather allows them to temporarily extend their range. For example, the black cutworm (*Agrotis ipsilon*) survives winters in the United States only in warm areas along the Gulf Coast and Florida, but annually this species disperses northward every spring, with some individuals reaching Canada. Reverse migrations occur in the late summer. From Mexico to Ecuador, the ducktail moth (*Urania fulgens*) makes annual across-land migrations from the Pacific side of the country to the Gulf of Mexico. The triggers for the migrating populations of the ducktail moth appear to be changes in palatability of host plants that grow in coastal areas.

FIGURE 16-27
The black cutworm, *Agrotis ipsilon*, is a migratory moth that regularly flies into the United States and Canada from areas where it survives winter in Mexico and Central America. Photograph courtesy of Jim Kalisch/University of Nebraska.

The best known moth migrations have evolved to avoid periods of hot, dry weather—a summer vacation of sorts. Over a broad area of the high plains and Rocky Mountain region, the army cutworm (*Euxoa auxiliaris*) makes an annual late spring flight to the mountains from the plains. These flights can attract great attention and much annoyance as the moths settle into various cracks to hide during the day. This behavior incidentally causes them to appear in high numbers in doors, cars,

and houses. During the summer the moths, known locally as the "miller moth," are in reproductive diapause, not laying eggs but getting fat by feeding on plant nectar interspersed with long rests under cool rocks. Only in late summer and early fall do they take flight to return to the lower elevation fields to lay eggs.

FIGURE 16-28
The army cutworm, *Euxoa auxiliaris*, is a moth that migrates from the Plains to the mountains in spring. The migrations of this insect often are large and cross major population centers that line eastern range of the Rocky Mountains. During these migrations many moths enter homes and automobiles and are popularly referred to as a type of "miller moth." Photograph by Whitney Cranshaw/Colorado State University.

Half a world away, almost the exact same type of migration occurs with the bogong moth (*Agrotis infusa*). Hot temperatures of summer cause them to leave the pastures and lowland areas of New South Wales and Queensland, Australia where they develop as a common cutworm caterpillar. Mass springtime migrations (September through November) of the adults send them to the Bogong High Plains of Victoria and the Snowy Mountains of New South Wales where they remain dormant, clustered in caves and under rocks. The annual flight path to and from the high country crosses Canberra, Sydney, and many other populated areas where they can be a serious, but temporary, nuisance. The bogong moth came to some international attention when a good-sized flight coincided with the opening ceremonies of the 2000 Olympics held in Sydney.

More limited migrations occur in the eastern Mediterranean and involve the Jersey tiger moth, *Euplagia quadripunctaria*. The heat of the summer drives many of these insects from the mainland to seek temporary lodgings in wooded valleys located on several

(continued)

of the islands off the coasts of Greece and Turkey. These are brightly patterned, colorful moths, and their massing is a tourist attraction on the island of Rhodes during July and August. Despite the considerable attributes of the Jersey tiger moths, local promoters of this event tout it as the "Valley of Butterflies," presuming this to be more attractive verbiage than the truthful "Valley of Moths."

FAMILY LYMANTRIIDAE—TUSSOCK MOTHS AND RELATIVES

Moths in the family Lymantriidae tend to be fairly plain and dull in color, of medium size, and many resemble cutworm moth adults. The caterpillars are hairy, and those that have dense tufts of hairs in patches of their body are referred to as tussock moths. All of the 32 described species of North American lymantriids feed on the leaves of trees and shrubs. Several species are important pests that were introduced originally from Europe or Asia.

Among the native species that are more notable are the whitemarked tussock moth (*Orgyia leucostigma*), a common shade tree pest of the midwest, and the Douglas-fir tussock moth (*Orygia pseudotsugae*), which sometimes explodes in numbers and can be the most important defoliating insects of forests in the northwestern United States and British Columbia. The satin moth (*Leucoma salicis*) was introduced from Europe into both New England and British Columbia around 1920, and has recently emerged as an important new pest in Wyoming and Montana where it feeds on aspen, poplars, and cottonwoods.

FIGURE 16-29
Larva of the whitemarked tussock moth, *Orygia leucostigma*. Tussock moth caterpillars get their name from the bushy tufts of hairs on their body. Photograph courtesy of David Cappaert/ Michigan State University/ Bugwood.org.

The most notorious member of this family is the gypsy moth (*Lymantria dispar*). Unlike many insects that originated elsewhere and accidentally became established in the North America, the gypsy moth was purposely brought here. In what ranks high amongst the all-time most ill-conceived entomological ideas, Leopold Trouvelot brought the gypsy moth to Medford, Massachusetts in 1866 in hope of using it for breeding a silk-producing species adapted to the Northeast. (Among other problems with this original concept is that the gypsy moth produces very little silk to support its largely naked pupa.)

Some of the gypsy moth caterpillars escaped into the nearby woodlands and soon established themselves. Within a decade, it was clearly becoming a serious forest insect that has a taste for the leaves of almost every deciduous tree in the northeast United States. By the late 1800s, the burgeoning populations of gypsy moths finally started to attract enough attention that efforts were attempted to control it. But it was far too late, particularly with the very limited insect control options of that era. Over the next century, the gypsy moth became one of the most important forest insects east of the Mississippi.

Since the original introduction into Massachusetts, the gypsy moth has steadily increased its distribution across the continent and currently the leading edge occurs in Wisconsin to the west and Georgia to the south. It has always been slowed somewhat by a central fact of its biology; the female, although winged, is incapable of flight. As a result, dispersal to new areas largely occurs by very young caterpillars that are light-weight but with long hairs; they can disperse to considerable distances being carried by wind. When caterpillars are fully developed, they usually wander and pupate a short distance away from the trees and shrubs on which they earlier fed. On later emergence from the pupa, gypsy moth females lay their eggs very close to the pupation site. If pupation and subsequent egg laying occurred on a trailer, nursery plant container, piece of backyard furniture, or other item that is then moved, the gypsy moth can be, as it frequently has been, transported with human aid almost anywhere.

FIGURE 16-30
Late instar gypsy moth caterpillars defoliating an oak tree. This insect was introduced into North America with the ill-conceived idea that it could be bred with silk moths. They escaped and subsequently the gypsy moth spread throughout much of the northeastern quadrant of the United States where it has become a major forest pest. Photograph courtesy of Tim Tigner/Virginia Department of Forestry/Bugwood.org.

FIGURE 16-31
Adults of the gypsy moth, *Lymantria dispar*, a species that shows considerable sexual dimorphism. The smaller, dark brown males can fly, while the larger, white and black speckled females cannot. Photograph courtesy of USDA APHIS PPQ/Bugwood.org.

FIGURE 16-32
Gypsy moth laying a mass of eggs. A typical egg mass contains about 200 eggs and these are covered with the golden scales from the abdomen of the mother. Photograph courtesy of Steven Katovich/USDA Forest Service/Bugwood.org.

A tremendous amount of effort has been expended to help manage gypsy moth populations, and in the process, several insect management techniques have been greatly advanced. The sex pheromones of gypsy moth have been used not only to detect its presence but to attempt to control this pest through mass releases of the chemicals to confuse males and prevent mating (mating disruption). Several novel insecticides, including those that act on insect hormones (insect growth regulators), largely owe their development to the gypsy moth as does the idea that insect pathogens (e.g., the nuclear polyhedrosis virus of gypsy moth) have potential to control insects. Also, a great many

natural enemies of the gypsy moth have been purposefully transported from Europe and established in North America to help control the insect, although with limited success—and some negative impacts on native moths. Somewhat ironically, there finally now is a natural enemy of gypsy moth that seems to be having a very big impact on gypsy moth populations, but it is one that itself was accidentally introduced. It is a fungus (*Entomophthora maimaiga*) that produces a lethal disease in gypsy moth and has proved to greatly reduce their numbers in the past couple of decades, being particularly effective following wet spring weather.

FIGURE 16-33
An early effort to attempt eradication of gypsy. This picture was taken in Malden, Massachusetts. The insect was originally introduced into the nearby town of Medford. Photograph courtesy of the USDA Forest Service Archives/Bugwood.org.

FIGURE 16-34
A virus-killed gypsy moth caterpillar. Caterpillars killed by this virus die in this characteristic position, hanging by the prolegs. The body later ruptures and the virus particles are released. Gypsy moth caterpillars become infected when they feed on leaves contaminated by the virus. Photograph courtesy of Steven Katovich/USDA Forest Service/Bugwood.org.

FAMILY ARCTIIDAE—WOOLLYBEARS/ TIGER MOTHS

The family Arctiidae is closely related to the Noctuidae (some taxonomists now place them within the Noctuidae family), but individual species of arctiids are usually far splashier. Adults often are attractive moths, sometimes with bright colorful patterning. A few that mimic wasps even fly during the day. Its caterpillars are hairy, sometimes very hairy, and may also be brightly colored.

(a)

FIGURE 16-35
A virgin tiger moth, *Grammia virgo*. Many tiger moths have strong banding and bright coloration of the hindwing. Photograph courtesy of Jim Kalisch/University of Nebraska.

(b)

FIGURE 16-36
A yellow woollybear (a), *Spilosoma virginica*; the adult (b) is known as the Virginia tiger moth. Photograph of the adult courtesy of Ken Gray/Oregon State University. Photograph of the larva by Whitney Cranshaw/Colorado State University.

The term "woolly bear" is sometimes applied to the species that are most densely covered with hairs and are commonly seen as they wander about, particularly late in the season. Within the United States there are at least three species with these features that are commonly considered woolly bears: the salt-marsh caterpillar (*Estigmene acrea*), the yellow woollybear (*Spilosoma virginica*), and the banded woollybear (*Pyrrharctia isabella*). The adults are known as tiger moths: the banded woollybear transforms into the Isabella tiger moth; yellow woollybear into the Virginia tiger moth.

Perhaps the most conspicuous member of this insect family in much of North America is the fall webworm, *Hyphantria cunea*. The fall webworm creates large, loose tents of silk in cottonwoods, chokecherry, and many other trees during summer. Fall webworms survive between growing seasons as a pupa within a cocoon in protected sites around the base of previously infested trees, emerging as the egg-laying adult moths in late spring. They breed in areas in the southern part of their range and two generations are produced annually. (Note that the fall webworm is

often mistaken for a "tent caterpillar." The tent caterpillars are members of a different family of moths, Lasiocampidae, and make tightly constructed silken tents in the crotches of limbs in spring.)

One enduring insect legend involving woollybears is that the severity of winter can be predicted by the width of the bands on the body of a banded woollybear (*Pyrrharctia isabella*) caterpillar. As the story goes, the broader the reddish brown center band, the milder will be the upcoming winter. Bands of this species do vary; however, the band width of the caterpillar increases with the age. Rather than reflecting the average temperature of the future winter, it instead indicates how far along in its development it is before going into winter hibernation. A previous warm summer or extended Indian summer during fall will likely produce woollybears with the broadest bands.

(a)   (b)

FIGURE 16-37

The fall webworm (a), *Hyphantria cunea*, feeds on a wide variety of trees and shrubs and is one of the most commonly encountered caterpillars that produces a tent of silk. The adult (b) is a white moth that lays eggs as masses on leaves in late spring and early summer. Photograph of the larvae by Whitney Cranshaw/Colorado State University. Photograph of the adult courtesy of Ken Gray/Oregon State University.

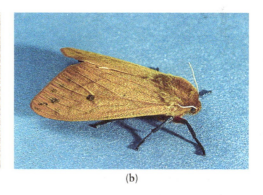

(a)   (b)

FIGURE 16-38

A banded woollybear, *Pyrrharctia isabella* (a). The caterpillars are seen wandering in fall and a common bit of folklore states that the width of the central band is a predictor of the severity of the upcoming winter. The adult (b) is known as the Isabella moth. Photograph of the larva by Whitney Cranshaw/Colorado State University. Photograph of the adult courtesy of Ken Gray/Oregon State University.

Nonetheless, woollybears remain a likable insect of interest. This has been parlayed into the Ohio Woollybear Festival, annually held in early fall in the town of Vermilion, a suburb of Cleveland. Originally started in 1973 as Ohio's answer to the groundhog Punxsutawney Phil, the event has become very popular and in 2006 attracted approximately 100,000 spectators. "Woolly worm" festivals are also held in Banner Elk, North Carolina and Beattyville, Kentucky.

FAMILY TINEIDAE—CLOTHES MOTHS AND RELATIVES

Very few insects have evolved an ability to feed on keratin, the tough protein that comprises much of animal hair and feathers. This ability occurs only among some beetles in the family Dermestidae (carpet beetles) and with caterpillars in the moth family Tineidae, the clothes moths. Woolen materials and furs may be extensively damaged by these insects. In natural systems, they are important macrodecomposers assisting in the breakdown of carrion.

In much of North America, the webbing clothes moth (*Tineola bisselliella*) and the casemaking clothes moth (*Tinea pellionella*) are the most common insects that damage fabrics. Well-cleaned fabrics are poor foods and fabrics are much more likely to be damaged if soiled by greases or other materials that can provide B vitamins and other needed nutrients. Raw wool is particularly favorable as food for clothes moth larvae.

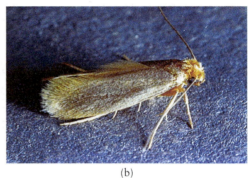

(a)   (b)

FIGURE 16-39

(a) Larvae of the casemaking clothes moth, *Tinea pellionella*. The larvae feed on wool, furs, and feathers, and are one of the few insects that can digest keratin, a major constituent of these animal fibers. Photograph by Whitney Cranshaw/Colorado State University.
(b) Adult of the webbing clothes moth, *Tineola bisselliella*. Photograph courtesy of Ken Gray/Oregon State University.

Larvae of both species produce webbing around their body, and in the latter species, this is formed into a tough silken case that is carried around as they feed. They are adapted to a totally dry diet and never take a drop of water. Instead, they extract what moisture they need from their food. Waste is converted into dry uric acid crystals, in a manner similar to other insects that do not have suitable access to free water.

Of course the adult moths do not feed on these materials; indeed adult clothes moths apparently do not feed at all. Clothes moths also are infrequently seen since they prefer darkened spaces and make only short flights, often preferring to crawl.

FAMILIES PYRALIDAE AND CRAMBIDAE—SNOUT MOTHS AND GRASS MOTHS

The snout moths and grass moths are generally small, rather delicate moths often marked with distinctive palps projecting from the base of the head giving them the appearance of the presence of a "snout." Together, between these two closely related families, there are over 1,375 described North American species with diverse life styles. They can be found in all manner of terrestrial environments.

Included in the Pyralidae are many of the "pantry pests" that feed on grains, seeds, dried fruits, and other stored foods. Among the more important species are the Angoumois grain moth (*Sitotroga cerealella*), Mediterranean flour moth (*Anagasta kuehniella*), almond moth (*Ephestia cautella*), and the Indianmeal moth (*Plodia interpunctella*). Some of these are normally found associated only with warehouses and grain storages, but the Indianmeal moth is well adapted to human households and is the most abundant and frequently encountered moth found breeding within homes.

Adult Indianmeal moths are rather small moths with a wingspan of about 16 mm. They can usually be distinguished from the myriad other "little brown moths" by having the distal half of the forewings substantially darker than the base. These insects may be seen flying around the home, particularly around dusk and during the night. Females lay eggs in areas where food is available and the larvae can develop on a wide variety of stored foods. Among the more common items that may be infested are seeds and nuts, dried fruits, vegetables or herbs, chocolate, crackers and other grain products, and dried pet foods. They also may feed on debris associated with the nests of mice, squirrels, and other rodents.

Indianmeal moth caterpillars are cream-colored or pale pink with a brown head and usually produce some silk that may thinly cover the food surface. When fully developed, the caterpillars usually disperse, often climbing walls and pupating near the ceiling. Indianmeal moths can produce several

(a) (b)

FIGURE 16-40

(a) Indianmeal moth, *Plodia interpunctella*, on infested grain. The Indianmeal moth is a very common pantry pest and infests materials such as nuts, dried fruits, and chocolate. (b) A larva and pupae of the Indianmeal moth. This insect produces a cocoon around the pupa, which has been removed on one of the pupae. Photographs by Whitney Cranshaw/Colorado State University.

(a)

(b)

FIGURE 16-41

A sod webworm larva (a) and adult (b). The larvae construct a tube of webbing around the base of grass plants and feed on the surrounding grass blades. The adults can be seen at dusk, hovering over lawns as they drop their eggs. Photographs courtesy of Jim Kalisch/University of Nebraska.

generations per year, but adults are most commonly seen during winter and early spring.

Most caterpillars in these two families feed on plants. Various species known collectively as sod webworms create silken tubes as larvae at the base of grass plants and will chew on the leaf blades of turf grasses. Adults are delicate, buff-colored moths that often are seen when disturbed to flight when walking across a lawn.

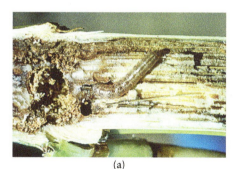

(a)

(b)

FIGURE 16-42

A European corn borer, *Ostrinia nubilalis*, larva burrowed in a corn stalk (a). An adult female of the European corn borer (b). Photographs courtesy of Jim Kalisch/University of Nebraska.

Other species tunnel into plants. Those in the genus *Dioryctria* develop in cones, trunks, and branches of pines, producing large ragged wounds marked with masses of pitch. The Zimmerman pine moth (*Dioryctria zimmermani*) is particularly damaging to pines in much of the central United States. Others attack important crop plants with the European corn borer (*Ostrinia nubilalis*) being particularly notorious in this regard. The European corn borer is not only one of the most important insect pests of corn grown east of the Rocky Mountains, but has also been reported to damage scores of other plants including peppers, beans, chrysanthemums, and even seedling apple trees.

● Biological Control of Weeds

A very high percentage of plants that develop as weeds become established by humans in new areas either by accident or through a misguided purposeful introduction. Released from many of the natural controls that suppressed them in their country of origin, they often do well, sometimes so well that they produce serious economic or environmental harm. In response to the problems posed by these invasive new weeds, biological control using herbivores may be considered. This involves first identifying the kinds of insects and/or mites that feed on and suppress the plant in its native region. If appropriate, these natural enemies may be considered for importation and release to help control the weed in the new location.

One of the earliest and most spectacularly effective demonstrations of this management method involved a small pyralid moth, the cactus borer (*Cactoblastis cactorum*), and prickly pear cacti in Australia. By 1900, six species of non-native prickly pear cacti (*Opuntia* spp.) had become established in Australia and one of these in particular, *Opuntia stricta*, spread over wide areas of the Australian continent. Within 25 years, the infested acreage increased six-fold, and plants were growing so densely that an area the size of Pennsylvania

FIGURE 16-43
Damage produced by the cactus borer, *Cactoblastis cactorum*, to a pad of prickly pear cactus. In addition to the tunneling produced by the insect, bacteria move into the wounds and rot the infested cacti. Photograph courtesy of Susan Ellis/USDA APHIS PPQ/Bugwood.org.

FIGURE 16-44
The copper leafy spurge flea beetle, *Aphthona flava*. This insect has been introduced into North America to help control leafy spurge, a serious invasive weed. Photograph courtesy of USDA ARS/Bugwood.org.

(continued)

was nearly impenetrable. In response, efforts were made to collect insects found to feed on *Opuntia* where it was native, including the United States, Mexico, and Argentina. As a result several insects were brought back to Australia and some became established. Only one, the cactus borer collected from Argentina, had a significant impact on the cactus.

A single introduction of 2,750 eggs was brought into Australia in 1925. From that original release populations of the insect were successfully established at several sites during the next 2 years. By the end of that time it was clearly evident that the insect was thriving, and, with a bit of redistribution, the cactus borer was established throughout the range of the alien cacti in Australia. By 1930, just 5 years after it was first introduced, the landscape was changing back dramatically as huge areas were cleared of prickly pear by the cactus borer. Since that time, prickly pear remains in Australia, but only in scattered patches, and its population densities are maintained at greatly reduced levels through the continued activities of the cactus borer.

Unfortunately, this story does not end here. Due to its success in Australia, the cactus borer was introduced to other areas of the world where exotic cacti are considered weeds. One such area included islands of the Caribbean. From here the moth was accidentally transported to Florida in 1989 and has become a serious invasive pest threatening native cacti in the southeast United States.

There have been many other releases of insects to control introduced plants that have become weeds. These have occurred across the country and in all habitats, coordinated by federal and state agencies. Among the weed biological control projects currently in progress within the United States are those to control water hyacinth clogging the waterways in the southeast, leafy spurge and toadflax damaging rangeland of the High Plains, tamarisk (salt-cedar) growing in dense thickets alongside irrigation canals and rivers of the arid west, and the purple loosestrife that impacts the ecology of wetlands in many northern and eastern states.

FAMILY TORTRICIDAE—LEAFROLLERS AND RELATIVES

The tortricid moths include several of the most important insect pests of fruit trees and forests of North America. They are relatively small moths, with a wing span usually less than 20–25 mm. The fore-wings often are rather square tipped, and overall coloration is usually dull with indistinct mottling so they rarely attract much attention. Most never reach a body length of more than 15–20 mm and they often are uniform in color with a dark head. Furthermore, the caterpillars often are hidden, either buried within the bud or fruit of plants, or sheltered within leaves; they are rarely seen.

The damage that some of these insects inflict attracts considerable attention. The single most damaging insect to apple and pear in North America is the codling moth (*Cydia pomonella*), whose caterpillar is the proverbial "worm" in the "wormy apple." Historically, a tremendous amount of insecticide has been directed against this pest, although in recent years substantial improvements have been made to more effectively manage it in large part with the assistance of pheromone-derived technologies (monitoring traps, mating disruption). Fruit and nut crops are also subject to injury by insects such as the oriental fruit moth (*Grapholita molesta*), orange tortrix (*Argyrotaenia citrana*), and grape berry moth (*Endopiza viteana*).

In coniferous forests, populations of spruce budworms periodically explode resulting in extensive defoliation of important forest trees such as spruce, Douglas fir, and balsam fir. The important species in western forests is the western spruce budworm (*Choristoneura freemani*) while in eastern forests the spruce budworm (*Choristoneura fumiferana*) is present. The young stages tunnel into the unopened buds and later stages consume entire needles. Related *Choristoneura* species sometimes produce outbreaks in aspen and pines.

The term "leafroller" applies to several species that feed on leaves. In the course of development, they pull together leaves and fasten them together to form a shelter within which they feed. Sometimes flower buds or developing fruit are incorporated into this shelter and also are injured. Among the important leafrollers are the fruittree leafroller (*Archips argyrospila*), omnivorous leafroller (*Cnephasia longana*), and the light brown apple moth (*Epiphyas postvittana*).

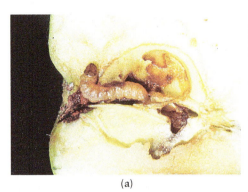

(a)   (b)

FIGURE 16-45
Larva (a) and adult (b) of the codling moth, *Cydia pomonella*. This insect is the most important pest of apples and pears in North America. It develops by feeding on the developing seeds of the core and is the culprit involved with "wormy apples." Photographs by Whitney Cranshaw/Colorado State University.

FIGURE 16-46
Larva of the western spruce budworm, *Choristoneura freemani*. This is one of the most important forest pests in the western United States and Canada. Photograph courtesy of USDA Forest Service Archives/Bugwood.org.

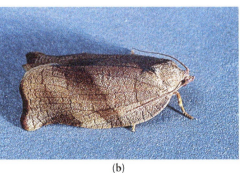

(a)   (b)

FIGURE 16-47
Larva (a) and adult (b) of the filbert leafroller, *Archips rosana*. Photographs courtesy of Ken Gray/Oregon State University.

● The Bounce Behind the Mexican Jumping Bean

Jumping beans have been a popular product sold in comic books and other kid-directed publications for over 80 years, and they were featured prominently in cartoon story lines from the 1930s through the 1950s. The secret of these odd novelties is an insect that lives within.

Cydia deshaisiana (Lepidoptera: Tortricidae) is a small moth that lays its eggs amongst the flowers of the shrub *Sebastiana pavoniana*, a native plant of the desert areas of northern Mexico. On hatching, the young larvae tunnel into the developing seeds. Throughout late spring and summer, they feed within the seeds, excavating a chamber. They further secure themselves to the inside with silken threads. Later, when the seeds drop, the insects remain secure within the tough seed coat.

FIGURE 16-49

A larva of the Mexican jumping bean moth, *Cydia deshaisiana*. The larva consumes the inside of the seeds of its host plant and then remains dormant within the seed coat. When warmed, it moves, causing the seed to move until it settles in a cooler, shadier location. Photograph by Whitney Cranshaw/Colorado State University.

FIGURE 16-48

Mexican jumping beans have been a novelty stage for decades. The "beans" twitch when warmed due to the movements of an insect larva that lives within the seed. Photograph by Whitney Cranshaw/Colorado State University.

When suddenly warmed, as from the heat of a hand, the caterpillar inside the "bean" begins to twitch, causing the seed to move. Presumably this behavior is useful for preventing the insect from overheating on the desert floor. Movements of the seed, with the insect inside, can shift the seed to shaded areas out of direct sunlight. The caterpillar remains within the seed throughout the rest of the year, but goes into a semi-dormant state (diapause) when full grown and ceases to feed.

In winter the insect finally transforms to a pupa, after leaving a small circular opening cut through the seed coat that is then covered with silk. Later, the adult moths emerge from the pupal case, pushing through the opening to the outside. They subsequently fly to new *Sebastiana* flowers and renew the cycle.

The jumping beans have long been a source of entertainment in their native area of Mexico where they are known as brincadores. As legend tells, it was a 12-year-old, Joaquin Hernandez, who first had the bright idea of selling the insect-infested seed. This led to an ongoing industry which still exports millions of jumping beans per year. The town of Alamos, Sonora calls itself the "Jumping Bean Capital of the World."

Technically the jumping bean is not a bean (seed of a legume) as it is the seed of a plant in the poinsettia family Euphorbiaceae. And to say they jump is a bit of a stretch; it is more of a bit of modest twitching. Regardless, the successful promotion of the "Mexican jumping bean" is a triumph of insect marketing.

FAMILY BOMBYCIDAE—THE SILKWORM AND RELATIVES

Several insects produce commercial products, notably the honey bee, cochineal scales, lac insects, and the silkworm. Among these, the silkworm (*Bombyx mori*) is the most thoroughly domesticated. It is a species no longer found in the wild and is now incapable of living without attending human care.

FIGURE 16-50
A mating pair of silkworm moths, *Bombyx mori*. This is the species that is the source of most cultivated silk and is reared commercially in many countries. Photograph courtesy of Jim Kalisch/University of Nebraska.

The origin of silkworm culture, known as **sericulture**, is lost in time. A common legend dates it to 2640 BC when Lady Si-Ling, wife of the Chinese emperor Huang-ti, noticed that a silkworm cocoon that accidentally fell into hot tea produced a mass of lustrous fibers. Regardless of origin, the words for silk used worldwide are derived from her name, and Lady Si-Ling has been considered the Goddess of the Silkworms, celebrated with annual festivals on emergence of mulberry leaves.

Mulberry, particularly white mulberry (*Morus alba*), is the food used by developing silkworm caterpillars, which have a prodigious appetite for the leaves. The caterpillars grow 25 times longer and become 12,000 times heavier during the 4 week period between egg hatching and pupation. Just prior to pupation, the caterpillars are transferred to trays with individual compartments; here they are allowed to spin their cocoons. Like other Lepidoptera, silk is produced from paired labial glands that open near the mouth. In the case of the silkworm, these glands are enormous and may comprise almost half the

mass of the insect. They are primarily used to construct the cocoon that will surround the pupa, and silk for the pupa is spun as a double-stranded dry fiber. A silkworm can spin a continuous strand of silk at a rate of 15 cm per minute. From each cocoon, a 300–1,000 m strand can be extracted for production of silk thread.

FIGURE 16-51
A full-grown larva of the silkworm. The silkworm is the only completely domesticated insect and is no longer found in the wild. Commercial rearing facilities feed the caterpillars leaves of mulberry on which the caterpillars develop over the course of a month. Photograph courtesy of Jim Kalisch/University of Nebraska.

To harvest the silk, the cocoons are boiled, which removes a gummy material known as sericin. The fiber is then painstakingly unreeled by hand and combined with the filaments from several other cocoons. Usually about six to eight filaments are used to produce silk thread. The overall production process is highly time consuming and expensive. To produce a 1-kg spool of silk thread involves the sacrifice of about 3,750 cocoons. Rearing this many caterpillars involves feeding 125 kg of mulberry leaves and the

FIGURE 16-52
Cocoons of the silkworm. To harvest the silk the cocoons are briefly boiled, which loosens the threads so that they can be unreeled for production of silk thread. Photograph by Jim Kalisch/University of Nebraska.

highly skilled process of reeling off the fibers from the cocoons may take another 10 h of labor.

The end result is silk thread that can be woven into fabric, a material with an extraordinary range of desirable characteristics. The density of silk fabric is low, so it is not bulky and is light. It combines high strength with elasticity (making it crease resistant) and insulates well. It is lustrous and has a high affinity for dyes, allowing it to produce brilliantly colored fabrics. Silk also lasts well, resisting microbial degradation, fabric-destroying insects, and fire. This combination has made silk the "most sought after, most jealously guarded, most furiously defended fiber in history."

Because of its properties, silk fabrics have been highly cherished; in AD 270 Europe silk was worth its weight in gold. Because of its high value, it made numerous contributions to human history, most notably the development of the **Great Silk Road**. Its beginnings date back over 2,300 years, and it originally ran along the northeast coast of the Mediterranean, through present-day Iran, Turkmenistan, and Kyrgyzstan. Ultimately it led to China, where the secret of silk production was guarded by pain of death. As the Great Silk Road developed and branched it was, by far, the longest road in the world at that time, extending 8,000 km, and provided the primary link between Europe, the Middle East, India, and China for over 1,000 years.

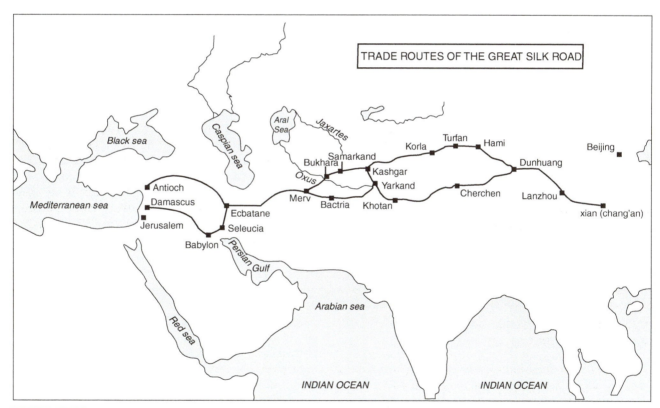

FIGURE 16-53

A map of the Great Silk Road. These were the main trade routes between Europe and Asia for many hundreds of years. Although many commodities were carried on these trade routes, it was the trade in silk originating from China that produced most activity. Figure by Matt Camper.

Silk Production in Insects

The fibrous protein known as silk is widely associated with spiders and with the cocoons of the silkworm. It is also used among a great many other arthropods. Some millipedes, symphylans, pseudoscorpions, collembolans, and diplurans have developed the ability to produce silk. The ability to produce silk also is common to a great many kinds of insects in addition to the cocoon-producing moths.

Among the insects there is a wide variety of silks produced and these are formed via a similarly diverse array of body structures. Special silk-producing glands have evolved in some insects. Among the orders Lepidoptera, Trichoptera, and Hymenoptera these glands are found in the labium, whereas the webspinners of the Embiidina have unique silk glands on the forelegs. All of these produce tough silk in sheet form. Highly elastic silks of helical structure are produced by the accessory glands of adult bees and the labial glands of flea larvae. The Malpighian tubules or peritrophic membrane may be the source of silk used by insects in the orders Neuroptera, Coleoptera, and a few Diptera. These silks are in the form of a polypeptide chain that reconfigures its structure when stretched.

Some insects produce different types of silk during different life stages. For example, honey bee larvae produce a silken cocoon for pupation from labial glands while adult honey bee workers produce silk from accessory glands that is used to strengthen newly constructed wax cells. Similarly, green lacewings pupate within a cocoon of silk, but this silk originates from the Malpighian tubules. Adults lay their eggs on a stalk from silk produced in the accessory glands of the reproductive tract.

There is also a wide array of wide functions served by silk among insects. These include some of the following:

Protection of eggs. Psocoptera may use sheeted silk to help protect their eggs. Some mantids produce a parchment-like cover of silk to protect the ootheca.

Protection of the pupa. Cocoons of silk are used to protect the pupal stage of many insects in the holometabolous orders Trichoptera, Lepidoptera, Siphonaptera, Hymenoptera, and Neuroptera. A few Coleoptera also use silk to protect the pupa.

Shelter. Many kinds of shelter are produced with aid of silk. Larvae of some moths and sawflies form silken tents. Other moths and the weaver ants construct shelters by fastening together leaves with the aid of silk. Caterpillars and larvae of caddisflies may construct silken cases that they further reinforce with pieces of plant debris or small pebbles. Both nymphs and adults of the webspinners (Embiidina) use extensive amounts of silk to line their burrows.

Food gathering. The netspinning caddisflies produce silken nets to capture floating food particles in flowing water. The unusual fly larva known as the New Zealand glowworm uses sticky silk threads to capture flying insects attracted to its glow.

Sperm transfer. The primitive wingless insects in the orders Thysanura and Microcoryphia fertilize eggs with indirect mating. Silk is used to allow the spermatophore to be picked by the female or for sperm to be transferred as droplets to the genital opening via silk threads. Silk stalks also support the spermatophores of the Collembola and other entognathous hexapods.

FAMILY SATURNIIDAE—GIANT SILK MOTHS AND ROYAL MOTHS

The family Saturniidae includes the largest and heaviest moths in North America, and, indeed, the world. The Atlas moth (*Attacus atlas*) of southeastern Asia has a wingspread of over 25 cm, and some North American species, such as the cecropia moth (*Hyalophora cecropia*), reach wing spans of 15 cm. In addition to their large size, these animals are often colorful, and many have unusual markings, notably eyespots on the wings; the luna moth (*Actias luna*) attracts particular attention because of its attractive light green coloration and long-tailed hindwings. Among silk moths differences also exist between the sexes, with males possessing very large plume-like antennae that they use to detect the sex pheromone emitted by the female. Typically, adults do not feed and live for only about a week; they survive on the stored fats accumulated as larvae.

Giant silk moth caterpillars can be brightly colored and possess curious outgrowths of the body. For example, the hickory horned devil (*Citheronia regalis*) that feeds on nut trees in the southern and eastern United States is characterized by bizarre fleshy extensions. A few of the more colorful species, such as the caterpillars of the Io moths (*Automeris* spp.) and buck moths (*Hemileuca* spp.), also have venomous hairs that can produce a very painful sting.

FIGURE 16-54
Adult of the cecropia moth, *Hyalophora cecropia*. Photograph courtesy of Jim Kalisch/University of Nebraska.

FIGURE 16-55
Adult of the luna moth, *Actias luna*. Photograph courtesy of David Cappaert/Michigan State University/ Bugwood.org.

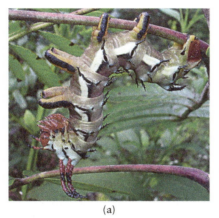

(a)

(b)

FIGURE 16-56
Larva (a) and adult (b) of the hickory horned devil, *Citheronia regalis*. The adult is also known as the regal moth. Photographs courtesy of Tom Coleman/USDA Forest Service/Bugwood.org.

(a)

(b)

(c)

FIGURE 16-57
Full-grown larva (a), pupa (b) and adult (c) of the polyphemus moth, *Antheraea polyphemus*. Like other giant silk moths they produce large cocoons, which have been cut away in image (b) to show the pupal stage within. Asian relatives of the polyphemus moth are cultivated for silk. Photographs courtesy of John Capinera/University of Florida, Whitney Cranshaw/Colorado State University, and David Leatherman, respectively.

Because of their large size, some giant silk moth caterpillars have been used as food, notably the mopane worm, *Gonimbrasia belina*. The larva of this moth is considered a delicacy in Southern Africa where a small industry surrounds its collection and processing

Most giant silk moths produce a large silken cocoon made of tough, durable silk. In parts of Asia several species are semi-domesticated and used in production of silk fabrics. Tussah silk is produced in China from the cocoons of *Antheraea pernyi*, and tensan silk in Japan from *Antheraea yamamai*. At least three species are used in silk production in

India: *Antheraea assamensis* (muga silk), *Antheraea paphia* (tasar silk), and *Samia eri* (eri silk).

Populations of several species of giant silk moths have declined greatly in the United States over the past century. Several explanations for this have been proposed and include loss of moths to street lighting and loss of habitat from development. Perhaps the most important factor in the northeastern region is the incidental effect of a parasitic fly, *Compsilura concinnata*, which was introduced from Europe to help control gypsy moth but also commonly attacks caterpillars of silk moths.

● Look but Don't Touch—Urticating Hairs

Among the many defenses that insects may deploy are **urticating hairs** that produce stinging and itching. Some caterpillars have envenomating hairs that are hollow and connect to a venom gland at the base. When brushed, the hairs easily break, releasing the fluids onto the skin. Others have dart-shaped or barbed hairs that produce irritation by mechanical means. Such **urticating** hairs can retain their activity even off the animal, when broken and air-borne, on old larval skins, or when incorporated into the cocoon.

Normal reactions to these stinging hairs include pain and redness; however, there is a tremendous range of individual reactions to the stinging hairs of caterpillars. Sometimes there are extensive rashes and persistent painful itching lasting for days. Rarely, more serious systemic reactions may occur, and barbed hairs that lodge in eyes can cause severe injury. Changes in reactions can also develop following repeated exposure due to changes in one's allergic response.

(a)

(b)

FIGURE 16-58

Larva (a) and adult (b) of the Io moth, *Automeris io*. The caterpillars of this species of giant silk moth have stinging spines that can produce a painful rash upon contact. The adult also has large eyespot markings that can be used in startle displays to deter predators. Photograph of the adult courtesy of Ronald Billings/Texas Forest Service/Bugwood.org. Photograph of the larva by Whitney Cranshaw/Colorado State University.

TABLE 16-4 Some North American caterpillars with stinging hairs that produce painful reactions in humans.[*]

COMMON NAME	SCIENTIFIC NAME	LEPIDOPTERA FAMILY
Hag moth	*Phobetron pithecium*	Limacodidae
Saddleback caterpillar	*Sibine stimulea*	Limacodidae

(continued)

COMMON NAME	SCIENTIFIC NAME	LEPIDOPTERA FAMILY
Spiny oakslug	*Euclea delphinii*	Limacodidae
Stinging rose caterpillar	*Parasa indetermina*	Limacodidae
Io moth	*Automeris io* and others	Saturniidae
Buck moths	*Hemileuca* spp.	Saturniidae
Puss caterpillar (asp)	*Megalopyge opercularis*	Megalopygidae
Puss moth	*Megalopyge bissesa*	Megalopygidae
White flannel moth/Hackberry leafslug	*Norape ovina*	Megalopygidae
Mesquite stinger	*Norape tenera*	Megalopygidae

*Hairs that can produce itching or development of allergic reactions are also found among some hairy caterpillars in the families Lymantriidae (tussock moths), Noctuidae (dagger moths), Arctiidae (tiger moths), and Zygaenidae (leaf skeletonizer moths).

(a)   (b)   (c)

FIGURE 16-59
Three North American caterpillars that possess stinging hairs and can produce a painful reaction when contacted: saddleback caterpillar (a), *Sibine stimulea*; caterpillar of the hag moth (b), *Phobetron pithecium*; caterpillar of the puss moth (c), *Megalopyge bissesa*. Photographs courtesy of Tom Murray, Jerry Payne/USDA ARS/Bugwood.org, and Sturgis McKeever/Georgia Southern University/Bugwood.org, respectively.

FAMILY SPHINGIDAE—SPHINX OR HAWK MOTHS

In North America, the sphinx or hawk moths rival the giant silk moths for size. The largest is the western poplar sphinx, *Pachysphinx occidentalis*, with a wing span of about 15 cm, but several other species are impressively large. Despite being rather heavy-bodied most are excellent fliers capable of long-distance migration.

Many sphinx moths also are able to hover in flight, an ability use as they visit flowers and feeding while still in flight. They possess very long tube-like mouthparts that probe to reach nectar deep within flowers, and sphinx moths are critical to the pollination of many plants. Plants dependent on pollination by sphinx moth typically have deep flowers accessible for pollination only by an insect with the very long mouthparts of a sphinx moth or a hummingbird.

FIGURE 16-60
A whitelined sphinx, *Hyles lineata*, feeding at a flower. This is the most common "hummingbird moth" in western North America. Photograph by Whitney Cranshaw/Colorado State University.

FIGURE 16-61
A hummingbird clearwing, *Hemaris thysbe*, feeding at a flower. This is one of the clearwing sphinx moths that fly during the day and have a general resemblance to bumble bees. Photograph courtesy of David Cappaert/Michigan State University/Bugwood.org.

Most sphinx moths fly at night, but a handful are active during the day-time, and these latter species are sometimes popularly known as "hummingbird moths" as they may closely resemble a hummingbird in both size and activity. The most common day flying species throughout most of North America is the whitelined sphinx, *Hyles lineata*. Also flying

during the day are sphinx moths of the genus *Hemaris* with clear wings making them resemble large bumble bees.

The caterpillar stage of a sphinx moth is typically known as a hornworm because most possess a prominent (but harmless) horn-like projection from the hind end. All hornworms consume leaves and a few cause enough injury to be ranked as garden or agricultural pests. Most notorious among the pest species are the larvae of the tomato hornworm (*Manduca quinquemaculata*) and the tobacco hornworm (*Manduca sexta*), both of which feed on tomatoes and other nightshade family plants (Solanaceae). As with most groups of insects, the great majority of hornworms produce insignificant plant injuries and some even feed on plants that are normally considered to be weeds.

FIGURE 16-62
A full-grown larva of the tobacco hornworm, *Manduca sexta*, a common insect pest of vegetable gardens, where it feeds on tomatoes and related plants. The closely related tomato hornworm, *Manduca quinquemaculata*, is also common in gardens. Photograph by Whitney Cranshaw/Colorado State University.

● Yucca Moth and the Yucca Plant—A Story of Mutualism

Yuccas are some of the most widespread and important plants populating the high deserts of the southwestern United States and northern Mexico. They have a long history of use by native Americans who have used the roots for producing soap, the leaf fibers for ropes and clothing, and the fruits for food. It is also now widely planted as a distinctive ornamental plant with its spiked leaves and prominent flowering stalk.

The successful reproduction of yucca is tightly bound to that of an insect, a yucca moth (Prodoxidae family). Some 13 species of these white moths are now recognized, which are extremely similar in appearance and collectively referred to as the *Tegeticula yuccasella* complex. Each species of moth is associated with its own species of yucca plant.

Yucca flowers have a unique vase-like form that requires careful placement of the pollen for pollination to occur. Only yucca moths can successfully pollinate these flowers resulting in successful fruit and seed production. Pollination is performed by the female moths after they mate and begin foraging on the flowers in search of pollen. With unique and highly specialized mouthparts, the moths form a ball from the sticky yucca pollen. On visiting a second plant they then move to the stigma of the flower and intentionally brush it with pollen from the ball, allowing fertilization to occur.

This behavior is not pure charity as the yucca moths benefit equally. After pollinating the flower, the

FIGURE 16-63

A yucca moth, *Tegeticula* sp., in a yucca flower. Photograph courtesy of Jillian Cowles.

female deposits one or more eggs within the developing yucca fruit. Eggs will hatch within the fruit and the developing caterpillar feeds on the still-developing seeds within the yucca fruit. The caterpillar never consumes all of the seed and only consumes approximately 20 seeds from a fruit that typically produces 300 seeds.

Relationships between two species that necessarily involve both mutual benefit and dependence are known as obligate mutualistic relationships. The yucca moth and the yucca plant well illustrate this; neither can survive without the other.

● Agave (Mezcal) Worm

The presence of a caterpillar or two in the bottom of a bottle of mezcal is one of the most unusual forms of insect consumption by humans. Mezcal is an alcoholic drink of ancient origin that is produced by fermenting the juice extracted from various species of agave plants (maguey). Most production occurs in the southern Mexican state of Oaxaca. It is related to tequila, a drink produced from a different species of agave and produced in other regions of Mexico.

Very often a caterpillar is added to the bottom of a bottle of mezcal. Usually this is *Hypopta agavis* (Cossidae family), a reddish caterpillar ("gusano rojo") that develops as a borer in the base of the agave plant. This caterpillar is also considered a local delicacy and is often eaten as the key ingredient in various cooked

dishes. Eating the agave worm has a long history, and in legend, it reportedly enhances strength and virility. The presence of an intact caterpillar also can indicate the alcohol content as it is an excellent preservative at sufficient concentration. However, the addition of these caterpillars in mezcal bottle is fairly recent, first appearing sometime after 1940.

The presence of the worm in alcoholic beverages is now strongly associated with mezcal. An effort by the Mexican government a few years ago to ban the caterpillar as a prohibited adulterant was met with a huge public outcry. Arguments that the caterpillar gives to mezcal much of its distinction, both as a novelty and perhaps as a bit of flavoring, caused the government to soon retreat from the idea and the caterpillar remains.

(continued)

FIGURE 16-64

Mezcal, an alcoholic drink produced by fermenting the maguey agave, often is sold with caterpillars in the bottom of the bottle. Often this is the reddish larva of *Hypopta agavis*, the "gusano rojo" that bores into the base of the maguey plant.

Order Trichoptera— Caddisflies

Entomology etymology. The name Trichoptera is derived from Greek meaning "hair wing," referring to the hairs that cover the wings of the adults. The common name caddis derives from English peddlers who advertised ribbon and yarn goods by fastening them to their clothing. The ribbons were known as "cadace," pronounced "caddys," and the peddlers were known as "caddice men." Many caddisfly larvae similarly attach pieces of leaves, stick, or pebbles to form a characteristic case. Collectively, members of this order are sometimes referred to as trichopterans.

Along with Ephemeroptera, Plecoptera, and Odonata, Trichoptera are among the few truly aquatic insect orders. Larvae of all species develop in water, and one family found in Australia and New Zealand (Chathamiidae) even develops in seawater, a phenomenon that is only duplicated among the insects by a few chironomid midges and seafaring water striders. In terms of species number, the order

Trichoptera is of moderate size among the insects, containing about 12,000 described species.

Caddisflies have two pairs of membranous wings that are covered with short hairs. The wings are held over the abdomen in a roof-like fashion. Adult caddisflies may superficially resemble moths and their flight behavior may even be quite moth-like. Long, segmented antennae and the presence of hairs (vs. scales) on the wings can be used to distinguish caddisfly adults. Caddisfly larvae have chewing mouthparts and a dark-colored head, and in addition to the normal three pairs of thoracic legs, there is also a pair of prolegs on the tip of abdomen. Each proleg bears a single claw. All caddisfly larvae are aquatic and acquire oxygen using gills located primarily on the abdomen.

Silk production (from labial glands) is an important feature of caddisfly larvae and is used for many purposes. Most caddisfly larvae live within portable cases constructed by webbing together small pebbles, leaf fragments, pieces of twigs, or other debris. The type of materials used in constructing the case and the form of the case are characteristic of different species. The type of case produced may also

FIGURE 16-65
Examples of caddisflies representing three families: Limnephilidae (a), Hydropsychidae (b), and Phryganeidae (c). Photographs courtesy of Tom Murray.

FIGURE 16-66
Three examples of cases produced by caddisfly larvae. Photographs courtesy of Tom Murray.

reflect the habitat of the caddisfly. Species that develop in fast-moving water often have a streamlined but relatively heavy case constructed of small pebbles. Caddisflies that sprawl on the bottom of calm water may have a light-weight, broad case that helps prevent them from sinking into soft silt. Irregular cases that form a retreat and are attached to solid surfaces are typically used by net-spinning species. One family of caddisflies (Rhyacophilidae), known as the free-living caddisflies, does not produce a case but uses silk threads attached to solid objects to help tether them as they crawl on the bottom of fast-flowing streams.

Among the caddisflies all manner of feeding behaviors can be found. Many shred and scavenge dead plant matter while others scrape algae or feed directly on aquatic plants. The net-spinning species

FIGURE 16-67
A netspinning caddisfly larva. These caddisflies construct a silken web that captures food particles filtered from the flowing water. The netspinning caddisflies (Hydropsychidae family) do not construct a larval case. Photograph courtesy of Tom Murray.

construct tiny nets that capture floating debris. A few caddisflies are predators of other aquatic invertebrates.

In quiet areas of streams and ponds, caddisflies pupate within a silken cocoon attached to the stream or pond bottom. Prior to emergence, the pupa cuts through the cocoon and then floats and swims to the surface. Adults then emerge from the pupa, either on the water surface or along the water's edge. As with mayflies, emergence of caddisfly adults may occur in synchronized events involving enormous numbers of individuals.

On emergence adults usually fly to standing vegetation along the edge of the water where they can be found during the day resting on the under sides of leaves. The mouthparts of adult caddisflies only allow them to sponge fluids such as water and honeydew; adults live for only about a month.

Eggs are typically dropped along the water surface, and the females appear to bob and bounce across the water surface during oviposition. Other species deposit eggs directly on solid surfaces along the shallow bottom areas at the edges of lakes and streams. For these caddisflies, females enter the water by walking or diving into it and then cement their eggs to various substrates below the surface. They can remain underwater for upwards of a half an hour, acquiring oxygen through a film of air trapped within the hairs on their body.

Depending on the species there may be one or two generations produced annually, and a few have a life cycle spanning 2 years. Anglers eagerly anticipate the annual "caddis hatch" of certain local species, as this can be a favorable time to fish for trout as they gorge on the newly emerged adults or pupae rising to the surface. As they are among the insects most sensitive to water pollution, the presence, or absence, of certain larval caddisflies also can be used as an indicator of water quality.

FIGURE 16-68
Trout fishermen are well tuned to the development of insects in waters where they fish so that they can match their fly patterns to the insects present on which the fish are feeding. Photograph courtesy of Tim Wood.

17

"Gift" Bearers of Plague—or a Plump Insect Wedding Present

Fleas and Scorpionflies

Although small in size, the peculiar insects known as fleas (Siphonaptera) have had a disproportionate effect on human history through their involvement in the transmission of plague-producing bacteria. In their own right these blood-feeding parasites cause tremendous discomfort to all mammals and birds on which they feed. Their closest relatives appear to be another odd group of insects known as scorpionflies (Mecoptera), which have their own peculiarities of habit and appearance.

Order Siphonaptera—Fleas

Entomology etymology. The order name—siphon (sucking) and aptera (wingless)—reflects the blood-feeding behavior of these wingless insects. The common name, flea, is derived from Old English "fleon," meaning "to flee," a reference to the activities and behaviors of the adults.

Siphonaptera, the fleas, is one of the smaller insect orders with only about 1,900 known species worldwide. Adults of all species are blood-sucking parasites, the great majority of them feeding on mammals. A small number (about 100 species in the genus *Ceratophyllus*) are known to feed on birds.

0.5 mm

FIGURE 17-1
Cat flea, *Ctenocephalides felis*.
Photograph courtesy of the Pest and Diseases Images Library (Australia)/
Bugwood.org.

Adult fleas have a truly unique appearance. Their overall body form is well adapted to life as a parasite of vertebrates. All are wingless, and the body is compact and, uniquely among insects, flattened laterally, a feature that allows them to move easily through the hairs or feathers of their animal host. The body of the flea is also dark and very heavily sclerotized, affording them protection from crushing due to biting and scratching of their host animals.

Their small size is also adaptive for a parasite; the most commonly encountered fleas are only 2–3 mm long. The largest flea on record is *Hystrichopsylla schefferi*, found only on the mountain beaver in the northwestern United States. Typical females are about 8 mm, but some exceed 1 cm.

The head of a flea is blunt and somewhat rounded in the side profile. Many species possess rows of stout hairs behind the head, known as a **pronotal comb**, and some also have a row of spines above the mouthparts (**genal comb**). The eyes are tiny and of the simple type; some species may lack eyes altogether. The mouthparts are downward-projecting stylets, used to suck blood from their hosts. The hindlegs of fleas are greatly enlarged and spiny, allowing them to jump onto and cling to their hosts. Overall, the entire body form of fleas is so unique that it prompted the pioneer insect morphologist R. E. Snodgrass to provide a special comment regarding this order of insects:

> No part of the external anatomy of an adult flea could possibly be mistaken for that of any other insect. The head, the mouthparts, the thorax, the legs, the abdomen, the external genitalia, all present features that are not elsewhere duplicated among the hexapods.

FIGURE 17-2
Close-up of the head of a cat flea, *Ctenocephalides felis*. Photograph courtesy of the Pest and Diseases Images Library (Australia)/Bugwood.org.

LIFE HISTORY AND HABITS

Fleas undergo complete metamorphosis. Although the adults live on their hosts, eggs are scattered around the resting areas and nest of the host animal (e.g., where your pet dogs and cats sleep). The larvae do not live on the host but usually can be found in the host's nest as well. Flea larvae are pale colored, legless, and wormlike with chewing mouthparts. In most species, the larvae feed primarily on the partially digested blood excreted by the adults found in the host nest or den. Flea larvae are quite sensitive to drying, a restriction that limits their distribution in drier climates to high humidity areas such as belowground animal dens.

Pupation occurs within a sticky, silken cocoon, which often has bits of soil or other debris incorporated in it. The cocoon, spun from the labial glands of the larva, protects the pupa from predation by other insects or cannibalism by other flea larvae. On transformation to the adult form, the larva may remain quiescent within the cocoon for an extended period and becomes stimulated to activity when exposed to the carbon dioxide and warmth of a potential host. Such behaviors can help fleas to survive extended periods without feeding, a useful adaptation when animal hosts are only periodically available.

FIGURE 17-3
Larva of a cat flea, *Ctenocephalides felis*. This is the most common species that also bites humans and dogs in North America. The larvae develop at sites where their animal hosts rest and feed on debris from the animal and waste blood excreted by adult fleas. Photograph courtesy of Ken Gray/Oregon State University.

Adults of both sexes feed on animal blood. Most fleas feed repeatedly, leaving the host between blood meals. A few species remain on the host for extended periods until they become fully engorged, a behavior similar to that found in some ticks. This behavior occurs with the sticktight fleas (*Echidnophaga gallinacea* is the most common species found in North America) that are normally

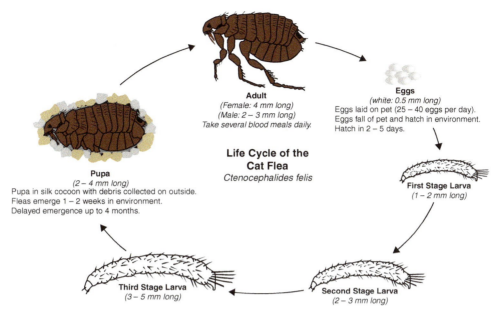

FIGURE 17-4
Generalized life cycle of a cat flea. Drawn by Scott Charlesworth and provided courtesy of Purdue University.

FIGURE 17-5
Pupa of a cat flea. Fleas produce a sticky, silken cocoon and it is usually well camouflaged with particles of soil and debris attached to it. Photograph courtesy of Ken Gray/Oregon State University.

associated with birds and occasionally found on dogs, cats, and some other mammals. Female sticktight fleas may feed for extended periods, up to 19 days, while males feed intermittently and move about to seek mates. Extended feeding also occurs with the chigoe flea (*Tunga penetrans*) found in coastal areas of South America, tropical Africa, India, and the West Indies. It is the tiniest flea species (1 mm), but the adult female burrows into the skin of its host and continues to feed until it is distended to the approximate size of a small pea. (The female lays eggs while still embedded within

the host.) In humans, chigoe fleas are found most often associated with the feet since bare feet and wearing open sandals allow the flea easy access. Skin lesions and infections commonly develop around the feeding sites of sticktight and chigoe fleas.

FIGURE 17-6
A feeding cat flea. Photograph courtesy of Ken Gray/Oregon State University.

Carbon dioxide, light (or the presence of overhead shadows), and heat are cues that stimulate fleas to actively seek their hosts. Once these cues are detected, the adults begin to jump toward the source of the stimulus. Fleas are arguably the most adept jumping animals on the planet. Despite their small size, they have been known to jump 20–30 cm. Flea jumping is greatly assisted by the presence of a highly elastic protein, **resilin**, found in critical points of their exoskeleton. Prior to a jump, the flea pulls the hindlegs into a ridge in the abdomen, compressing an area of resilin. The flea then makes a rapid downward swing of the hindlegs, releases the stored energy, and shoots into the air, twirling as it goes. It has been estimated that during the jump a flea may have to absorb g-forces of up to 149. (A jet pilot may experience a g-force of approximately 9; anything above a g-force of 15 can be fatal to humans.)

It is not only the distance of their jumping but also the ability of fleas to repeat the process that is remarkable. Perhaps the record is held by the oriental rat flea, *Xenopsylla cheopis*. When suitably stimulated in the presence of other fleas, this insect has been shown to be capable of jumping on average once every second for 3 days! Should a suitable animal host be nearby, sooner or later one of the jumps will almost certainly bring the flea onto its body, and its spiny legs will allow the flea to grasp the hairs of the host.

Adult fleas bite with specialized mouthparts that form a three-part stylet bundle, composed of a highly modified labium and two stylets originating from an area of the maxilla (lacinium). The bundle is serrated, allowing it to saw through skin and locate a blood vessel. Saliva is injected into the bite to inhibit clotting, and the flowing blood is sucked rapidly into the mouth with muscular pumps in the head. A red, itchy area develops at the bite site, largely due to a reaction to proteins in the saliva. Often, there are multiple attempts to bite, which produce small clusters of small red sores. Reaction to flea bites in humans varies tremendously from individual to individual and through time. In some humans, flea bites are uncomfortably very itchy, but people who are chronically exposed often develop immunity to the effects of flea bites.

There are official common names for many of the flea species that cause the greatest problems to humans. Unfortunately, these common names do not always adequately describe their hosts. For instance, the human flea (*Pulex irritans*) will incidentally bite humans, but they are much more commonly associated with denning animals such as foxes, skunks, or badgers, and even occur in the nests of birds such

as owls and ducks. The most commonly encountered flea in the United States is the cat flea (*Ctenocephalides felis*), which sometimes does infest cats and often incidentally bites humans, but it is primarily known as the bane of dogs wherever it occurs. (It has been said, by one who clearly was not a big dog lover, that "dogs have fleas to keep them from thinking about being dogs.")

PLAGUE

Plague is a terrible disease that has transformed human history in a profound manner. Plague is produced by infection with the bacterium *Yersinia pestis* and is normally a disease of rodents. It is primarily spread from host to host by fleas. Originally it was found in regions around China, and humans have been moving *Yersinia pestis* (via incidental transport of plague-infected rodents and their fleas) in the course of trade and exploration for centuries, perhaps millennia. Presently, plague is endemic in much of Asia, the Middle East, parts of South America, southern Africa, and the western United States.

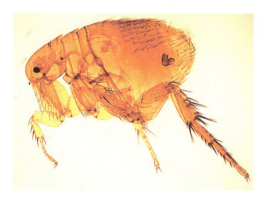

FIGURE 17-7
Oriental rat flea, *Xenopsylla cheopis*. This species is an efficient vector of the bacterium that produces plague and was a central figure in the Black Death pandemic, which periodically devastated Europe for almost three centuries beginning in 1347. Photograph courtesy of the Centers for Disease Control and Prevention Images Library.

Three different forms of the disease are recognized in humans. The most common is **bubonic plague** that manifests symptoms about 6 days after the flea bite. The lymph glands become tender and swollen, a symptom described as "buboes." The area of the bite often becomes gangrenous, and hemorrhaging occurs under the skin, producing purpling and discoloration. If untreated, mortality typically averages 50%–60%. In rare cases, the bacteria move directly into the blood from the site of bite or from a lymph node and rapidly reproduce in the liver and spleen. This condition is known as **septicemic plague**; it is almost invariably lethal and can kill within a day. Plague bacteria can also be spread without a flea bite as in **pneumonic plague**, where it is spread in sputum coughed out by infected hosts (human or animal) and inhaled directly into the lungs. The mortality for pneumonic plague can be as high as 90% with death occurring in 1–6 days from acquisition.

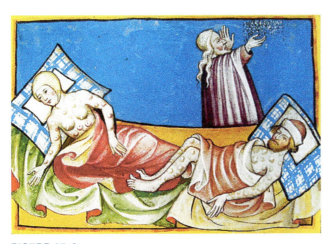

FIGURE 17-8
Illustration of the effects of bubonic plague during the second pandemic. The large bumps on the bodies of the victims are buboes, enlarged nymph glands that are a characteristic symptom of bubonic plague. Illustration from the Toggenburg Bible (1411).

Plague is sustained among rodents, with some species of rodents unaffected but still hosting the bacteria (symptomless hosts). Mice, rats, gerbils, ground squirrels, and prairie dogs are among those mammal species that are susceptible to the bacterial pathogen. Some predatory mammals are also susceptible, such as cats and ferrets.

Not all species of fleas are effective at transmitting plague. For those species of fleas that are particularly important as vectors, the bacteria grow well in the proventriculus of the flea foregut, ultimately packing it to create a blockage of the food canal. The blockage causes the flea to starve and ultimately kills it. Prior to death, *Yersinia pestis*—infected fleas will repeatedly attempt to feed during which they may regurgitate the bacteria into the

feeding wounds, often then infecting the animal host (including humans). Worldwide, the oriental rat flea is considered to be the most important and effective vector of plague in humans, but several other flea species can be significant in the maintenance and spread of plague in animal hosts.

Periodically, plague has exploded among human populations with devastating effects. Three extensive plague outbreaks were recognized as pandemics (i.e., epidemics occurring over a wide area) with the first plague pandemic often referred to as the Plague of Justinian. The origin of this pandemic is often proposed as having originally spread from Ethiopia to Egypt and then moving to the eastern Mediterranean on grain ships carrying plague-infected rats. However, transport of plague infested rodents from the east via the Great Silk Road is also considered a possibility. Regardless, in 541 plague exploded in Constantinople (present day Istanbul), capital city of the Byzantine Empire ruled by the Emperor Justinian. During its peak between 5,000 and 10,000 people were dying each day, and the disease subsequently spread throughout most of the entire Mediterranean region, ultimately reaching as far as parts of Ireland and Denmark. In some areas, up to a fourth of the population is estimated to have died from plague, and it sapped the power of the Byzantine Empire at a critical period in the midst of a major military campaign that had nearly consolidated the area of present day Italy under its sole control. The subsequent collapse of the empire in the face of invading armies—and the subsequent shaping of the entire political geography of the region for centuries—is thought to have been greatly affected by the extent and timing of the first plague pandemic. New outbreaks of plague reappeared periodically, about once every generation, until it disappeared around 750.

The second pandemic hit Europe in 1347 and resulted in such terrible effects over the next three centuries that it later became known by historians as The Black Death. Its beginnings have been traced to the trading post of Kaffa on the Crimean shore of the Black Sea where Genoese traders were besieged by Tartars. During the siege, an outbreak of plague occurred that ultimately devastated the attacking Tartars and spread into the Genoese trading post. (One source suggests that the disease was spread to the Genoese by the Tartars by catapulting the dead, flea-infested bodies of their comrades into the town.) The traders, escaping back to Europe on ships,

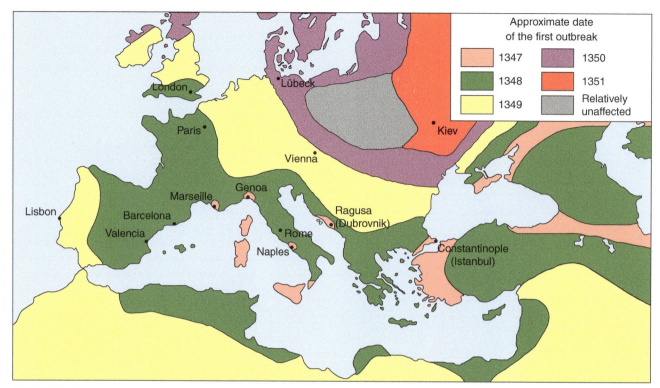

FIGURE 17-9

Course of the first wave of plague during the second pandemic. The outbreak decimated human populations and was later referred to as Black Death. Figure drawing by Matt Camper.

carried plague-infested rats with them. The withdrawing Tartars helped spread the disease to areas that now include Russia, India, and China.

The effects of plague on Europe are best documented. Within 5 years of its establishment, one-third of the European population had died, dropping the estimated population from 75 million to 50 million. The dramatic impact was reported by Francesco Petrarca, Italian scholar and poet, who wrote about the effects of this first wave of plague as it swept through southern Europe:

> … future generations would be incredulous, would be unable to imagine the empty houses, abandoned towns, the squalid countryside, the fields littered with dead, the dreadful silent solitude that seemed to hang over the whole world…

Popular conceptions of "Black Death" involved events around the Great London Epidemic of 1665–1666, which occurred at the very end of the second pandemic. At its peak, as many as 7,000 people died in a week in London, with the corpses carted away to be buried in mass graves known as plague pits.

Whereas the death toll in Europe was great, even greater loss of life was estimated to have occurred in China, which may have lost half of its population during the peak of the outbreak. Furthermore, plague outbreaks associated with the second pandemic periodically recurred every couple of decades in many areas of Europe and Asia for over 300 years.

The third pandemic emerged in China during the 1850s. Over the next 50 years, its impact was greatest in India and China where perhaps 12 million people died. It was widespread in Southeast Asia and spread on shipping routes to the Americas. In 1899 it reached Honolulu, leading to panicked disease management decisions (burning of buildings in an effort to control rats) that resulted in the notorious Great Chinatown Fire in Honolulu of

World Distribution of Plague, 1998

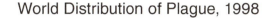

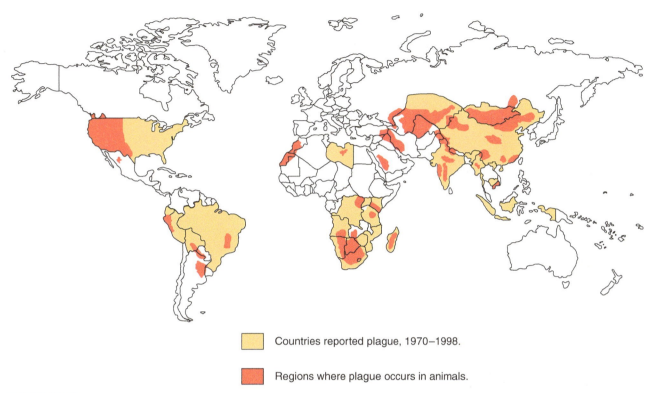

■ Countries reported plague, 1970–1998.

■ Regions where plague occurs in animals.

FIGURE 17-10
Current worldwide distribution of plague. Figure courtesy of the Centers for Disease Control and Prevention.

1900. Later that year it spread into San Francisco, apparently via stowaways who jumped from a plague-quarantined ship in the San Francisco harbor. It became established among rats and later moved into wild rodents, where it remains today as **sylvatic plague**.

The last rat-related plague cases in the United States occurred in 1924–1925 in Los Angeles. Since then all have been associated with wild rodents, and plague presently occurs endemically among ground squirrels, prairie dogs, and chipmunks. During the period 1971–1995 there was an average of 13 human plague cases a year in the United States, with about one in seven proving fatal even with the availability of effective antibiotics. The great majority of US cases have occurred in north-western New Mexico and north-eastern Arizona where plague is primarily maintained among rock squirrels and spread through bites of the rock squirrel flea (*Oropsylla montana*).

FIGURE 17-11
Removal of plague victims during the Great London Epidemic of 1665. The bodies were subsequently buried in mass graves of plague pits.

● Plague, the Black Death, and European History

The catastrophic effects of the second plague pandemic shaped many features of European history. The disease struck down all ranks and classes of people with no clear pattern, and the idea that microbes (e.g., bacteria) were the source of disease was some 500 years in the future. Society collapsed.

A grim acceptance of imminent doom often took over, leading to an intense pursuit of pleasure. Even many clergy of the time were involved in the moral breakdown, giving themselves up to "gayness and gluttony." One of the major social impacts of the plague was on clerical influences, resulting in the rise of various religious sects. One group that challenged the Church authority was the Lollards, started by John Wyclif. They demanded such changes as a vernacular order of service, translation of the Bible to English, and refusal to worship images and relics. The Lollards proved to be ahead of their time and were subsequently later suppressed after the Church reestablished its shaken authority. They periodically reemerged and ultimately combined with the Protestants of Martin Luther.

The flagellants were one of the more conspicuous groups that were strengthened during the early stage of the pandemic in central and northern Europe. They believed plague to be a form of "divine chastisement" and attempted to divert this by chastising themselves through public and private whippings. Some groups, such as the Brothers of the Cross, wore a special uniform, lived under strict discipline, and conducted flagellations according to a set ritual. Although originally embraced by the Church, the flagellant movement ultimately became a challenge to the rich upper classes and the Church, and so it too was later brutally suppressed.

It was also a time of terrible intolerance as ignorance of plague caused many to seek culprits for blame. Sometimes the targets were the nobles or those with some noticeable crippling; however, it was the Jews who

FIGURE 17-12
The rise of the cult of flagellants was one of many reactions to the social breakdown that followed the incomprehensible human suffering of the plague pandemic.

were most commonly attacked. In Basel and Freiburg (currently in Switzerland), all known Jews were herded into buildings and burned. In Strassburg over 2,000 were hanged at the Jewish cemetery. This persecution initiated a Jewish migration to the north and east into Germany, Poland, and Russia.

With the huge loss of life, there were shortages of workers, and as a result, workers became empowered for the first time. Wages went up to allow a new measure of prosperity, and the aristocracy never fully recovered its former strength and influence. Within 150 years, the feudal system of England collapsed and allowed the rise of the tenant farmer. Several popular uprisings during this period also helped establish concepts of democracy that later became gradually institutionalized in Europe and America.

Order Mecoptera— Scorpionflies and Hangingflies

Entomology etymology. The order name is based on the Greek words "meco" (long) and "ptera" (wings), a feature that is reasonably characteristic, although other orders (e.g., Odonata) have proportionally longer wings. The common name of the order is derived from the bizarrely enlarged genitalia of males in the family Panorpidae (common scorpionflies), which superficially resemble the stinger of a scorpion.

The scorpionflies are a small order of insects with about 600 known species (ca. 75 species in North America). They are an ancient group of insects that were formerly far more abundant and retain many features that have changed little since the lower Permian age. Adults are medium sized (9–25 mm) and notably possess an elongated beak that gives them a sort of a horse-faced appearance. Wings are of approximately equal size and with similar venation. The rarely seen larvae have an elongate body form and resemble caterpillars or certain types of active beetle larvae.

Mecopterans are typically found in forested areas. Most of the adult and larval stages are

(a)

(b)

FIGURE 17-13

(a) Female and (b) male scorpionflies, *Panorpa* sp. Photograph courtesy of Tom Murray.

scavengers of dead insects. Predation of live insects occurs among larvae of some species, and adult hangingflies (Bittacidae family) may actively prey on live soft-bodied insects. The latter spends a lot of its time hanging on plants by its long forelegs, allowing the hindlegs to sweep out and capture flying insects that pass nearby.

Mating rituals of both common scorpionflies (Panorpidae family) and hangingflies involve males using freshly captured prey or a dead insect as an enticement to induce females to mate. Such "enticements" are known as **nuptial gifts**. Females select mates, in part, based on the quality of these offerings, and mating occurs while the females feed on the offering.

FIGURE 17-14

A hangingfly. Photograph courtesy of Tom Murray.

Male scorpionflies will steal nuptial gifts from one another to present offerings to females. Another method of acquiring a nuptial gift is removing a dead insect from a spider web. Many scorpionflies are killed by spiders in their efforts to pick out a suitable nuptial gift. On failing to acquire a suitable insect, a male may regurgitate a salivary secretion as a substitute gift that may, or may not, be acceptable to the female.

Snow scorpionflies (Boreidae family) display many behaviors unique within the order. These insects feed on mosses, and adults are active in the middle of winter. The females are wingless and are

FIGURE 17-15
A female snow scorpionfly. Photograph courtesy of Tom Murray.

sometimes called "snow fleas" (a name perhaps better saved for the springtails that periodically mass on the snow surface). They also have their own mating peculiarities, which include the male carrying the female *in copula* for hours and "reciprocal intromission" where the female inserts her genitalia into the male.

The Mecoptera is an insect group of particular interest to insect systematists. They appear to be ancestors of several currently recognized orders, including Diptera. Some families (snow scorpionflies in particular) have many physical features common with fleas (Siphonaptera). Future taxonomic revisions may likely see combinations of at least some members of Mecoptera with the fleas.

True Flies and the Twisted-Wing Parasites

Two pairs of wings are the norm for most insects with the ability to fly. But there are two orders of insects with only a single pair—the true flies (order Diptera) and the peculiar twisted-wing parasites (order Strepsiptera).

Order Diptera—The True Flies

Entomology etymology. The name of the order is derived from the unique wings of these insects; they possess only a single functional pair, "di" (two) and "pteron" (wing). These are the "true flies." Collectively members of this order are sometimes referred to as dipterans.

When using a common name for an insect in this order the word "fly" is separated from the descriptor. House fly, deer fly, and fruit fly would be examples. Insects from other orders that are described as a "fly" properly have the name combined, such as damselfly, mayfly, butterfly, or scorpionfly.

The order Diptera is a large one, with over 150,000 described species. Flies are also very ecologically diverse, with representatives found filling essentially every type of ecological role: feeding on plants in diverse ways, scavenging on all types of organic material, preying on and parasitizing other arthropods, or living as internal parasites of mammals.

The defining feature of the order is the single pair of wings, attached to the mesothorax. The hind pair of wings is reduced to small knob-like organs, known as **halteres**. The halteres function as flight stabilizers, as demonstrated when their experimental removal cause flies to go into uncontrolled flight. (Attachment of a thread to the abdomen has been shown to help some insect re-establish a measure of stability, apparently acting somewhat in the manner of a kite's tail) The fly mesothorax is greatly enlarged and packed with large flight muscles. These are not directly attached to the wings but instead function as **indirect flight muscles** that compress plates of the thorax to which the wings attach. Furthermore, each contraction of the thorax (and wing beat) need not be stimulated by a single nerve firing. **Asynchronous muscles**, which can contract multiple times from a given nervous impulse, power the thoracic compressions that produce wing beats. This neuromuscular arrangement can allow some Diptera to achieve astonishingly high wing beat frequencies; a tiny biting midge (*Forcipomyia* sp.) is reported to be able to beat its wings over 1,000 times/s.

highly variable. In many species of Diptera, the larval mouthparts are extremely reduced and function as small hooks that slash tissues of plants, prey on insects, or eat other foods.

FIGURE 18-1
Side view of a house fly, *Musca domestica*. The small yellow knob is the haltere, the highly modified hindwing found in the order Diptera. The haltere acts as a balancing organ that helps stabilize the insect in flight. Photograph courtesy of Brian Valentine.

Adults of all Diptera have mouthparts designed to suck fluids, but there are many variations within this type of feeding behavior. Although most of them siphon or sponge existing fluids, others have evolved various mechanisms for the mouthparts to cut or pierce, allowing them to feed on the blood of insects, birds, mammals, and other animals. Larval mouthparts also are

FIGURE 18-3
Many flies, such as this blow fly, feed on fluids by means of sponging mouthparts, a type of modified labium. Photograph courtesy of Brian Valentine.

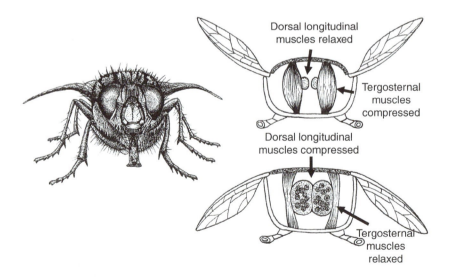

FIGURE 18-2
Indirect flight muscles are used to power the wings of Diptera and some other insects. With this method of wing movement the muscles are connected to points on the mesothorax, rather than wings. Two major groups of muscles (dorsal longitudinal muscles, tergosternal muscles) alternately compress the mesothorax. The wings are extensions of the mesothorax and beat with the thoracic flexing. (Smaller muscles groups fine tune wing flexing.) The flight muscles of flies are also *asynchronous*, which allow multiple muscle contractions from a single nerve impulse, helping to create very high wing beat frequency. Drawings by Matt Leatherman.

Conversely, quite elaborate mouthparts occur among some Diptera larvae, particularly those that filter particles out of water.

There are two major subdivisions (suborders) of Diptera: Nematocera and Brachycera. Those in the suborder Nematocera typically have adults with long antennae and are sometimes referred to as "long-horned flies." Many have a narrow, sometimes even delicate, body type. Larvae have a distinct dark head with mandibles that move laterally. The larvae lack thoracic legs, but during the pupal stage the developing appendages appear as if glued to the body (**obtect** pupal form). Various "midges" and "gnats," as well as mosquitoes, black flies, and crane flies, are placed within this suborder.

FIGURE 18-5
Life stages of the house fly, *Musca domestica*, a typical member of the dipteran suborder, Brachycera. Photograph courtesy of USDA ARS/Bugwood.org.

FIGURE 18-4
A fungus gnat is a fairly typical representative of Diptera in the suborder Nematocera. This suborder includes insects such as mosquitoes, midges, gnats, and crane flies. Photograph courtesy of Brian Valentine.

The suborder Brachycera includes the "shortened-horn flies" characterized by having inconspicuous antennae, sometimes reduced to a small knob with a protruding bristle (aristate antennae). The body form of the adult is often compact. Larvae are typically of the maggot form, with the head end narrowed; no conspicuous "head" is present. Feeding occurs with mouth hooks that often contract within the head area. The last larval skin covers the pupa so it has a smooth, rounded appearance (**coarctate** pupal form) and is known as a **puparia**. House flies, blow flies, flower flies, and horse flies are Brachycera representatives.

FAMILY TIPULIDAE—CRANE FLIES

Crane flies are slender insects with long, brittle legs and have a superficial appearance to giant mosquitoes. Indeed, "mosquito hawk" is a name commonly given to adult crane flies, although this name is entirely inappropriate because adults are not predators, do not bite, and feed only on a little nectar, if they feed at all. Crane flies are common insects that occur almost everywhere but the driest locations, from the northernmost Arctic to the tropics and from sea level to the top of high mountains. They are also the largest family among the Diptera, with some 15,270 described species, over half described by a single individual, C.P. Alexander. (Dr. Alexander during his career at the University of Massachusetts described approximately 11,000 species of crane flies, a feat that ranks him as the most prolific taxonomist of all time.)

There is wide range in size, with the largest North American species (*Holorusia hespera*) having a wing span of 85 mm. The halteres are also unusually large among crane flies, making them quite easy to see. Prominent halteres are also present even on wingless species, such as the *Chionea* spp. found walking across snow as they search out nests of small mammals in which they live. Almost all crane flies occur within two

FIGURE 18-6
A mating pair of crane flies. Photograph courtesy of Tom Murray.

subfamilies that can be usually be distinguished by whether they hold their wings out to the side (Tipulinae subfamily) or fold them over the back (Limoniidae subfamily) when at rest.

Crane fly larvae have an elongate, cylindrical body form, with fleshy lobes on the hind end (adjacent to the spiracles) with the head area partially sunken into the thorax. Color ranges from dirty cream to brown and larvae of the larger species (typically within the genus *Tipula*) are often known as "leatherjackets" because of their thick skin. Most crane fly larvae develop as scavengers that feed on decaying plant matter, but others eat mosses, liverworts, fungi, or roots of living plants. Crane flies tend to be most common in moist soil and many occur along the edges of and sometimes within ponds and streams. A few are adapted to drier sites, and a species accidentally introduced into the Pacific Northwest, the European

crane fly (*Tipula paludosa*), is a pest of pastures and lawns where it feeds on the roots of grasses. Adults of this large crane fly, and several others, also frequently enter homes incidentally when attracted by lights, where they occur as particularly noticeable, but innocuous, temporary nuisance invaders.

FAMILY CHIRONOMIDAE—NON-BITING MIDGES

The non-biting midges are the most abundant family of insects that develop in water. They include over 1,000 species in North America but, because of their relatively small size, innocuous habits—and because the adults are often mistaken for mosquitoes—they are frequently overlooked.

Chironomid midge larvae that develop in water typically feed on fine particles of organic matter, algae, and other small organisms they filter from the water or gather from the surface of the bottom sediment. Where conditions are favorable, their populations can become extremely dense, and as many as $50,000/m^2$ have been observed. Although small in size, larvae have a fundamentally important role in the breakdown and recycling of nutrients in waters. Additionally, all stages can serve as a highly significant food source for various predatory insects, fish, and waterfowl.

FIGURE 18-8
Larva of a chironomid midge. Chironomid midges are the most abundant kind of insect found in many lakes and waterways. Photograph courtesy of Tom Murray.

Adults emerge from water and fly to nearby land. Sometimes adult emergence is synchronized, triggered by temperature or other changes in the water. Large numbers periodically swarm cities around the Great Lakes, where they are sometimes called

FIGURE 18-7
A crane fly larva. These insects are usually found in moist soil and sometimes referred to a "leatherjackets." Photograph courtesy of Tom Murray.

"muffleheads" because of the prominent feathery antennae of the males. One of their more memorable appearances was during the 2007 American League baseball playoffs, when they temporarily halted play in a game that the Cleveland Indians hosted against the New York Yankees. (The Indians won the game.)

Adults are short-lived and do not feed, but sometimes they attract attention by producing

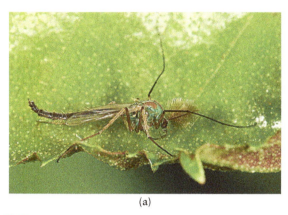
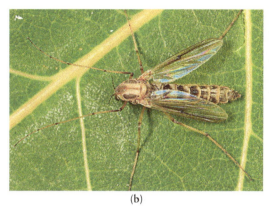

(a)   (b)

FIGURE 18-9

Male (a) and female (b) chironomid midges. The males can be distinguished by their very large plumed antennae. Photographs courtesy of Tom Murray.

conspicuous mating swarms. Mating swarms of midges often only include a few individuals but occasionally can involve thousands that may suddenly come together during periods of calm weather. These mating swarms typically form at particular sites, such as small sun-lit clearings, but often they occur as "top swarms" over some landscape marker. These markers can be all manner of objects: a soccer ball on the lawn, a parked car, a church steeple—or the head of a human.

Mating swarms are comprised entirely of males, with females quickly darting in for only a brief period to find a mate. The males are highly attuned to the sounds of a nearby female and midge swarms often respond to other sounds. For example, a sharp hand clap may cause a swarm to become temporarily disarrayed. Specific notes may cause midge swarms to rise or fall in height.

Despite their rather fragile appearance, the chironomid midges include some species that show impressive resilience to adversity. The blood-red midges or "bloodworms" (*Chironomus* species) are among the most tolerant insects to polluted water. They are able to withstand such conditions because they produce hemoglobin in their blood, which allows oxygen to be stored when it becomes

available. *Polypedilum vanderplanki*, a species that inhabits temporary water pools in West Africa, is capable of dehydrating to <3% of its body weight, allowing it to survive during the hot, dry seasons. In this incredibly dehydrated condition it also can survive immersion in liquid helium (−270°C).

A few genera of chironomid midges are represented among the very small number of insects that have adapted to marine life. One example is *Pontomyia natans*, a marine midge that occurs in the waters off Samoa. All stages are submarine and the adults lack functioning wings. Males move through the water by swimming with their front legs while the adult females remain in place after pupation, lacking antennae, mouthparts, legs, wings, and halteres. The chironomid midges also have notable claim in including the largest insect restricted to the Antarctica land mass, *Belgica antarctica*, a wingless midge that is about 12 mm in length.

FAMILY CULICIDAE—MOSQUITOES

Mosquitoes are all-too-familiar insects to almost anyone on the planet, and various representatives among the 3,500 known species can be found almost anywhere that pooled water is periodically present. Mosquitoes are rather slender-bodied flies with narrow wings and a long proboscis containing the stylets used for feeding. Fine scales cover their wings and their bodies, producing elegant patterning on many species. Both males and females have a long proboscis, although in the males it is adapted only to suck nectar. Females also may feed on nectar but are blood feeders as well, using blood meals to nourish developing eggs. The sexes are often most easily separated by looking at the antennae; those of males are typically much larger and more feathery than those of the female. Males use their antennae largely as sound detectors cued into the hum produced by the wings of a flying female.

Larvae of all mosquitoes develop in water and are called "wrigglers"; they are active insects that can rapidly swim through the water. They have an elongate body with a rounded head and bulbous thorax. Among the most commonly encountered North American species, a snorkel-like breathing tube is present on the tip of the abdomen allowing them to acquire oxygen with periodic visits to the surface. Almost all feed by filtering small particles out of the water, although a few are predators of other mosquito larvae.

The four larval stages are usually completed within 10 days, but development time varies with

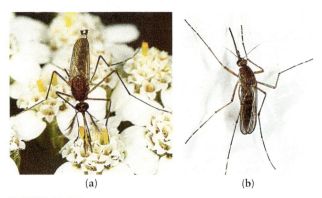

(a)   (b)

FIGURE 18-10

Male (a) and female (b) mosquitoes, *Ochlerotatus stimulans*. Only the females feed on blood; males are primarily nectar feeders. Photographs courtesy of Tom Murray.

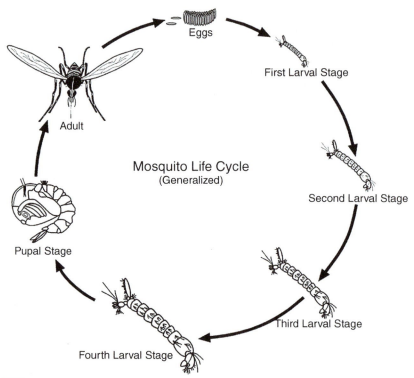

Mosquito Life Cycle
(Generalized)

Eggs

First Larval Stage

Second Larval Stage

Third Larval Stage

Fourth Larval Stage

Pupal Stage

Adult

FIGURE 18-11

Generalized life cycle of a mosquito. Scott Charlesworth artwork/design. Drawing courtesy of Purdue University.

Note:
Each larval stage is larger than the previous one. Molting occurs between each larval and pupal stage. Larval and pupal stages are aquatic.

water temperature. Pupae also live in water and as appear C-shaped, with a pair of trumpet-like horns behind the head through which they breathe at the surface. They are unusually active for a pupal stage and are known as "tumblers."

FIGURE 18-12
Mosquito larvae (*Culiseta* sp.) known as "wrigglers" develop in water and breathe through a spiracle that extends from the tip of the abdomen. Photograph courtesy of Tom Murray.

FIGURE 18-13
A mosquito pupa, known as a "tumbler." Unlike most insects the pupal stage of mosquitoes are quite active. Periodically, they return to the surface to acquire air through the paired spiracle on the thorax. Photograph by Tom Murray.

Most mosquito bites of humans in North America can be blamed on species in the genera *Aedes* and *Ochlerotatus*. These are "floodwater mosquitoes" that lay their eggs on moist soil along the edge of receding water. Eggs can remain dormant for long periods but will hatch when flooding waters rise to again cover the eggs. Heavy rainfall or spring snow-melt are typical events that trigger eggs to hatch, with adults

appearing about 2 weeks later. In areas with cold winters, adults of *Aedes* and *Ochlerotatus* species die out in cold weather, but the eggs can survive this harsh season. In tropical areas, all stages can be found the year round.

FIGURE 18-14
Aedes vexans is the most important species involved in biting humans in most of North America. It is a floodwater type of mosquito that lays its eggs at the edge of temporary pools of water. The eggs hatch with a later flooding event that covers the eggs. Photograph courtesy of Tom Murray.

Most other mosquitoes (e.g., species in the genera *Anopheles*, *Culex*, *Culiseta*) that occur in North America spend winter as mated females hiding in protected areas while in a dormant condition (diapause). All of them lay their eggs directly on the surface of water either as massed egg rafts (*Culex*, *Culiseta*) or as individual floating eggs (*Anopheles*).

Although all mosquitoes develop in water of some type, different species have preferred places. Most *Culex* mosquitoes lay eggs in stagnant pools of water that have algae growth. Others prefer small pools, such as the "tree hole" mosquitoes that breed in water that collects in hollows within trees. Containers of water produced by humans are particularly favored by the notorious yellow fever mosquito, *Aedes aegypti*. Most floodwater mosquitoes usually lay eggs in low-lying areas that periodically flood with fresh water from rainfall or flooding. Along coastal areas some mosquitoes are adapted to brackish water and are known as "saltmarsh mosquitoes." Only rarely will mosquitoes be found developing in running water or permanent ponds where predatory fish are present.

FIGURE 18-15
Culex pipiens laying an egg mass on the surface of a water pool.
Photograph courtesy of Susan Ellis/USDA APHIS PPQ/Bugwood.org.

No other family of insects has so consistently and profoundly affected the course of human activity than the mosquitoes. Where mosquitoes are abundant, nuisance problems at dusk may make outdoor activities unbearable. Tens of millions of dollars are spent annually on window screening, insect repellents, and other attempts to control bites from these insects. The nuisance aspects of biting aside, the most important effect of mosquitoes on humans is the spread of important diseases caused by various viruses, protozoans, and filarial worms. Several of these diseases have long affected human history, and continue to do so to the present.

TABLE 18-1 Some of the more important disease agents transmitted by mosquitoes. Natural hosts are animal hosts of the pathogen that are regularly infected, can allow sustained development and transmission of the pathogen, and serve as reservoirs of the pathogen. Tangential hosts are those that are occasionally infected but cannot sustain the pathogen ("dead-end hosts").

| DISEASE AGENT | HOSTS | | IMPORTANT VECTORS |
	NATURAL	TANGENTIAL	
Viruses			
Eastern equine encephalitis	Bird	Human	*Coquilletidia perturbans*
Venezuelan encephalomyelitis	Small mammal	Human, horse	*Culex pipiens*, etc.
Western equine encephalomyelitis	Bird	Human, horse	*Culex tarsalis*
Dengue	Human		*Aedes aegypti, Aedes albopictus*
Japanese encephalitis	Swine	Human	*Culex tritaeniorhynchus*
Rift Valley fever	Sheep, goats, cattle	Human	*Aedes* spp.
St. Louis encephalitis	Bird	Human	*Culex pipiens, Culex nigripennis*
Yellow fever	Primate	Human	*Aedes aegypti, Aedes africanus*
La Crosse encephalitis	Rodent	Human	*Aedes triseriatus*
West Nile virus	Bird	Human, horse	*Culex tarsalis, Culex pipiens*
Protozoans			
Malaria	Human		*Anopheles* spp.
Malaria	Bird		*Culex* spp.

(continued)

TABLE 18-1

| DISEASE AGENT | HOSTS | | IMPORTANT VECTORS |
	NATURAL	TANGENTIAL	
Filarial nematodes			
Wuchereria bancrofti	Human		*Culex* spp. *Mansonia* spp.
Brugia malayi	Cats, canids	Human	*Culex* spp., *Mansonia* spp., *Aedes* spp.

Source: Adapted from Eldridge, B.F. 2003. Medical entomology: A textbook on public health and veterinary problems caused by arthropods. Springer.

Yellow Fever. Viruses transmitted by mosquitoes and other arthropods are often referred to as **arboviruses** (**ar**thropod-**bo**rne **viruses**) and the most serious of these produces yellow fever. Originally restricted to Africa, where it occurred primarily among primates, it was soon spread through the Americas with the slave trade and has long caused disease outbreaks in coastal cities. It is a terrible disease that causes excruciating pain, hemorrhaging, vomiting of tarry black material (blood), and jaundice—a yellowing skin symptom that gives the disease its name. There are no effective treatments for the disease and mortality often exceeds 10%.

Yellow fever virus is transmitted by the yellow fever mosquito, *Aedes aegypti*, a mosquito that is uniquely adapted to living among humans. It prefers to breed in small pools of rainwater that collects in rain barrels, flower pots, drains, old tires, gutters, and

FIGURE 18-16

Aedes aegypti is the vector of the viruses that produce yellow fever and dengue fever. It is unusually well adapted to living among humans and breeding occurs in small pools of water, often located in materials created by humans (e.g., old tires, bird baths). Photograph by John Gathany/Centers for Disease Control and Prevention.

other sites provided by human activities. Fortunately for much of North America, *Aedes aegypti* is not adapted to cold winters and can only remain a permanent resident in warmer climates.

One of the biggest impacts of yellow fever was on the development of the Panama Canal. The idea for a short-cut across the isthmus between the Atlantic and Pacific Oceans was long a dream; shipping was otherwise forced to take the roundabout and treacherous route around the Cape of Good Hope. The first attempt at building a canal was made by the French under the guidance of Ferdinand de Lesseps, who earlier had successfully overseen the building of the Suez Canal. The French effort in Panama turned into a disaster and was abandoned in 1889 after 20,000 lives had been sacrificed in the effort. On average 176 out of every 1,000 workers involved in the project died, most from either malaria or yellow fever.

Fifteen years later the project was renewed by the United States. By this time the cause of yellow fever had been discovered through the collective efforts of such pioneer medical entomologists as Walter Reed, James Carroll, Aristides Louis Agramonte, and Jesse Lazear. (Lazear died during studies of the disease following his purposeful exposure to yellow fever – bearing mosquitoes.) Furthermore, methods to manage mosquitoes had been developed, and in the Canal Zone these were rigorously and effectively enforced by Colonel William Gorgas so that yellow fever was eliminated within a year. Construction began in 1907 and the Panama Canal was completed in 1914, during which time only 0.6% of the workers died from mosquito-borne disease.

Historically, yellow fever has been present in the United States. The most notorious outbreak occurred in Philadelphia during the summer of 1793 and killed

FIGURE 18-17
Discarded tires are a notorious site for breeding of *Aedes aegypti* and the tiger mosquito, *Aedes albopictus*. Photograph courtesy of the Centers for Disease Control and Prevention Images Library.

FIGURE 18-18
Yellow fever strongly shaped the history of the construction of the Panama Canal. The original French effort was disastrous as the insect-borne diseases, yellow fever and malaria, killed over 1/6th of the workers and sickened many more. By the time the Americans resumed the work, the role of mosquitoes in yellow fever was understood. A massive effort was made to control mosquitoes prior to resumption of building, which was highly successful in reducing incidence of yellow fever. This historical photograph was taken in 1907 during the Culebra Cut, the largest project involved in the building of the Canal.

about 5,000 people, approximately 10% of the population. At that time Philadelphia was serving as the temporary center of the fledgling American government, but its reputation for pestilence associated largely from yellow fever caused it to be dropped from consideration as the permanent US Capitol. For many decades yellow fever would periodically occur during the warmer months in port cities, with subsequent outbreaks extending up riverways, which originated from infested ships arriving from tropical areas. During the last major US outbreak in 1878, some 20,000 people died in the areas bordering the southern Mississippi River, with particularly devastating effects on New Orleans and Memphis.

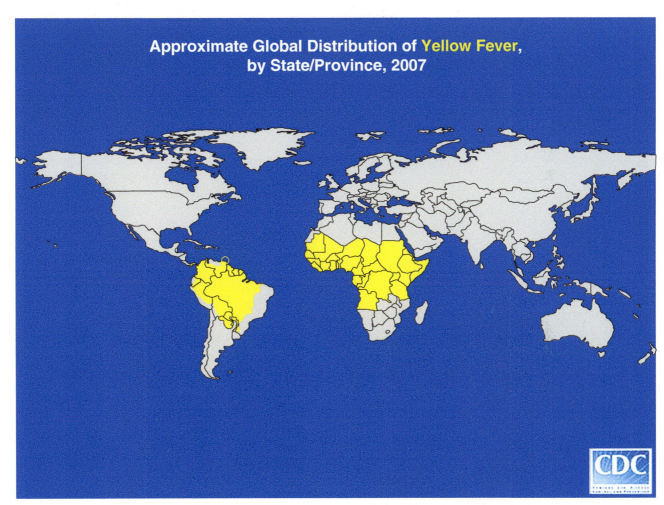

FIGURE 18-19
Worldwide distribution of yellow fever as of 2007. Figure provided by the Centers for Disease Control and Prevention.

West Nile virus and other arboviral encephalitides. For over a century, yellow fever has been eliminated from the United States, and the other major arthropod-borne virus (arbovirus) disease of the tropics, dengue fever, has yet to take hold. Nonetheless, several encephalitis-producing viruses are present in North America that are spread by mosquitoes, including eastern equine encephalitis, western equine encephalomyelitis, and St. Louis encephalitis. A recent addition to this group is West Nile virus, which has long occurred in parts of Africa and the Middle East but was first identified in North America in New York City during 1999. It has now spread to most areas of the United States.

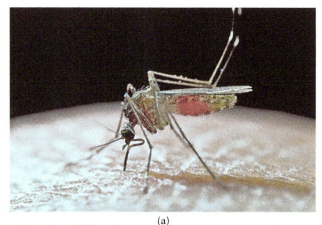

(a)

(b)

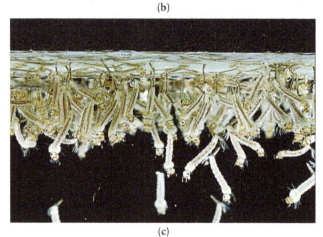

(c)

FIGURE 18-20

Culex tarsalis (a) is the primary vector of the virus that produces West Nile Virus disease. It lays eggs in the form of rafts (b) on the surface of still water and the larvae (c) feed within these pools. Photographs by Jack Kelly Clark and provided courtesy of the University of California IPM Program.

These are all primarily diseases of wild birds. Normally they are transmitted from bird to bird by bird-feeding mosquitoes, such as various *Culex* species and *Coquilletidia perturbans*. Humans, and

sometimes horses, are uncommon hosts of these mosquitoes; however, when they do bite after previously feeding on an infected bird, they can transmit the virus to mammals.

After West Nile virus became established among bird populations, it made a steady march westward across the country. Bird deaths presaged outbreaks with some species of birds being highly susceptible to the virus, most notably the family Corvidae (crows, magpies, ravens). The number of human infections is unknown; fully 80% of all infected people show no symptoms. Mild symptoms occur in about 20% of the cases, but in about 1 in 150, serious complications develop when the virus invades the nervous system. Deaths do occur and peak mortality in the United States occurred in 2003 and 2004 when there were 264 and 284 West Nile virus deaths reported, respectively. In general the pattern has been that number of infections peak in an area during the year that follows first appearance of the disease and then it declines. However, annual incidence varies and 2012 saw the highest number of early season cases ever, concentrated in Texas, Mississippi, Louisiana, Oklahoma, South Dakota, and Michigan.

Malaria. Malaria is, by far, the most important arthropod-borne disease worldwide. Annually, it produces between 350 and 500 million new infections and results in the deaths of over 1 million people, most commonly in Sub-Saharan Africa. (This is about two people per minute, mostly children, who currently die of this disease.) Over 40% of the world's population lives where malaria is endemic and regularly occurs. It is the fourth leading cause of death among children in developing countries.

It has also had a long history of causing human disease. The intermittent fevers and chills typical of malaria are mentioned in Assyrian and Babylonian medical texts and the disease "burning ague" appears in Leviticus in the Old Testament; a more complete description of the disease was provided by Hippocrates (460–377 BC). The word malaria is derived from the Latin words for "bad air" since it was long thought to be caused by inhaling air from swampy or marshy areas. The discovery of the causal organism, and the swamp-associated mosquito vector, was not achieved until the early 1900s.

Malaria is produced by an infection by the protozoans in the genus *Plasmodium* with *Plasmodium vivax* and *Plasmodium falciparum* being most important. Various *Anopheles* mosquitoes are the vectors of the disease, and the protozoan depends on the mosquito not only for transportation but to

FIGURE 18-21

Anopheles gambiae, one of the *Anopheles* species that is an important vector of malaria. Photograph by John Gathany/Centers for Disease Control and Prevention.

complete parts of its life cycle as well. The *Plasmodium* life cycle is complicated, and sexual reproduction only occurs in the mosquito; other stages of the organism complete the life cycle in humans. Haploid stages known as sporozoites form later within the mosquito and these move into the insect's salivary gland where they can be introduced into humans during a subsequent blood meal.

Within humans initial infection by the sporozoite occurs among cells of the liver. Six days later a different form of the *Plasmodium* (trophozoite) moves into blood cells beginning multiple infective cycles involving forms known as merozoites. Most effects of malaria result from the waves of merozoites bursting from blood cells and invading new ones. In some forms of malaria, emergence of

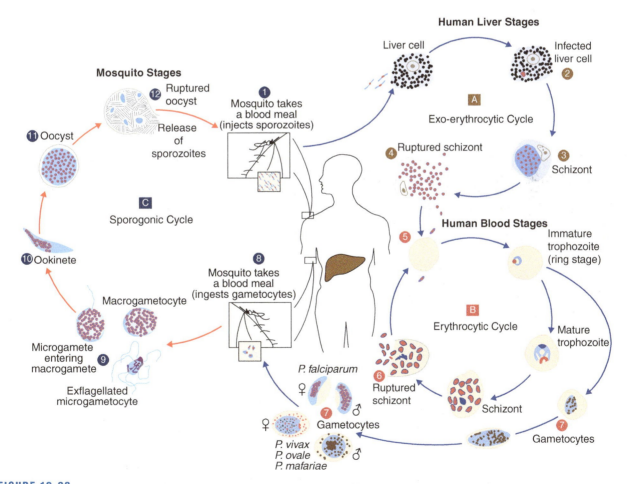

FIGURE 18-22

The life cycle of the *Plasmodium* that produces malaria in humans. The life cycle is extremely complex and involves stages that are completed in both the human and *Anopheles* mosquito. Drawing courtesy of Centers for Disease Control and Prevention.

the merozoites from cells is regular, producing alternating symptoms of fever and chills every 48 h (*Plasmodium vivax*) or 72 h (*Plasmodium malariae*). Most of the debilitating effects of malaria are associated from loss of blood cells. Death, usually limited to infections with *Plasmodium falciparum*, most often results as a byproduct of capillary blockage of the brain and other tissues.

Unlike most mosquito-borne diseases, malaria is restricted to humans and there is no wild animal host reservoir. (Some birds suffer from malaria produced by different *Plasmodium* species, but these diseases are restricted to birds.) The basic transmission cycle involves: (1) infected human who is fed on by an appropriate *Anopheles* mosquito; (2) maturation of the *Plasmodium* sexual stage within a mosquito over the course of several days; and (3) introduction of sporozoites by the mosquito into a new human host with the next feed. A sequence of different developmental stages of Plasmodium occur in the vertebrate host.

Efforts to combat malaria have been applied to all aspects of the transmission cycle. It has proved to be a frustratingly difficult task, in part because of the large number of different *Anopheles* species that may be vectors and the different species and strains of the malaria-producing organism. For a period of time, following the discovery of effective insecticides such as DDT, some thought that using insecticide could eradicate malaria. Unfortunately, strains of mosquitoes soon developed resistance to these chemicals, rendering them ineffective. Drugs that can kill the *Plasmodium* have long had important value, notably quinine and chloroquine, but strains of the disease resistant to these drugs have also evolved. Intensive efforts have been made to attempt development of an effective vaccine, but this remains elusive.

Nonetheless, progress certainly has been made and large areas of the world are now malaria-free through a combination of efforts. In the United States malaria once ranged into upstate New York and Minnesota and was formerly very common in the southern states. Eradication was achieved by the early 1950s and cases that now do occur within the country have originated from areas of the world where it remains endemic.

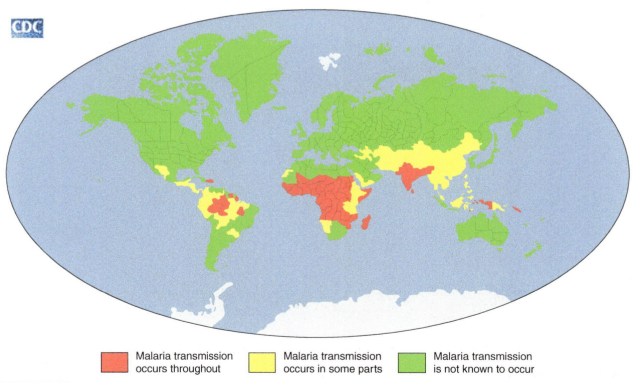

| | Malaria transmission occurs throughout | | Malaria transmission occurs in some parts | | Malaria transmission is not known to occur |

FIGURE 18-23

Current worldwide distribution of malaria. Figure provided by the Centers for Disease Control and Prevention.

● Mosquito Repellents

Mosquito protection efforts have likely been going on since man and mosquitoes first came together, and they continue unabated. The importance of mosquitoes as vectors of numerous diseases, not to mention their considerable nuisance issues, has drawn particular attention of researchers associated with public health agencies and the military. It also draws considerable commercial attention, because the demand and hope for solutions to mosquito bites drives a commercial market worth hundreds of millions of dollars. The result is an onslaught of products, some with value and some without.

Most fundamental has been the search for the ultimate effective repellent. Out of this work only a very small number of materials have been found to provide both broad activity against many kinds of mosquitoes and acceptable persistence. Long atop the short list of products that are recommended by the Centers for Disease Control and Prevention (CDC) is DEET (*N,N*-diethyl-*m*-toluamide), a material that apparently acts by blocking chemical receptors of the mosquito used to identify odors of potential host animals (e.g., lactic acid).

There has also been extensive evaluation of natural products for mosquito repellents and interest remains high in this area, in no small part because natural products are perceived by many potential consumers as somehow being safer. Thousands of plant materials have been examined

and from this it is evident that some mosquito repellent materials can be found in many plants including thyme, garlic, rosemary, cedar, pine, verbena, lemon balm, lavender, pennyroyal, geranium, cinnamon, catnip, Osage orange, basil, and peppermint. Unfortunately, effectiveness of these all tends to be very short-lived and requires constant re-application making them poor repellents in actual practice. However, a few natural products have shown good activity as mosquito repellents: a combination of geranium oil/soybean oil/coconut oil product (Bite Blocker[R]) has shown good persistence and the Centers for Disease Control (CDC) recently added lemon oil of eucalyptus on its short list of most effective repellents.

Sometimes mosquito repellent claims veer into less credible areas. Annually appearing in Sunday magazines and some garden publications are various "mosquito repellent plants" that purportedly will keep mosquitoes away if placed around the yard. All of these plants do share one feature that appears as the center of their claims; these plants contain oil of citronella, which has modest, but short-lived mosquito repellent activity. Unfortunately the amount of citronella these plants contain is often insufficient to provide any noticeable mosquito repellency and it is only released when plants are bruised. It is likely that about the only time these "mosquito repellent plants" would be effective for

FIGURE 18-24
Representative products containing mosquito repellents that are considered to be effective. Active ingredients, from left to right: DEET, picaridin, oil of lemon eucalyptus, soybean oil/geranium oil, and permethrin. Permethrin is permitted for use on clothing but not for direct application to skin. Photograph by Whitney Cranshaw/Colorado State University.

(continued)

repelling mosquitoes for even a brief period is if one placed the plant in the lap and slapped it.

Perhaps the most widespread and persistent scams related to mosquito repellents are various ultrasonic or other sound production devices. Associated claims may include testimony about their ability to mimic the wing beat frequency of dragonflies or sometimes that of the male mosquito, allegedly causing female mosquitoes to avoid the area. These claims are utterly false at many levels (female mosquitoes do not hear or respond to either male mosquitoes or dragonflies) and have been repeatedly and consistently repudiated, but they arise annually based on these testimonials and marketing. Such sales are illegal in the very few states where pest control product claims must be proved prior to in-state sales, but are allowed elsewhere. Ultrasonics, be they for use in control of mosquitoes, flea, cockroaches, rats, or whatever pest, prove to be among the best ways to learn the pest control lesson that "if looks to be too good to be true—*it is* too good to be true."

FAMILY SIMULIIDAE—BLACK FLIES

Black flies are tiny, humpbacked flies that develop in water. Adult females of most species are blood feeders and, along with mosquitoes, black flies can often be such annoying and persistent pests that they can preclude outdoor activity when they are active.

Black flies are blood pool feeders, similar to horse flies. They possess mouthparts designed to cut a tiny slice through the skin to open up capillaries. Concurrently, they introduce saliva with anesthetic and anticlotting activity that allows them to continue to feed but which may later cause reactions in the host animal. "Black fly fever" is a generalized condition that can be produced by large numbers of bites, marked by fever, nausea, headaches, and swollen lymph nodes. On occasion, large numbers of black flies have been known to even kill livestock. This can be due in part to blood loss and extreme annoyance, but also to a reaction to the effects of the saliva introduced during biting that produces a toxic shock (simuliotoxicosis).

After about 2–3 min of feeding, a black fly is fully engorged and leaves the host to convert the blood meal into 200–800 eggs. These are laid as masses on, or immediately adjacent to, running water. On hatching the larvae attach themselves to some solid surface, such as a rock, using a sucker appendage at the hind end. Black fly larvae feed with large fan-like combs projecting from the labium that strain passing food particles and move them into the mouth. Pupation also occurs underwater within a silken cocoon that somewhat resembles a slipper.

Although some black flies can produce many generations a year, most US species that are serious human biters produce only one generation a year. *Simulium venustum* is the key pest in the north central and northeastern United States where "black

(a)

(b)

fly season" typically occurs around the period between Mother's Day and Father's Day. Black flies are day active insects, and most biting occurs just after dawn and around dusk. The adults are attracted to carbon dioxide expelled during breathing and some dark colors (e.g., navy blue) increase attraction.

FIGURE 18-26
Larvae and pupae of a black fly exposed from underneath a rock in a small stream. Photograph by Whitney Cranshaw/Colorado State University.

Unlike mosquitoes, they do not fly at night and do not attempt to enter tents to feed on sleeping campers.

Most other species of black flies primarily bite birds and may only be a minor annoyance to humans. In the United States, problems with black flies affecting birds often have involved the turkey gnat (*Simulium meridionale*), which has been known to kill poultry and drive emus to their death. (Particularly the dark headed males that attract most black flies.) Black flies also can transmit species of the protozoan *Leucocytozoan*, producing malaria-like diseases in birds.

In parts of Africa and South America, black flies are much more serious problems as vectors of *Onchocerca volvulus*, a filarial worm that produces human onchocerciasis. Also known as "river blindness" the disease is estimated to currently infect nearly 18 million people, with incidence particularly high in some areas of western Africa.

The course of onchocerciasis may vary, with about a third of infected humans developing skin diseases that can involve intense itching, depigmentation, and loss of skin elasticity. More severe are cases where the filarial worms damage vision, and about 350,000 people are now blind from this disease. Intensive efforts have been made to control river blindness, and it has been eliminated from many countries. Nonetheless, it remains a serious problem in many savannah areas of Africa, where 10%–15% of the population in some villages may be blind from this cause.

The most extensive effort to manage black flies in the United States occurs in Pennsylvania where 45 rivers and streams are routinely monitored for the presence of *Simulium jenningsi*, the most serious black fly pest species in the mid-Atlantic states. The treatment of choice is the bacterium *Bacillus thuringiensis* var. *israelensis* (Bti) that is introduced into the waterways and rapidly kills black fly larvae if they ingest it. This bacterium has selective effects on larvae of certain flies that develop in water or moist soil and Bti is also widely used to control mosquitoes in ponds and pools where they breed.

● *Bacillus thuringiensis*—A Microbe for Managing Insect Pests

Insects can succumb to diseases caused by all manner of animal pathogens: viruses, bacteria, fungi, and protozoans. Natural outbreaks of these insect diseases have long intrigued scientists as a possible means of managing insect pests, and some of these disease organisms have been developed specifically for that purpose. No microbe has been more thoroughly studied and exploited than the bacterium *Bacillus thuringiensis*.

Bacillus thuringiensis (known as Bt) is a rather peculiar organism to have emerged as the leading example among microbial insecticides. Widely present in soils and many other environments, myriad Bt strains have been identified that kill insects through the production of various toxic proteins. (Most Bt toxins are produced within individual cells, often in the form of crystals, and are known as **endotoxins**.) Bt toxins can kill many types of insects without the bacteria producing an infection; indeed, dead bacteria or just the toxins alone by the microbes may be insecticidal. As Bt-toxins may kill many kinds of insects that normally would not contact them, the insecticidal activity of many of these strains seems to be an incidental characteristic of the bacteria that has proved beneficial to humans.

(continued)

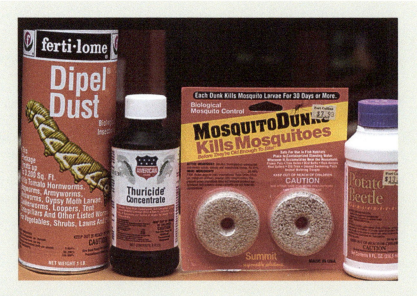

FIGURE 18-27

Bacillus thuringiensis-containing products displayed on the shelves of a nursery. On the left are dust and liquid formulations of *Bacillus thuringiensis* var. *kurstaki* that controls certain leaf-feeding Lepidoptera larvae. In the center is a "doughnut" formulation of *Bacillus thuringiensis* var. *israelensis* used to control mosquito larvae in water. On the right is a formulation of *Bacillus thuringiensis* var. *tenebrionis*, which is active against leaf beetles. Photograph by Whitney Cranshaw/Colorado State University.

The insecticidal activity of *Bacillus thuringiensis* was first observed in 1901. Interest in Bt insecticides resumed in the 1950s with the commercial development of a strain that was marketed under the trade name Thuricide. When applied as a spray or dust to foliage and ingested by caterpillars, Bt would stop their feeding within hours. Initial infection destroys midgut epithelial cells, which allows midgut fluids to enter the hemolymph. This raises the hemolymph pH, resulting in paralysis and termination of feeding, ultimately resulting in death from the combination of starvation and cell damage. In some cases the bacteria will grow within the hemolymph of affected insects, killing them by blood poisoning.

These early Bt products were quickly recognized for many advantages. Most immediately, they were essentially non-toxic to humans and showed no obvious effects on non-target species of wildlife such as birds, mammals, or fish. Later, the selective effects on target insects, sparing beneficial species such as pollinators, predators, and parasites (e.g., lady beetles, parasitic wasps), became a valuable trait. These selective characteristics were in sharp contrast to most insecticides in use at the time that had a broad spectrum of ecological effects including being toxic to humans

The past 50 years have seen sustained interest in Bt research, and many new strains of the bacteria have been identified, producing a wide array of toxic proteins targeting many different insects. Although the first strains just affected caterpillars (larvae of Lepidoptera), a new

subspecies emerging in the late 1970s (*israelensis*) was found to be effective on larvae of certain Diptera larvae—including mosquitoes, black flies, and fungus gnats. Another subspecies (*morrisoni*, strain *tenebrionis*) that had insecticidal activity against some beetles was also identified.

The use of Bt took a different turn with development of genetic engineering technologies. Since the insecticidal component activity of Bt involved the toxins, and did not require the whole organism, it proved relatively easy to identify specific genes involved in production of various toxins (usually endotoxins). These genes could then be inserted into plants through genetic engineering techniques and, if successful, result in **transgenic plants** that had the ability to produce the Bt toxins.

Genetically modified crop plants that produce Bt toxins effective against certain caterpillars and beetle larvae, particularly field corn and cotton, are now widely grown in the United States. The practice is controversial and it is in the forefront of the debate on genetically modified organisms (GMOs). Advocates point out the reduction in use of broad-spectrum insecticides when Bt-modified crops are used, and there has been a tremendous reduction in insecticide use following adoption of such crops in some important cropping systems (e.g., reduction in use of corn rootworm and European corn borer insecticides in the Corn Belt). Opponents argue that use of such crops will allow insects to become resistant to Bt, possibly affect other non-target insects, and may have unknown health hazards.

(continued)

TABLE 18-2 Sources of some toxins used in agriculture that are derived from *Bacillus thuringiensis*. Note that multiple toxins with important insecticidal activity often are produced by an individual Bt subspecies.

SUBSPECIES	ENDOTOXINS	TARGET PEST GROUPS
aizawi	Cry1Aa, Cry1Ab*, Cry1Ca, Cry1Da, Cry1Fa2	Many lepidopteran larvae
israelensis	Cry4Aa, Cry4Ab, Cry11Aa, Cyt1Aa	Larvae of mosquitoes, black flies and fungus gnats
japonensis	Cry8Ca	Scarab beetle larvae
kurstaki	Cry1Aa, Cry1Ab*, Cry1Ac*, Cry2Aa	Many lepidopteran larvae
morrisoni	Cry3Aa, Cry 3Bb*	Leaf beetle larvae, including corn rootworms
—	mCry3Aa* (modified from Cry3Aa from *morrisoni*)	Corn rootworm larvae
—	Cry1F* (modified from Cry1Fa2 from *aizawi*)	Many lepidopteran larvae
—	Cry34Ab1* and Cry35Ab1* (used only in combination)	Corn rootworm larvae
—	VIP3Aa20* (Vegetative Insecticidal Protein, not an endotoxin)	Many lepidopteran larvae

Source: Modified from Federici, B.A., Park, H.-W., and Bideshi, D.K. 2010. Overview of the basic biology of *Bacillus thuringiensis* with emphasis on genetic engineering of bacterial larvicides for mosquito control. *The Open Toxicology Journal* 3: 83–100.
*Used in transgenic crops.

FAMILY BIBIONIDAE—MARCH FLIES

March flies are dark colored, hairy flies that normally draw little attention. (The heyday of the March flies was in the middle to late Tertiary, when they apparently were among the most abundant of all insects.) Larvae of most species lead unobserved lives developing in decaying organic matter and only a very few species cause any damage to living plants. Adults usually emerge sometime in spring, do not feed, and live only for a few days. The adult period is dedicated entirely to mating and egg-laying.

In Florida and parts of some Gulf Coast states, one species is very well known, *Plecia nearctica*. Known locally as the "lovebug," this insect period- ically occurs in huge swarms, and on mating,

FIGURE 18-28
A mating pair of March flies, *Bibio slossonae*. Photograph courtesy of Tom Murray.

the insects usually remain coupled until death. Lovebug season occurs twice a year, in late April and May then again in late August and September. Often, they meet their end as a splatter on the grill or windshield of a passing automobile or truck. Swarming tends to concentrate around areas of vehicle traffic, because chemicals formed from irradiation of diesel exhaust (heptaldehyde, formaldehyde) are highly attractive to the swarming insects. Radiators and air intakes sometimes become clogged with lovebugs, and bacteria consuming their remains can etch paint.

FIGURE 18-29
An automobile splattered with "lovebugs," *Bibio nearctica*. Photograph courtesy of Herbert A. "Joe" Pase III/Texas Forest Service/Bugwood.org.

The lovebug phenomenon in Florida is actually fairly recent and some old timers may remember when this was not a semi-annual event. The species is apparently native to coastal areas of Texas and Louisiana and gradually extended its range, reaching panhandle Florida around 1949. There the lovebug found ideal conditions. Extensive pasture lands full of fresh cattle manure kept moist with rains provided ideal sites for the breeding larvae.

FAMILY CECIDOMYIIDAE—GALL MIDGES

Adult gall midges are inconspicuous insects, tiny, fragile flies with long antennae. Feeding activities produced by their larvae produce more attention and several gall midges damage crops or produce plant galls.

Of the 1,200 known North American species about half produce some sort of gall. Most cause fairly simple deformities such as a thickening and curling of leaves, needle stunting, or changes in fruit shape. Others produce galls of more bizarre appearance. For example, the terminal buds of willows are turned into pine cone shaped "willow cone galls" by various *Rhabdophaga* species and the juniper tip midge (*Oligotrophus betheli*) transforms the ends of branches so that they resemble flowers.

Historically, the most important agricultural pest among the gall midges has been the Hessian fly (*Mayetiola destructor*). The larvae tunnel the stems of wheat plants producing stunted growth and weakening so that plants may break. Fortunately, effective management methods have been identified, including attention to planting date and the development of resistant varieties, which have greatly reduced damage by this insect. Other crop pests in localized areas include the sorghum midge

(a) (b)

FIGURE 18-30

Two galls of woody plants produced by gall midges. (a) The juniper tip midge, *Oligotrophus betheli*, produces a stunting and swelling of juniper tips. (b) The pinyon spindlegall midge, *Pinyonia edulicola*, induces a swelling in the base of pinyon needles and larvae live within the chamber produced. Photographs courtesy of David Leatherman and Whitney Cranshaw, respectively.

(*Stenodiplosis sorghicola*), which prevents kernal formation of sorghum, and the sunflower midge (*Contarinia schulzi*), which can so severely distort sunflower heads that it has limited the ability to produce the crop in some locations. In gardens, the growing tips of roses are often killed by rose midges (*Dasineura rhodophaga*). The European species known as the swede midge (*Contarinia nasturtii*) has recently become established in Ontario and parts of the northeastern United States where it is a concern because of its potential to damage the new growth of various cabbage family plants.

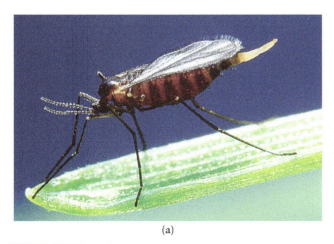 

(a) (b)

FIGURE 18-31

Adult Hessian fly (a), *Mayetiola destructor*, and larvae in the base of a wheat plant (b). Feeding by Hessian fly can cause serious stunting and distortion of wheat plants. Photograph of an adult by Scott Bauer/USDA ARS. Photograph of the larvae courtesy of Jim Kalisch/University of Nebraska.

FAMILY TABANIDAE—HORSE FLIES AND DEER FLIES

The family Tabanidae includes moderate- to large-sized flies (6–30 mm) that have a stout body and large head. Their prominent eyes are usually brightly colored, even iridescent, and sometimes patterned. Their most noticeable behavior is that females of most of the 317 known North American species are vicious biters of warm-blooded animals. Some popular names of representatives include "greenheads" found around salt marshes near East Coast beaches and "bulldogs" in northern Canada.

Horse flies and deer flies occur around wetlands. Eggs are laid in tidy layered masses either on vegetation above water or directly on moist soil. The larvae live within shallow areas of ponds or saturated soils alongside water bodies. Larvae feed on worms, insect larvae and other small animals. Pupation usually occurs in somewhat drier areas of soil above the water table.

Females often can produce an initial mass of eggs from the energy stored as a developing larva; however, most species then must feed on blood to

FIGURE 18-32
Head of a deer fly, *Chrysops flavidus*. Many deer flies have brightly patterned compound eyes. Photograph courtesy of Sturgis McKeever/Georgia Southern University/Bugwood.org.

FIGURE 18-34
A horse fly, *Tabanus proximus*, feeding. Both horse flies and deer flies are "pool feeders," making wounds that produce considerable bleeding. The flies then quickly lap up the blood and leave the host. Photograph courtesy of Sturgis McKeever/Georgia Southern University/Bugwood.org.

(a)

(b)

FIGURE 18-33
Egg mass of a horse fly (a). Egg masses are usually laid on vegetation over damp areas along the edge of water. Larvae of horse flies and deer flies (b) live in moist soil as predators, feeding on other arthropods. Photographs courtesy of Sturgis McKeever/Georgia Southern University/Bugwood.org.

nourish production of additional eggs. Mouthparts are designed to quickly cut flesh, producing a pool of blood that is subsequently lapped up. All males, and females of a few species, sustain themselves solely on nectar and pollen.

FAMILY ASILIDAE—ROBBER FLIES

Move a bit beyond a lake or pond where dragonflies reign and you often will find robber flies to be the insect rulers of the airways. Although there is a wide range in size (3–50 mm) among the 1,000 plus species that occur in North America many are large flies with a distinctive long, slender body. With very few exceptions, all are predators of other insects. Their large eyes are widely spaced and their unique needle-like mouth is very effective for stabbing prey. Some robber flies mimic the appearance of various local bees and wasps.

FIGURE 18-35
A robber fly, *Proctocanthus* sp., feeding on a wasp mantidfly. Robber flies catch a wide range of flying or jumping insects that come into range. They have piercing mouthparts that can inject a paralyzing toxin, which quickly incapacitates their prey. Photograph by Whitney Cranshaw/Colorado State University.

Robber flies are most often seen in open, sunny areas, perched on vegetation that provides good views of passing prey, usually some flying or jumping insect. When potential prey is detected the robber fly darts out, grabs it with the front legs and injects a paralyzing saliva, rapidly incapacitating the insect. The saliva also has components that begin to liquefy the tissues which are subsequently sucked out.

FIGURE 18-36
A robber fly, *Diogmites angustipennis*, emerging from its pupal skin. Robber fly larvae live in the soil and are thought to be predators of other soil-dwelling arthropods. Photograph by Whitney Cranshaw/Colorado State University.

Robber flies can frequently be seen coupled in mating and sometimes flying together for short distances. Males of the more common species can be distinguished by having a large bulbous area at the tip of the abdomen; the tip of the abdomen of females is slender, and is used to insert eggs into soil, cracks of rotting wood, or flower heads. The habits and biology of the immature stages are poorly known but they are thought to be soil-dwelling predators of other insects, such as the white grub larvae of scarab beetles.

FAMILY GLOSSINIDAE—TSETSE FLIES

Few insects have had such a sustained and devastating impact on a region of the world as have tsetse flies (*Glossina* spp.). They are vectors of the protozoan parasites that produce **sleeping sickness** in humans and **nagana** in livestock, diseases that have long devastated sub-Saharan Africa. Sleeping sickness has not only caused direct human misery by inflicting sickness and death on countless numbers of people, it has crippled the development by local economies by denying the use of livestock for food and transport.

These diseases are produced by at least four species of *Trypanosoma*. Most important is *Trypanosoma brucei* that affects humans and a wide range of domestic animals (cattle, horses, goats, camels). Other species of *Trypanosoma* can affect different specific types of livestock including *T. congolese* (cattle), *T. simiae* (pigs), and *T. vivax* (cattle). These organisms naturally occur among a wide variety of native mammals, but these non-domesticated hosts have developed immunities to the parasite such that serious illness does not develop.

Sleeping sickness of humans, produced by infection with *Trypanosoma brucei*, produces a range of unpleasant symptoms beginning with high fever, headache, and a rubbery feeling of the glands of the neck. The name of the disease results from some of its effects on sleep and activity. Infected individuals may experience alternating periods of lethargy followed by extreme, sometimes violent activity. Sleeping patterns are reversed so that one sleeps during the day and has insomnia at night. Tremors and gradual paralysis may develop and a loss of appetite contributes to a gradual wasting. Stresses on the heart ultimately may lead to death.

There are estimated to be 16 species of tsetse flies that can transmit *Trypanosoma* spp. to livestock; six of these also transmit the pathogen to humans. The individual fly species are each found in different habitats including forested riverways, high rainfall coastal forests, and wooded savanna. Together, the various tsetse flies range through much of Central Africa.

Reproduction of tsetse flies is unique. Females give birth to live, full-grown larvae. All of the first three larval instars develop within the mother, who nourishes them with highly nutritious secretions of their specialized accessory glands. Only when the young are sufficiently grown will she give birth; the larvae immediately burrow into soft soil where they promptly pupate. Adults emerge from the pupa about a month later. Adult females reproduce continuously and produce about 8–10 young.

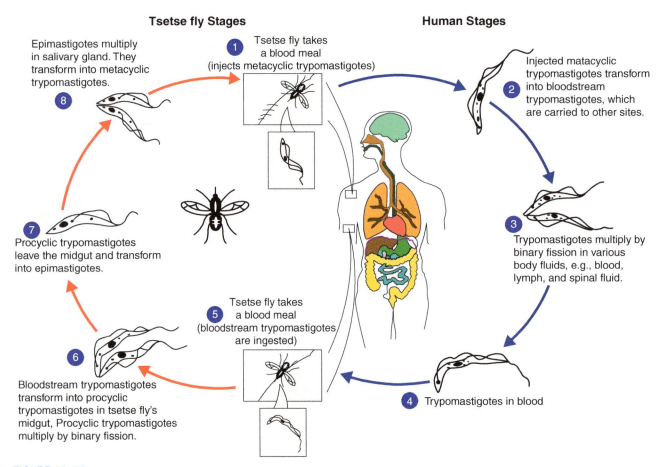

Tsetse fly Stages

Human Stages

Epimastigotes multiply in salivary gland. They transform into metacyclic trypomastigotes.

8

1 Tsetse fly takes a blood meal (injects metacyclic trypomastigotes)

Injected matacyclic trypomastigotes transform into bloodstream trypomastigotes, which are carried to other sites. **2**

7 Procyclic trypomastigotes leave the midgut and transform into epimastigotes.

3 Trypomastigotes multiply by binary fission in various body fluids, e.g., blood, lymph, and spinal fluid.

5 Tsetse fly takes a blood meal (bloodstream trypomastigotes are ingested)

6 Bloodstream trypomastigotes transform into procyclic trypomastigotes in tsetse fly's midgut, Procyclic trypomastigotes multiply by binary fission.

4 Trypomastigotes in blood

FIGURE 18-37

Life cycle of African trypanosoma, often known as sleeping sickness. The disease is produced by infection with a protozoan parasite (*Trypanosoma* spp.) that is transmitted to animals, including humans, by the bite of an infected tsetse fly. Drawing courtesy of the Centers for Disease Control and Prevention.

Tsetse flies spend most of their time resting in shaded, cooler wooded areas. They are active before and after the hottest period of the day and may forage for host animals up to 200–330 m from resting sites. Wild mammals (primarily suids and bovids) are their primary hosts, but tsetse flies are known to feed on reptiles and birds. When they do bite it involves a rapid penetration of the skin to produce a pool of blood. The blood is then ingested rapidly so that they usually spend only about a minute feeding.

Feeding may occur every day and females need at least three meals to nourish each larva they produce. During this time only a small percentage of individual tsetse flies will incidentally acquire *Trypanosoma* protozoans from the blood of an infected host. Regardless, should they acquire the organism they subsequently retain it for a long period and may transmit it to new animals for a period exceeding 3 months. A combination of effective drugs, selective insecticide treatments applied to fly resting areas, and forest clearing have eliminated or suppressed tsetse fly borne diseases in many areas. Nonetheless, they remain elsewhere, and *Trypanosoma* protozoans remain endemic among the many wild mammals that can support the organisms without producing symptoms of disease.

FAMILY SYRPHIDAE—FLOWER FLIES OR HOVER FLIES

Flower flies are often the most common and conspicuous flies found around flowering plants. Many of them are such effective mimics of bees or wasps that their presence may be overlooked. The 6,000 species of flower flies known to occur worldwide range from small to large (4–25 mm) and from slender to heavy-bodied. Often they have orange or yellow markings simulating stinging insects and some have dense hairs covering the body, further mimicking a fuzzy bee.

Adults of most flower flies feed on nectar. They are excellent fliers and as they move about plants they are capable of rapid darting flights in all directions. Additionally, they can hover in place, leading to the other common name "hover flies." Their activities among flowers produce substantial pollination and they are considered second only to bees in value as pollinators.

Habits vary among flower fly larvae. Many are predators, particularly of aphids, and these have piercing stylets to slash their prey. Plant injuries are

FIGURE 18-38
Mimicry of bees and wasps is common among the flower flies (Syrphidae family). A flower fly is collecting nectar at the top of the picture and a honey bee is foraging beneath it. Photograph by Whitney Cranshaw/Colorado State University.

FIGURE 18-39
Flower flies are sometimes known as "hover flies" due to their ability to hover for sustained periods. Photograph courtesy of Brian Valentine.

produced by the "bulb flies" that grind into the base of flower bulbs and other plants. (Adult bulb flies are excellent mimics of bumble bees.) Other species of flower flies feed as scavengers of decaying plant matter. One conspicuous group of scavengers is the rattailed maggots (*Eristalis* sp.) that live in watery ooze of plants or in shallow polluted water. They breathe through a very long tail-like spiracle that can reach the water surface like a periscope. Adults of these species mimic honey bees and are known as drone flies.

FIGURE 18-40
A flower fly larva feeding on an aphid. Flower fly larvae feed almost exclusively on aphids and are among their most important natural enemies. Photograph courtesy of Brian Valentine.

FAMILY CALLIPHORIDAE— BLOW FLIES

Blow flies are moderate-sized flies common around yards and many have vivid metallic coloration with the most familiar often "greenbottle" or "bluebottle" flies. Adults feed on nectar or fluids sucked from the surface of carrion, dung, cut fruit or various types of garbage. They are among the most important of the "filth flies" and their incidental movements can allow them to cross-contaminate foods with the organisms that produce dysentery and other food-borne illnesses.

The great majority of blow fly larvae develop as scavengers and adults readily visit a recently killed

FIGURE 18-41
A greenbottle type of blow fly, *Lucilia* spp. Photograph by Whitney Cranshaw/Colorado State University.

animal or fresh animal droppings. Eggs are laid as masses and young maggots rapidly develop, breaking down tissues with digestive enzymes. Their appearance coincides with a bloating corpse, which is sometimes then described as "fly blown," hence the term "blow fly."

The period from egg deposition to a fully developed larva can occur in a little more than a

FIGURE 18-42
Blow flies are among the very first insects attracted to fresh carrion. Masses of eggs are being laid at this time and the young larvae rapidly develop on the decaying flesh. Photograph by Whitney Cranshaw/Colorado State University.

FIGURE 18-43
The blow fly larvae feeding on these chicken parts are about 4 days old. They develop rapidly and will be full grown in a few more days, after which they will usually wander from their food in search of a place to pupate. Photograph by Whitney Cranshaw/Colorado State University.

week. The larvae usually then migrate a short distance away and pupate. Pupation occurs within the last larval skin and results in a dark pupal form (**puparia**) that somewhat resembles a smooth seed. Development rate is well-correlated with temperature and some species are important indicators in the science of forensic entomology, where they can be used to accurately determine time of death.

By their association with dead flesh blow flies are often considered repulsive. However, they have been used in a unique medical treatment known as **maggot therapy**. This developed from the observations by physicians during war times in the 1800s that gravely wounded soldiers considered beyond saving often survived and recovered more quickly than those provided the medical treatment of their time. These observations were expanded further during World War I when an orthopedic surgeon, William S. Baer, observed that certain species of blow flies found in wounds consumed only the dead tissues, leaving the healthy tissue intact and allowing it to heal more effectively. Later he followed up with experiments conducted at the Johns Hopkins Institute where it was learned that the maggots also produced an antibiotic waste product, allantoin. These studies ultimately led to the practice of maggot therapy where certain blow fly maggots, aseptically reared, are introduced to clean out deep infected wounds that resist other treatment. Its practice has resumed in recent years with the appearance of antibiotic-resistant strains of bacteria and diabetic-induced gangrene.

A particularly nasty member of the family is the screwworm, *Cochliomyia hominivorax*. Unlike other members of this family it is an obligatory parasite, developing on the flesh of warm-blooded animals. It has been particularly important as a pest of cattle and in 1935 there were 1.2 million cases reported in Texas alone. Although the screwworm cannot penetrate skin directly (as can bot flies), sores and open wounds of any type (dehorning, castration, cuts, insect/tick bites) are attractive to the egg-laying adult females. Masses of eggs are laid in the wound and the larvae develop rapidly, becoming full-grown in less than a week. They then drop to the ground, pupate, and emerge as an adult about a week later. Continuous re-infestations can occur and screwworms have been known to kill an animal in 10 days, usually due to shock.

FIGURE 18-44
Head of a screwworm larva, showing the mouth hooks with which it feeds. Other blow fly larvae have similar mouthparts which allow them to cut tissues on which they feed. Photograph by Scott Bauer/USDA ARS.

The screwworm is native over a broad area of the Americas and originally ranged as far north as Texas, New Mexico and Arizona. It was also accidentally introduced into Florida and some Caribbean islands.

FIGURE 18-45
A cluster fly, *Pollenia pediculata*. Cluster flies attract attention when they move into the upper areas of buildings in late summer, where they cluster in wall voids through the winter. Periodically, cluster flies may work their way into living areas and are considered a significant nuisance pest. Unlike other blow flies that develop on carrion or decaying plant matter the cluster flies develop as parasites of earthworms. Photograph by Whitney Cranshaw/Colorado State University.

Through a particularly elegant management approach, known as the Sterile Insect Technique (SIT), this insect has been eliminated over a wide area and is now confined south of the Isthmus of Panama.

Other blow flies have substantially more innocuous habits. One group that commonly enters homes during winter is the cluster flies (*Pollenia* spp.). These flies often move behind the sun-exposed south and west walls of buildings in late summer, infiltrating behind walls and concentrating in upper stories. Although dormant during winter, they may become active during warm days. During such episodes they may incidentally emerge into living areas where they may lazily fly about in a zombie-like flight. Cluster flies do not reproduce in homes nor are they considered filth flies, because they develop during the warmer months as parasites within certain species of earthworms.

● Sterile Insect Technique—Turning Insect Biology against Itself

Some of the most novel techniques developed to manage insects involve **autocidal controls,** by which the insect can be turned against itself in some manner. The most widely used autocidal method is the sterile insect technique (SIT) where insects are reared, sterilized, and allowed to mate with wild insects leading to either no egg production or infertile/non-viable individuals. If successful, ultimately the pest population crashes.

SIT was developed and demonstrated spectacularly in a program to eliminate the New World screwworm (*Cochliomyia hominivorax*) from North America. This insect formerly ranged into many states of the southern United States where it caused terrible injuries to livestock and wildlife. United States Department of Agriculture (USDA) researchers identified a key aspect of its biology that provided a weak link: females mate only once. This made the insect particularly vulnerable to the use of SIT. With this information in hand, E.F. Knipling, R. Bushland, and coworkers developed a program to overwhelm the wild insect populations with sterilized insects and provided the theoretical framework of how SIT could bring about their local extinction.

After demonstrating its potential in 1954 on an isolated screwworm infestation on the Caribbean island of Curacao, eradication efforts moved to the Florida peninsula. Rearing facilities produced up to 50 million sterile flies/week and spanning an 18-month course over 2 billion flies were released by air over an area the size of Utah. It was a complete success that completely eradicated the insect from the southeast United States.

New goals were soon set to further eliminate screwworm from the remainder of the infested areas of the United States, and this was achieved in 1966. Re-infestations from Mexico necessitated the development of cooperative efforts between the United States and Mexico to eradicate the insect over a broader area. In 1977 the screwworm rearing

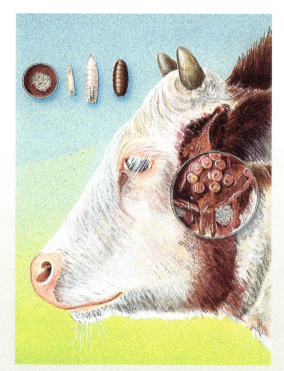

FIGURE 18-46
An illustration of the life stages of the screwworm, *Cochliomyia hominivorax*. The screwworm is one of the only blow flies that readily colonizes living tissue, although eggs are originally laid at small wounds. The screwworm was formerly a serious pest of livestock and wildlife in the southern United States, but has been eradicated through the sue of SIT. Illustration by Art Cushman. Figure courtesy of USDA/Smithsonian Institution Entomology Laboratory and Bugwood.org.

facility was moved to Tuxtla Gutierrez in Chiapas, Mexico and intensive efforts were put in place to eradicate it from that country. The program showed sustained success and ultimately all the countries of

(continued)

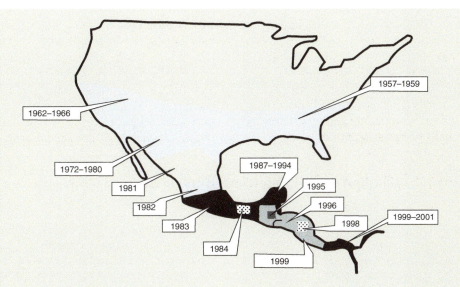

FIGURE 18-47
The screwworm has been progressively eradicated from the southern United States through most of Central America using the Sterile Insect Technique (SIT). The dates indicate when the SIT was initiated. The insect is presently eliminated from North America through the Isthmus of Panama. Figure from the UN Food and Agriculture Organization.

Central America became participants, allowing screwworm to be progressively eradicated by SIT throughout Central America. Mexico was declared free of screwworm in 1991, Belize and Guatemala in 1994, El Salvador in 1995, Honduras in 1996, Nicaragua in 1999, and Costa Rica in 2000. Efforts now are in place to sustain a screwworm fly-free barrier zone along the narrow Darien Gap of Panama, preventing reinvasion from South America where the insect still remains a very important livestock pest.

Since the original demonstration of the SIT there have been several successful uses against other insects as well. It has been particularly effective for eradication of some newly introduced insects that are still in low populations, such as various tropical fruit flies (e.g., the oriental fruit fly, *Bactrocera dorsalis*) and the pink bollworm (*Pectinophora gossypiella*). For some eradication efforts the SIT has been combined with other techniques, such as insecticide baiting, to control introduced populations of the Mediterranean fruit fly (*Ceratitis capitata*) in California and Florida.

● Use of Insects in Forensics

Insects can be used as forensic indicators in many kinds of criminal investigations and they have a history that far predates the current wave of interest in the subject stimulated by CSI-type television shows. For example, the presence of insect species that have a restricted geographic range can determine if a vehicle has been driven in a certain area or where an infestation of some insect found in food or furniture may have originated. Much of the science of forensic entomology, however, is used to help determine the time of death.

Two fundamental concepts underlie using insects in crime scene investigations involving a death. One is that, in any area, at any time there is a fairly predictable **faunal succession** of insects that come to carrion to feed

directly on the carcass or to prey on and parasitize those insect that do utilize carrion. One can therefore determine the progress of decay (approximate time since death) based on the animals feeding on or within the body. The specific insect species involved may vary with geographic location and time of year, but the faunal succession associated with different stages of decay can be summarized as follows:

Fresh carrion may be almost immediately visited by various blow flies (Calliphoridae), flesh flies (Sarcophagidae), muscid flies (Muscidae), and sphaerocerid flies (Sphaeroceridae). Among the most common of these first visitors are the

(*continued*)

black blow fly (*Phormia regina*), various green-bottle flies (*Lucilia* spp.) and bluebottle flies (*Calliphora* spp.). These lay eggs or living larvae that rapidly develop. Scavenging species of yellowjackets are also common at fresh carrion during summer.

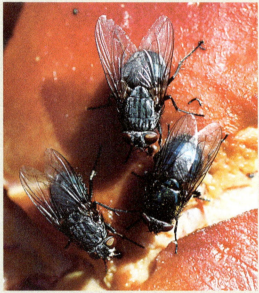

FIGURE 18-48
Bluebottle flies, *Calliphora* spp., and the black blow fly, *Phormia regina*, are among the first insects to discover and lay eggs on fresh carrion. Photograph by Whitney Cranshaw/Colorado State University.

FIGURE 18-49
Hairy rove beetles are one of the predators that are attracted to carrion after a few days, when the blow fly larvae are present on which they feed. Photograph by Whitney Cranshaw/Colorado State University.

Bloat stage carrion, which typically develops within 24 h, continues to be attractive to egg-laying adults of many of the same species, but rove beetles (Staphylinidae) that prey on fly larvae also are attracted. Skipper flies (Piophilidae), hump-backed flies (Phoridae), and sepsid flies (Sepsidae) also may begin to visit and lay eggs. Carrion beetles start to arrive.

During the **active decay** stage, the larvae of previously visiting blow flies and flesh flies have often done extensive feeding, and the bloated body deflates, often oozing fluids into the soil. Muscid flies and skipper flies continue to visit and lay eggs, but often different species than were originally present. Predators such as rove beetles and hister beetles (Histeridae) increase in numbers.

Following this is a period of **advanced decay** when the remains dry rapidly and most of the flesh has been consumed. Some rove beetles continue to be present along with piophilid flies. Certain carrion-feeding checkered beetles (Cleridae), such as the redlegged ham beetle (*Necrobia rufipes*) are attracted at this stage. As the carcass continues to dry, certain carrion-feeding sap beetles (Nitidulidae) and dermestid beetles (Dermestidae) come to predominate.

The second important concept is that the rate of insect development is predictable based on temperature. For investigative purposes, most important are immature stages of blow flies and flesh

FIGURE 18-50
Insects such as the redlegged ham beetle, *Necrobia rufipes*, and dermestid beetles are attracted to carrion after it has dried and are in the last wave of insects that feed on and help decompose animal matter. Photograph courtesy of Jim Kalisch/University of Nebraska.

(*continued*)

flies collected at the site. Meticulous care must be made in collecting the insects and subsequently handling them, often requiring that some be reared to determine the exact insect species and life stage present. When the forensic entomologist is able to determine the species of insect present, literature on the insect's habits and development can be referenced. With this information in hand, combined with weather records that provide local temperatures at the site, an informed estimate can be made of the "time since colonization." This figure can be considered when determining the postmortem interval, or time since death. Insects can also be used to determine the presence of toxins or drugs associated with the corpse, because they can incorporate them into their body in detectable quantities after consuming tissues.

FAMILY MUSCIDAE—HOUSE FLIES AND RELATIVES

God in His wisdom
made the fly
and then forgot
to tell us why.
- Ogden Nash

Adults of the family Muscidae are usually moderate sized, dull gray/brown flies covered with bristles. Although some 620 species occur in North America alone, one species dominates human attention—the house fly, *Musca domestica*.

Like many muscid flies, the maggot-form larvae of the house fly develop as scavengers. It is a classic "filth fly" that breeds readily in manure and in rotting vegetable waste, and thus it is heavily contaminated with bacteria. Their life cycle is extremely rapid, well adapted to these ephemeral food sources, and larvae may be full-grown within a week after eggs are laid.

The incredible reproductive capacity of the house fly has long been obvious and occasionally has been quantified. One famous calculation was done a century ago by the entomologist L.O. Howard who estimated the potential population of house flies that theoretically could be produced if starting with a single fertilized female in Washington, DC during the beginning of "fly season" (April 15 through September 10). Making a series of assumptions, most notably that all progeny survived and developed normally, he estimated that 5,598,720,000,000 (5.5 trillion+) house flies could result by the end of the 5 month period.

Adult house flies feed using a prominent proboscis comprised of an enlarged labium that is tipped with small pores allowing it to sponge fluids. Only liquids can be ingested and if feeding on solid material (e.g., sugar cube) they regurgitate the fluids in their foregut to help dissolve the food. As it dries, this regurgitated material darkens and appears as "fly specks."

FIGURE 18-51
House flies, *Musca domestica*, commonly breed in decaying vegetation, such as this overripe melon. Photograph by Whitney Cranshaw/Colorado State University.

Heavy contamination with bacteria is a constant with the house fly. The larvae live amongst decay and indeed require the presence of decay bacteria to develop. Adults visit all manner of materials and provide an excellent means of cross contamination. House flies have been implicated in transmission of a great many disease-producing bacteria, notably those that involve food or water-borne diseases such as typhoid and dysentery. Related to this is one of history's little ironies. When the automobile was first introduced, among its many features that were promoted was that it could serve as a non-polluting alternative to the horse. At that time of the introduction of the automobile, there were massive problems with accumulating horse manure, and associated swarms of house flies were very common in cities and towns.

Several other muscid flies are important as pests of livestock. The face fly (*Musca autumnalis*) develops in fresh cattle manure, and adults visit the heads of cattle and horses, feeding on fluids around the eyes and mouth. In addition to their nuisance effects that disturb the animals, they also can transmit pathogens that produce pink eye disease. Even more bothersome is the bush fly (*Musca vetustissima*) that can make life miserable for humans and livestock in rural areas of Australia. Other flies have the mouthparts modified so that they can draw blood such as the horn fly (*Haematobia irritans*) which can be a seriously irritating pest of pastured cattle. In areas where straw waste mixes with water and soil, stable flies (*Stomoxys calcitrans*) thrive, and are known as the "biting house fly."

FIGURE 18-53
Illustration of the life cycle of a horn fly, *Haematobia irritans*, and the injury it causes to cattle. Adults are blood feeders and most commonly feed on the legs of the animal. Larvae develop in moist manure. Illustration by Art Cushman. Figure courtesy of USDA/Smithsonian Institution Entomology Laboratory and Bugwood.org.

FIGURE 18-52
Face flies, *Musca vetustissima*, breed in moist animal manure. The adults often feed on fluids around the head of animals and can transmit pathogens that produce pink eye. Photograph courtesy of Clemson University/Bugwood.org.

FIGURE 18-54
A stable fly, *Stomoxys calcitrans*. Sometimes called the "biting house fly," the adults have mouthparts designed to pierce and feed on blood. Moist straw mixed with manure is a favored breeding site for the larvae. Photograph courtesy of David Shetlar/The Ohio State University.

FAMILY OESTRIDAE—BOT FLIES AND WARBLE FLIES

Bot flies are notorious insects that develop as internal parasites of mammals, producing a condition known as **myiasis**, where organs and/or tissues are infested by their large grub-like larvae. In North America 41 species occur and some of the more prominent are summarized by genera as follows:

Gasterophilus species. Known as the horse bot flies, the adults resemble honey bees. Eggs of the most common species, *Gasterophilus intestinalis*, glue their eggs to the hairs of the inner foreleg where they are stimulated to hatch by the animal's warmth and saliva due to the horse licking the area. The larvae are then accidentally swallowed and develop in the stomach of the horse. The following spring or summer they become full-grown, detach from the horse's stomach and drop out with the feces, where they pupate. Two other species present in North America lay their eggs, respectively, near the nose and throat of the horse and the larvae crawl into the animal where they also develop within the alimentary tract.

FIGURE 18-55
An adult rabbit bot fly, *Cuterebra* sp. Photograph courtesy of Jim Kalisch/University of Nebraska.

Cuterebra species. Known as robust bot flies, the adults are large, stout bodied, hairy flies that resemble bumble bees. In North America there are 29 species and all are parasites of rabbits or rodents. These large insects can be huge in proportion to the size of a tiny host mouse or squirrel and often may kill it. Many species lay

their eggs at the entrance of the animal burrows and will sometimes incidentally parasitize dogs or other visiting animals.

FIGURE 18-56
Rabbit bot fly larva emerging from host. Photograph courtesy of Ben Whately/University of Nebraska.

Hypoderma species. Three species of the ox warble flies parasitize cattle, wild deer, and caribou in North America. They lay their eggs on the hairs of the legs and then penetrate the skin and undergo a long (4 month+) migration through the body of the host. Ultimately they lodge just under the skin of the back where their presence can be seen in the form of a large swelling (warble) with a small breathing tube penetrating the skin surface. When full-grown the larvae emerge through the skin and drop to the ground where they pupate.

Adults of these bot flies somewhat resemble bumble bees and they are fast fliers. They also make an audible buzzing noise in flight that causes nervousness in the host animals. Often they may trigger cattle to stampede in response to their buzzing. From such behavior that are often referred to as "heel flies." The effect of these flies on cattle is also known as "gadding" and has led to the term "gadfly" that refers to a persistent and annoying pest. (Note: Deer flies and horse flies of the family Tabanidae are also sometimes referred to as "gadflies.")

One of the most unusual bot flies is *Dermatobia hominis*, which occurs over a broad area from southern Mexico through northern Argentina. Sometimes known as the human bot fly (or torsalo), host animals include cattle, dogs, pigs, sheep, horses, cats, monkeys and, rarely, humans. It is a serious pest

FIGURE 18-57
Life cycle of the cattle grub. The larvae develop as an internal parasite of cattle. The adults lay eggs on the legs of their host, and their buzzing while in flight can cause distress that may cause cattle to stampede. Adult cattle grubs are sometimes known as "heel flies" or "gad flies." Illustration by Art Cushman. Figure courtesy of USDA/Smithsonian Institution Entomology Laboratory and Bugwood.org.

of livestock and is estimated to produce annual losses to cattle of over a quarter billion dollars in the form of reduced weight, hide damage, reduced milk production, and death of calves.

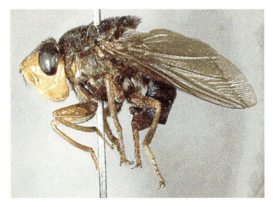

FIGURE 18-58
A human bot fly, *Dermatobia hominis*. Larvae develop as internal parasites of many mammals, including humans. Photograph by J. Eibl and provided courtesy of the Centers for Disease Control and Prevention.

Adults of the human bot fly never directly approach their host animal. Instead they use an intermediate insect to carry their eggs, a behavior known as **phoresy**. Adult females hover in the vicinity of an animal host and then seize visiting mosquitoes or biting flies onto which they will glue their own eggs then release it. When the mosquito (or other biting fly) then visits a mammalian host, the body warmth causes the bot fly egg to hatch, and the resultant larva then tunnels into the skin. Over the next 6 weeks the bot fly larva feeds and grows, producing considerable pain and discomfort from the spines of its body. When full grown, the larva exits the skin, drops to the ground to pupate. Usually the wound heals without complication.

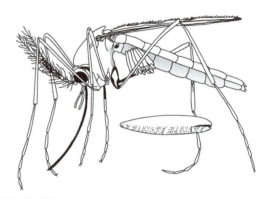

FIGURE 18-59
The human bot fly does not lay its eggs directly on its host but instead uses an intermediary insects, such as a mosquito. In this image bot fly eggs are attached to the underside of a mosquito. The bot fly eggs hatch when brought next to a warm body, as occurs when the mosquito finds an animal host on which to feed. Illustration by Art Cushman. Figure courtesy of USDA/Smithsonian Institution Entomology Laboratory and Bugwood.org.

FAMILY TEPHRITIDAE—FRUIT FLIES

Fruit flies are small- to medium-sized flies (2–12 mm), usually colorful and often with intricately patterned wings, leading to the common name "picture-winged flies." Larvae of almost all of the 4,450 known species develop in fruit, seeds, stems, or other plant parts, and the fruit flies are the most important family of flies affecting agricultural crops. About 70 species are considered to be major pests, with the majority found in tropical areas. In addition to direct crop losses and high associated control costs, the presence of certain fruit fly species in a region limit export markets, as fruit may be banned from entry to countries not infested with the insect. Despite efforts to limit their spread, many important fruit fly pest species are easily transported in fruit have been accidentally spread around the world.

In North America, important pest species include the apple maggot (*Rhagoletis pomonella*), European and eastern cherry fruit flies (*Rhagoletis cerasi* and *R. cingulata*, respectively), blueberry maggot (*Rhagoletis mendex*), and walnut husk fly (*Rhagoletis completa*). In addition, many millions of dollars are spent by state and federal agencies in eradication efforts whenever new fruit fly species become accidentally introduced into the country. This has happened repeatedly with some species, notably the oriental fruit fly (*Bactrocera dorsalis*) and the Mediterranean fruit fly (*Ceratitis capitata*). On the other hand, a few fruit flies have been purposefully introduced into the United States and Canada for biological control of invasive weeds, such as spotted knapweed and Canada thistle.

Many fruit flies exhibit elaborate courtship, mating, and oviposition (egg laying) behaviors. Males are known to release sex pheromones to attract females and engage in elaborate wing-fanning behaviors in an effort to disperse these chemicals. Males may also engage in stylized fights to control territories where females are likely to visit. Mating behaviors may also be complex and can involve

FIGURE 18-61
Injury to apple by the apple maggot, *Rhagoletis pomonella*. The larvae make meandering feeding tracks through the flesh and egg-laying punctures by the adults dimple the skin of affected apples. Photograph by Whitney Cranshaw/Colorado State University.

intricate courtship dances, wing flicking, and sometimes, presentation of nuptial gifts by the male to the female.

FIGURE 18-62
An oriental fruit fly, *Bactrocera dorsalis*, ovipositing on a fruit. Photograph courtesy of Scott Bauer/USDA ARS.

When female fruit flies oviposit they often also lay down a pheromone chemical signal designed to deter future egg laying at the site. Such "oviposition deterrent" or "spacing pheromones" help ensure that an individual fruit is not colonized such that serious competition for food occurs among the developing larvae. When full grown, larvae of many seed-feeding

FIGURE 18-60
Adult of the Mediterranean fruit fly, *Ceratitis capitata*. The "medfly" is a serious pest of fruit crops and it has been the repeated subject for eradication whenever it has become accidently introduced and established within the continental United States. Photograph courtesy of Scott Bauer/USDA ARS.

species will often pupate within the plant; those that develop within fruit exit and drop to the ground. To find a suitable site for pupation, larvae of some fruit flies can jump several centimeters, hooking their mouthparts to the end of the body, then releasing like a spring.

FAMILY DROSOPHILIDAE—VINEGAR FLIES/SMALL FRUIT FLIES/POMACE FLIES

Vinegar flies are tiny flies, typically only 1–2 mm, with dusky wings and they are most commonly encountered near a fruit bowl. Larvae feed on yeasts and other microorganisms associated with over ripe fruit, plant ooze, stale beer in the bottom of bottles, and various kinds of decaying plant material. Often they are known as "fruit flies," but as this name is best used to describe another fly family (Tephritidae) that damages developing fruit, they are alternately known as "little fruit flies." (This distinction should perhaps also be reflected in the oft-penned bit of graffiti—"Time flies like an arrow; (little) fruit flies like a banana.")

The occasional minor annoyance they produce aside, vinegar flies have made tremendous contributions to science. One species in particular, *Drosophila melanogaster*, could serve as the poster insect for the field of functional genetics. Despite their small size they have large chromosomes in the salivary glands that are easy to study. Because of their small size and enviable reproductive ability they are very easy to rear and maintain in a small space. About 25 generations may be produced in a single year and the occurrence of mutants has allowed discovery of such fundamentals of genetics as chromosomal sex linkage, crossing over, inversions, translocation, and duplication. *Drosophila melanogaster* also became the first insect to have its genes sequenced—all 13,600 of them.

Small fruit flies have also been widely used in studies of insect behavior and physiology. Among the more unusual findings occurred during a research project investigating competition effects of sperm size. The species *Drosophila bifurca* was found to have flagella on the sperm that, if teased out completely, stretch 6 cm—20 times the length of the insect's body.

In the isolation of the Hawaiian Islands a very interesting complex of *Drosophila* species has evolved. Many are found on only a single island and on that island they are associated with only a few

FIGURE 18-63
A vinegar fly/small fruit fly, *Drosophila* sp. These common flies develop on yeasts that grow on overripe fruit and other suitable sites, including unwashed beer and soft drink containers. Photograph courtesy of Brian Valentine.

FIGURE 18-64
A simple trap for small fruit flies. In the base of the glass is some fermenting material on which yeasts grow; a mixture of beer and overripe banana was used in this trap. A cone on the top can allow the flies to come into the container but have difficulty exiting. Photograph by Whitney Cranshaw/Colorado State University.

types of plants. Extensive destruction of native host plants through effects of introduced mammals (e.g., rats, pigs, cattle) and competition from invasive plants have prompted concerns about their imminent extinction, and in 2006 twelve Hawaiian species of *Drosophila* were listed as endangered or threatened species.

Order Strepsiptera— Twisted-Wing Parasites

Entomology etymology. The order name is derived from the Greek words for twisted (streptos) and wing, a reference to the unique fan-like hindwings (with only a few lateral veins) of the males. Insects of this order are commonly referred to as strepsipterans.

The twisted-wing parasites are a small insect order (596 species worldwide) that all develop as internal parasites of other insects. They are also among the most baffling orders to biologists, showing an array of unique physical features and behaviors, making them the "enigmatic order." Some systematic studies suggest that perhaps their closest relatives among the insects are the beetles.

Adult forms show extreme sexual dimorphism. Males are winged, with a single pair of large fan-like hindwings; forewings are small and clubbed, resembling the halteres found in Diptera. They have eyes with large facets resembling a raspberry and possess prominent branched antennae. Females lose their legs as they develop within their insect host and will not develop legs, antennae or eyes; they remain within the host for their entire life. An exception to this habit does occur among the most primitive stresipteran family, Mengenellidae, which parasitize silverfish. These species produce females that leave the host just before pupation and retain mobility, eyes, and small antennae.

(a)

(b)

FIGURE 18-66

Polistes paper wasps parasitized by strepsipterans. The female strepsipteran is immobile and embedded between segments of the wasp abdomen. Photographs courtesy of (a) David Shetlar/The Ohio State University and (b) A.D. Ali/The Davey Institute.

FIGURE 18-65

An adult male strepsipteran. Photograph courtesy of David Shetlar/The University of Ohio.

The female's head protrudes through the body wall of its host insect, and females feed directly on host tissues. Mating occurs via an external brood canal that allows the sperm to move into her body cavity. Subsequent to mating, the adult female's body will become packed with developing embryos. Ultimately, thousands of tiny (0.1 mm) larvae emerge from this brood canal.

These larvae are very active insects, sometimes known as triungulins (a term given also to the first stage larvae of blister beetles and mantidflies). As these larvae disperse they seek out passing host insects and, if successful, attach to them. The larvae then burrow through the host exoskeleton, forming a small pocket just above the cuticular epidermis and then molt to a legless second instar. They continue to molt and develop at this site and do not completely shed the old larval skin at each molt. In males, pupation occurs within the last larval skin in a manner similar to many Diptera.

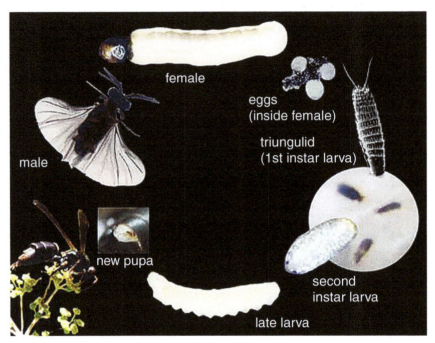

FIGURE 18-67
Life cycle of a strepsipteran parasite of a paper wasp. Figure courtesy of Elke Buschbeck and Brigit Ehmer/University of Cincinnati.

Developing Strepsiptera rarely kill their host insect but may adversely affect it so that it is sterilized. Several insect orders serve as hosts for Strepsiptera including Hemiptera, Hymenoptera, Orthoptera, Mantodea, and Hymenoptera. In one unusual family (Mymecolacidae) the two sexes develop in different types of hosts with males parasitizing ants and females parasitizing Orthoptera or Mantodea.

Appendix I: State Insects

State Insects, State Butterflies, or other arthropods that have been formally recognized in North America.

STATE	STATE INSECT*,†	SCIENTIFIC NAME
Alabama	Monarch butterfly	*Danaus plexippus*
	Eastern tiger swallowtail butterfly*	*Papilio glaucus*
Alaska	Fourspotted skimmer dragonfly	*Libellula quadrimaculata*
Arizona	Twotailed swallowtail butterfly	*Papilio multicaudatus*
Arkansas	Honey bee	*Apis mellifera*
	Diana fritillary butterfly*	*Speyeria diana*
California	California dogface butterfly	*Zerene eurydice*
Colorado	Colorado hairstreak butterfly	*Hypaurotis crysalus*
Connecticut	European praying mantid	*Mantis religiosa*
Delaware	Convergent lady beetle	*Hippodamia convergens*
	(Eastern) tiger swallowtail butterfly*	*Papilio glaucus*
District of Columbia	None	
Florida	Zebra butterfly	*Heliconius charithonius*
Georgia	Honey bee	*Apis mellifera*
	(Eastern) tiger swallowtail butterfly*	*Papilio glaucus*
Hawaii	None	
Idaho	Monarch butterfly	*Danaus plexippus*
Illinois	Monarch butterfly	*Danaus plexippus*
Indiana	None	
Iowa	None	
Kansas	Honey bee	*Apis mellifera*
Kentucky	Viceroy butterfly	*Limenitis archippus*

(*continued*)

425

STATE	STATE INSECT*,†	SCIENTIFIC NAME
Louisiana	Honey bee	*Apis mellifera*
Maine	Honey bee	*Apis mellifera*
Maryland	Baltimore checkerspot butterfly	*Euphydryas phaeton*

FIGURE I-1
Maryland was one of the first states to designate a state insect, choosing the Baltimore checkerspot. Photograph courtesy of David Cappaert/Michigan State University/Bugwood.org.

STATE	STATE INSECT*,†	SCIENTIFIC NAME
Massachusetts	Ladybug beetle	Unspecified species
Michigan	None	
Minnesota	Monarch butterfly	*Danaus plexippus*
Mississippi	Honey bee	*Apis mellifera*
	Spicebush swallowtail butterfly*	*Papilio troilus*
Missouri	Honey bee	*Apis mellifera*
Montana	Mourning cloak*	*Nymphalis antiopa*
Nebraska	Honey bee	*Apis mellifera*
Nevada	None	
New Hampshire	Twospotted lady beetle	*Adalia bipunctata*
	Karner blue butterfly*	*Lycaeides melissa samuelis*
New Jersey	Honey bee	*Apis mellifera*
New Mexico	Tarantula hawk wasp	*Pepsis formosa*
	Sandia hairstreak butterfly*	*Callophrys mcfarlandi*
New York	Ninespotted lady beetle	*Coccinella novemnotata*

(*continued*)

STATE	STATE INSECT*,†	SCIENTIFIC NAME

FIGURE I-2
The ninespotted lady beetle is the state insect of New York. In recent decades, its numbers have drastically declined and it was the catalyst for a national citizen science survey, "The Lost Ladybug Project." Photograph by Whitney Cranshaw/Colorado State University.

North Carolina	European honeybee	*Apis mellifera*
North Dakota	None	
Ohio	Ladybug beetle	Unspecified species, probably *Hippodamia convergens*
Oklahoma	European honeybee	*Apis mellifera*
	Black swallowtail butterfly*	*Papilio polyxenes*
Oregon	Oregon swallowtail butterfly	*Papilio oregonius*
Pennsylvania	Pennsylvania firefly	*Photuris pennsylvanica*
Rhode Island	None	
South Carolina	Carolina mantid	*Stagmomantis carolina*
	Eastern tiger swallowtail butterfly*	*Papilio glaucus*

(*continued*)

STATE	STATE INSECT*,†	SCIENTIFIC NAME

FIGURE I-3
Four states have chosen the eastern tiger swallowtail as either the state insect or the state butterfly. Photograph courtesy of Steven Katovich/USDA Forest Service/Bugwood.org.

FIGURE I-4
The Carolina mantid is the state insect of South Carolina. Photograph courtesy of Sturgis McKeever/Georgia Southern University/Bugwood.org.

STATE	STATE INSECT*,†	SCIENTIFIC NAME
South Dakota	Honey bee	*Apis mellifera*
Tennessee	Firefly beetle	Unspecified, probably *Photinus pyralis*
	Ladybug beetle	Unspecified species
	Honey bee †	*Apis mellifera*
	Zebra swallowtail butterfly*	*Eurytides marcellus*
Texas	Monarch butterfly	*Danaus plexippus*
Utah	Honey bee	*Apis mellifera*
Vermont	Honey bee	*Apis mellifera*
	Monarch butterfly	*Danaus plexippus*

(*continued*)

STATE	STATE INSECT*,†	SCIENTIFIC NAME
Virginia	(Eastern) tiger swallowtail butterfly	*Papilio glaucus*
Washington	Green darner dragonfly	*Anax junius*
West Virginia	Honey bee	*Apis mellifera*
	Monarch butterfly*	*Danaus plexippus*

FIGURE I-5

The monarch butterfly is the state insect or state butterfly of seven states and has been proposed as the national insect by the Entomological Society of America. Photograph courtesy of David Cappaert/Michigan State University/Bugwood.org.

Wisconsin	Honey bee	*Apis mellifera*
Wyoming	Sheridan's hairstreak*	*Callophrys sheridanii*

*State butterfly.
†State agricultural insect.

Notes on State Insects: California was the first state to have a proposal to establish a State Insect (1929), the California dogface butterfly, but this was not formalized until 1973. This was followed within a year by Maryland, Arkansas, and North Carolina. At present all but seven states and the District of Columbia have some official State Insect with Wyoming being the most recent (2009), selecting the Sheridan's hairstreak as the State Butterfly. West Virginia added the honey bee in 2002, moving the monarch to become the official State Butterfly. Arkansas established the Diana fritillary butterfly as its official State Butterfly in 2007, supplementing the honey bee as the State Insect.

Efforts have been made to establish the **monarch** as the National Insect of the United States. This has been a project originating from the Entomological Society of America, the largest professional entomology organization in the country.

The **white admiral butterfly** (*Limenitis arthemis*) has been proposed as the official insect of Quebec. Although not formally ratified, Quebec remains the only Canadian province to have moved toward recognition of any insect, to date, as a provincial symbol.

FIGURE I-6
The white admiral butterfly has been proposed as the official insect for the province of Quebec. Photograph courtesy of David Cappaert/Michigan State University/Bugwood.org.

Appendix II: Largest Arthropods

Examples of some of the larger living (extant) and extinct arthropods. Those marked with * are aquatic species.

COMMON NAME	SCIENTIFIC NAME	CLASS: ORDER	CURRENT STATUS	PHYSICAL FEATURES
*Sea scorpion	*Jaekelopterus* sp.	Euryptida: Pterygotoidea	Extinct	2.5 m body length
Griffinfly	*Meganeura monyi*	Hexapoda: Protodonata	Extinct	Wingspan up to 75 cm, body weight estimated at least 450 g
Millipede ancestor	*Arthropleura* spp.	Arthropleuridea: Arthropleurida	Extinct	Body length 2.5 m
Coconut crab	*Birgus latro*	Malacostraca: Decapoda	Living, SE Asia	Body weight 4 kg, legspan 1 m
*Japanese spider crab	*Macrocheira kaempferi*	Malacostraca: Decapoda	Living, North Pacific	20 kg weight, 60 cm body length, 4 m legspan
*North Atlantic lobster	*Homarus americanus*	Malacostraca: Decapoda	Living, North Atlantic	22 kg body weight, 1.18 m body length
Giant African millipede	*Archispirostreptus gigas*	Diplopoda: Spirostreptida	Living, Africa	Body length 30 mm
Amazonian giant centipede	*Scolopendra gigantea*	Chilopoda: Scolopendromorpha	Living, tropical South America	Body length 33 cm
Goliath birdeater	*Theraphosa blondi*	Arachnida: Araneae	Living, tropical South America	Body length across carapace 30 mm
Scorpion	*Heterometrus swammerdami*	Arachnida: Scorpiones	Living, India	Body length 29 cm, weight 57 g
Walkingstick	*Phobaeticus serratipes*	Hexapoda: Phasmatodea	Living, West Malaysia	Body length (legs outstretched) 555 mm
Giant longhorned beetle	*Titanus giganteus*	Hexapoda: Coleoptera	Living, tropical South America	Body length up to 16.7 cm

(*continued*)

COMMON NAME	SCIENTIFIC NAME	CLASS: ORDER	CURRENT STATUS	PHYSICAL FEATURES

FIGURE II-1
Giant longhorned beetle, *Titanus giganteus*.

COMMON NAME	SCIENTIFIC NAME	CLASS: ORDER	CURRENT STATUS	PHYSICAL FEATURES
Little Barrier Island giant weta	*Deinacrida heteracantha*	Hexapoda: Orthoptera	Living, New Zealand	Weight (pregnant female) 71 g, body length with ovipositor 11 cm
Elephant beetles	*Megasoma* spp.	Hexapoda: Coleoptera	Living, tropical South America	Body length (horned males) 13.7 mm

FIGURE II-2
Elephant beetle, *Megasoma mars*.

Goliath beetles	*Goliathus* spp.	Hexapoda: Coleoptera	Living, tropical Africa	Weight 45 g +, body length 11 cm

(*continued*)

COMMON NAME	SCIENTIFIC NAME	CLASS: ORDER	CURRENT STATUS	PHYSICAL FEATURES

FIGURE II-3
Goliath beetle, *Goliathus* sp.

| Queen Alexandra's birdwing butterfly | *Ornithoptera alexandrae* | Hexapoda: Lepidoptera | Living, Papua New Guinea | Wingspan may exceed 25 cm |

FIGURE II-4
Queen Alexandra's birdwing butterfly, *Ornithoptera alexandrae.*

(continued)

COMMON NAME	SCIENTIFIC NAME	CLASS: ORDER	CURRENT STATUS	PHYSICAL FEATURES
White witch	*Thysania agrippina*	Hexapoda: Lepidoptera	Living, Mesoamerica, northern South America	Wingspan may exceed 28 cm

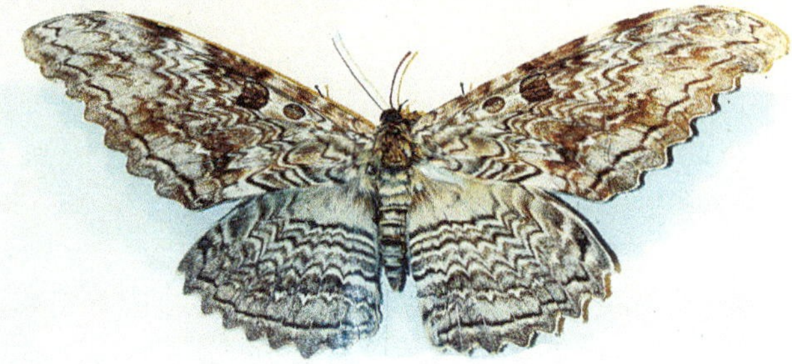

FIGURE II-5
White witch, *Thysania agrippina*.

Atlas moth	*Attacus atlas*	Hexapoda: Lepidoptera	Living	Area of wing up to c. 400 cm^2

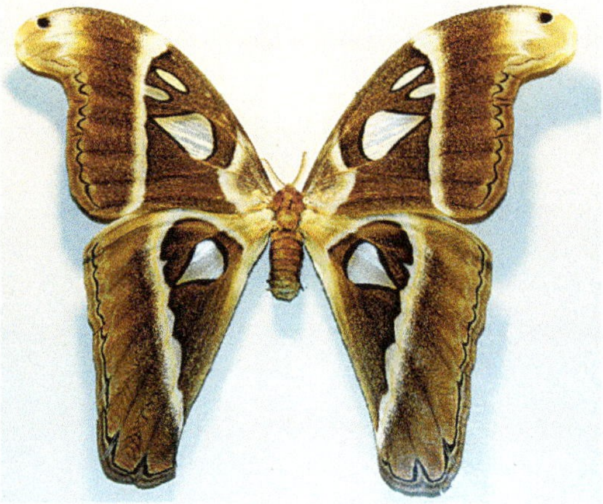

FIGURE II-6
Atlas moth, *Attacus atlas*.

Photo Credits. All photographs by Whitney Cranshaw/Colorado State University.

Subclass Protura (Proturans)

Approximate Number of Species: 500.
Development: Anamorphic (the number of body segments increases from 9 to 12).
Mouthparts: Small scraping mouthparts kept within a pouch (entognathous).
Wing Type: Wingless.
Leg Adaptations: The front pair of legs has sensory function. The remaining legs evolved for walking (cursorial).
Other Morphological Features: No eyes. No antennae. Very small (<1 mm).
Mating and Reproduction: Indirect sperm transfer via external spermatophores.
General Behavior and Ecology: Soil dwelling and primarily fungivores.
Unique or Unusual Features: The absence of antennae and the increase in the number of body segments are unique among Hexapoda.

Proturan

Subclass Collembola (Springtails)

Approximate Number of Species: 6,000.
Development: Ametabolous. Molting continues after adult stage.
Mouthparts: Small scraping mouthparts kept within a pouch (entognathous).
Wing Type: Wingless.
Leg Adaptations: Walking.

Other Morphological Features: Peg-like structure on the abdomen (collophore) used to absorb water. Taillike spring mechanism (furcula) for locomotion present in many species.
General Behavior and Ecology: Soil dwelling. Most are scavengers that feed on fungi and decaying plant matter. Some species are predators of nematodes and other minute animals.
Mating and Reproduction: Indirect reproduction. Males produce spermatophores on surfaces and females draw them into their genital opening. In some species, elaborate mating dances are used to guide females to the spermatophores.
Unique or Unusual Features: Furcula allows for short jumps. The most abundant terrestrial hexapods.

Entomobryid springtail

Globose springtail

Subclass Diplura (Diplurans)

Approximate Number of Species: 800.
Development: Ametabolous. Molting continues throughout life.
Mouthparts: Mouthparts allow scraping and chewing but have only a single point of attachment.
Wing Type: Wingless.
Leg Adaptations: Walking.
Other Morphological Features: Elongate body form. Long taillike cerci modified to grasp prey in predatory species. No eyes. Long unsegmented antennae.
General Behavior and Ecology: Found in moist sites. Feeding habits vary by suborder; some feed on live and decaying plant matter; others are predators.
Mating and Reproduction: Indirect. Males deposit spermatophores on the substrate that females seek. Females subsequently draw the spermatophores into their genital opening. Females care for eggs and stay with their young during early development.
Unique or Unusual Features: Able to regenerate lost appendages.

Dipluran

Japygid dipluran

Order Microcoryphia (Jumping Bristletails)

Approximate Number of Species: 350.
Development: Ametabolous.
Mouthparts: Base of mandibles with only single point of attachment, allowing rotation. Food is picked at rather than chewed.
Wing Type: Wingless.
Leg Adaptations: Walking and running (cursorial).
Other Morphological Features: Three long cerci extend from abdomen. Body with humped thorax and covered in scales. Long antennae and two long maxillary palps. Large compound eyes.
General Behavior and Ecology: Primarily feed on lichens and fungi. Usually found in grassy or wooded areas and are primarily nocturnal.
Mating and Reproduction: Indirect via production of externally deposited spermatophores. Silk may be used as guides to spermatophores.
Unique or Unusual Features: Capable of jumping, produced by arching thorax and sudden flexing of abdomen.

Jumping bristletail

Jumping bristletail

Order Thysanura (Silverfish)

Approximate Number of Species: Approximately 400.
Development: Ametabolous and insects continue molts in adult stage.
Mouthparts: Chewing.
Wing Type: Wingless.
Leg Adaptations: Running and walking (cursorial).
Other Morphological Features: Three long filaments extend from the tip of the abdomen. Flattened body covered with scales. Very long and thin antennae. Eyes reduced.
General Behavior and Ecology: Omnivorous scavengers. Normally found in moist environments (under bark and leaves and in animal burrows); a few adapted to human dwellings and may be household pests. Some species adapted to very dry conditions.
Mating and Reproduction: Indirect, involving use of externally laid spermatophores, often attached to silken threads.

Order Ephemeroptera (Mayflies)

Approximate Number of Species: >2,500.
Development: Hemimetabolous. Most immature stages (nymphs) are aquatic. The last immature stage (subimago) is winged.
Mouthparts: Immature stages have chewing mouthparts. Mouthparts of adults and subimago are vestigial.
Wing Type: Membranous, held vertically, forewing much larger than hind. Wings do not fold when at rest.
Legs Adaptations: Walking and crawling. The forelegs of adult male are often elongated and used to clasp the female while flying.
Other Morphological Features: Three, sometimes two, long abdominal tails. Nymphs acquire oxygen through their abdominal gills.
General Behavior and Ecology: Nymphs are aquatic with most feeding on algae, decaying plant materials, diatoms, and fungi. Nymphs of some species are carnivorous. Adults are extremely short-lived and do not feed.
Mating and Reproduction: Adult emergence is synchronized. Males form large swarms, into which females fly and select a mate. Oviposition occurs on or in the water.
Unique or Unusual Features: Subimago stage of mayflies are the only winged immature stages found among extant insects.

Firebrat

Common silverfish

Mayfly (female)

Mayfly (nymph)

Mayfly (subimago male)

Mayfly (nymph)

Order Odonata (Dragonflies and Damselflies)

Approximate Number of Species: 4,000+.
Development: Hemimetabolous.
Mouthparts: Chewing. The mouthparts of nymphs include an extensible, hinged labium that can be used to grab prey.
Wing Type: Straight and membranous. At rest, the wings are held out horizontally from the body in dragonflies and over the abdomen in damselflies. The wings do not fold when at rest.
Leg Adaptations: Walking or crawling in nymphs. In adults, evolved for catching prey in "basketlike" configuration while in flight and for perching.
Other Morphological Features: Very large compound eyes and head can pivot for a wide field of view. Abdomen long and thin. Many are colorful and often, there are substantial differences in coloration between sexes. Damselfly nymphs have three long flattened gills extending from the end of the abdomen; gills of dragonflies are internal in rectum.
General Behavior and Ecology: Nymphs and adults are predators. Nymphs are aquatic. Adults are terrestrial and excellent fliers.
Mating and Reproduction: Males are strongly territorial. Mating wheels are formed while in copula and males may continue to grasp female for an extended period through oviposition. Oviposition of damselflies involves attaching eggs to submerged vegetation or other objects. A direct oviposition into water during flight common with dragonflies.
Unique or Unusual Features: Adult males have secondary genitalia on the third segment of their abdomen where sperm is transferred. Dragonfly nymphs utilize jet propulsion for locomotion. Some dragonfly species make long annual migrations.

Damselflies (mating pair)

Damselfly (nymph)

Dragonfly

Dragonfly (nymph)

Order Orthoptera (Grasshoppers, Crickets, and Katydids)

Approximate Number of Species: 25,000.
Development: Hemimetabolous.
Mouthparts: Chewing.
Wing Type: Forewings thickened (tegmina) and straight and protect membranous hindwings. The hindwings can unfold for flight.
Leg Adaptations: Hindlegs of most species modified for jumping (saltorial). The other legs allow walking. Species with soil-burrowing habits have forelegs modified for digging (fossorial).
Other Morphological Features: Medium-sized insects with large compound eyes. A distinct ovipositor is present in many species.
General Behavior and Ecology: Most species are phytophagous.

Jerusalem cricket

Grasshopper

Mating and Reproduction: Males often attract females by calling or chirping. Many species lay eggs as pods in soil.

Unique or Unusual Features: Stridulatory organs are often present for sound production. Phase polymorphism is present in some species. Some species are capable of long-distance migration.

Field cricket

Katydid

Order Phasmatodea (Walkingsticks and Leaf Insects)

Approximate Number of Species: >3,000.
Development: Hemimetabolous.
Mouthparts: Chewing.
Wing Type: Most are wingless or brachypterous. Some with reduced front tegmina and rear functional membranous wings.

Leg Adaptations: Legs are extremely long and thin, used in slow walking (cursorial). Prominent defensive leg spines present in some species.

Other Morphological Features: Body form of walkingsticks is long and slender, often large insects. Mimic sticks and twigs. Leaf insects have broad leaflike body.

General Behavior and Ecology: All are phytophagous.

Mating Reproduction: Often by parthenogenesis. Sexually reproducing species display strong sexual dimorphism, with males much smaller than females.

Unique or Unusual Features: The use of crypsis and mimicry of foliage is the primary means of avoiding predation. Most capable of reflexive bleeding or squirting caustic fluid if disturbed. Nymphs can regenerate lost limbs to some extent. Order includes the longest insects.

Walkingstick

Walkingstick

Order Grylloblattodea (Rock Crawlers)

Approximate Number of Species: 34.
Development: Hemimetabolous.
Mouthparts: Chewing.
Wing Type: Wingless.
Leg Adaptations: Walking (cursorial).
Other Morphological Features: Elongated body, large head, and long antennae. Eyes often absent or reduced. Long abdominal cerci.
General Behavior and Ecology: Live in cold forested areas with very low optimum development temperature. Omnivorous feeding habit.
Mating and Reproduction: Sexual reproduction.
Unique or Unusual Features: Relatively rare insects that cannot tolerate warm temperatures and often found at high altitudes. Present only in Northern Hemisphere.

Rock crawler

Order Mantophasmatodea (Heelwalkers, African Rock Crawlers, and Gladiator Insects)

Approximate Number of Species: >15.
Development: Hemimetabolous.
Mouthparts: Chewing.

Wing Type: Wingless.
Leg Adaptations: Forelegs are modified to grasp prey (raptorial); the remaining legs are used for walking (cursorial).
Other Morphological Features: Lack ocelli.
General Behavior and Ecology: Predators of other arthropods. Found in arid habitats of southwest Africa.
Mating and Reproduction: Sexual reproduction.
Unique or Unusual Features: Relative newly discovered order, first described in 2002.

Heelwalker (nymph in alcohol)

Order Dermaptera (Earwigs)

Approximate Number of Species: 1,900.
Development: Hemimetabolous.
Mouthparts: Chewing.
Wing Type: Forewings are tegmina and reduced in size to cover only a small area of the abdomen. Hindwings are membranous and tightly folded up under the forewings.
Leg Adaptations: Walking (cursorial).
Other Morphological Features: Small- to medium-sized, often dark in color. Large pincerlike cerci present in both sexes, usually more bowed in males.
General Behavior and Ecology: As an order, they are predominantly omnivorous. Some are primarily predaceous, others primarily herbivorous. Nocturnal activity.
Reproduction: Sexual. Some species display maternal care for their young. A few species are live-bearers.
Unique or Unusual Features: Several species display thigmotaxis, favoring dark, humid, tight spaces in which they dwell during the day.

European earwig (male)

Doru earwig (female)

Order Plecoptera (Stoneflies)

Approximate Number of Species: 2,000.
Development: Hemimetabolous. Immature stages are aquatic, adults terrestrial.
Mouthparts: Chewing.
Wing Type: Membranous wings that fold flat over the abdomen when at rest.
Leg Adaptations: Adults walk and may run rapidly. Aquatic nymphs crawl.
Other Morphological Features: Nymphs often have a flattened body and a pair of long thin cerci. Nymphs acquire oxygen though thoracic gills associated with the legs.
General Behavior and Ecology: Immature stages are aquatic. Most develop on as shredders of living and dead plant matter or graze algae. Some are predaceous. Adults are usually found on vegetation next to streams and rivers and feed minimally.

Mating and Reproduction: Mating involves extensive communication by abdominal drumming. Eggs are laid as masses in water.
Unique or Unusual Features: Most are found in well-oxygenated, clear, cool running water. Presence/absence of species can be used to assess water quality. Adults of some species emerge in winter.

Stonefly

Stonefly (nymph)

Stonefly

Order Embiidina (Webspinners)

Approximate Number of Species: 360.
Development: Hemimetabolous.
Mouthparts: Chewing.
Wing Type: Males possess membranous wings; females are wingless. Wings are flexible so that they can be pulled through burrows.
Leg Adaptations: Walking type (cursorial), with modifications for running through tunnels. Enlarged silk glands present on tarsi of forelegs.
Other Morphological Features: Elongate body form adapted to living within tunnels. Body not heavily sclerotized.
General Behavior and Ecology: Live in silken tunnels found under rocks and debris. Most scavenge dead plant matter.
Mating and Reproduction: High level of maternal care, with females tending eggs and early instar nymphs.
Unique or Unusual Features: Specialized glands on front tarsi are unique silk-production organs.

Webspinner

Webspinner (winged male)

Order Zoraptera (Angel Insects)

Approximate Number of Species: 32.
Development: Hemimetabolous.
Mouthparts: Chewing.
Wing Type: Most are wingless. If present, wings are membranous and develop in generations undergoing crowding and food deprivation.
Leg Adaptations: Walking (cursorial).
Other Morphological Features: Most are pale-colored, blind, and wingless. Dark, winged forms occasionally are produced in response to overcrowding. Long bead-like antennae similar to termites.
General Behavior and Ecology: Primarily fungivores. Found in moist decaying wood.
Reproduction: Sexual reproduction, but little studied.

Zorapteran

Order Blattodea (Cockroaches)

Approximate Number of Species: 4,000.
Development: Hemimetabolous.
Mouthparts: Chewing.
Wing Type: Front pair are tegmina. Rear pair are membranous. Some species wingless or brachypterous.
Leg Adaptations: Walking or running (cursorial). Some capable of very fast movement.

Other Morphological Features: Head is concealed by shield-shaped pronotum. Body is dorsoventrally flattened. Adults with obvious abdominal cerci. Very long, thin antennae project forward from the head.

General Behavior and Ecology: Most species are omnivorous and are particularly abundant in forested tropical areas. Some that feed on wood show semisocial behaviors living as groups of mixed ages. Most often, they gather in groups (harborages) for cover. A small number are adapted to living within human structures and are considered major pests.

Mating and Reproduction: Sex pheromones are produced to attract mates. Eggs are laid as a group within a case (ootheca), which may be dropped or retained by the female until hatch.

Unique or Unusual Features: Extremely fast runners. Often used as an experimental animal.

Death's head cockroach

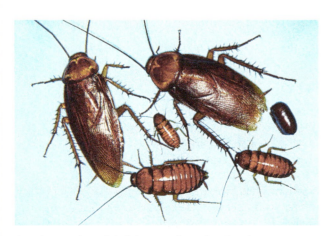

American cockroach (adults, nymphs and ootheca)

Order Isoptera (Termites)

Approximate Number of Species: 2,900.
Development: Hemimetabolous.
Mouthparts: Chewing.
Wing Type: Reproductive adults possess membranous wings but lose them soon after courtship. All other forms are wingless.
Leg Adaptations: Walking (cursorial).
Other Morphological Features: Antennae bead-like and straight. Most stages are pale-colored.
General Behavior and Ecology: Form very large social colonies that can consist of 100,000s of individuals. Depending on the family, colonies are classified as subterranean, mound-building, or nesting in dry wood. Distinct behavioral and morphological castes are present: workers, soldiers, and reproductives. Most feed on dead plant material or wood, utilizing gut symbionts to digest cellulose. Trophallaxis is common within a colony.

Mating and Reproduction: Sexual, with primary reproduction by a queen (fertile female) that is reproductively active and mates with a male throughout life. Reproductive forms with wings are produced that leave colonies during mating flights. Flightless secondary reproductives can be produced within colonies, which may lay some eggs.

Unique or Unusual Features: All species are eusocial. Reproductive male (king) and female (queen) pair for life.

Subterranean termites (soldier and workers)

Subterranean termite (winged reproductive)

Chinese mantid

European mantid (nymph)

Order Mantodea (Mantids)

Approximate Number of Species: 2,900.
Development: Hemimetabolous.
Mouthparts: Chewing.
Wing Type: Front pair of wings are thickened and protective (tegmina), rear wings membranous. Some individuals with brightly colored rear wings.
Leg Adaptations: Prothoracic legs enlarged and raptorial for prey capture. Meso- and metathoracic legs are usually long and thin and used for walking, cursorial.
Other Morphological Features: Greatly elongated prothorax. Large, widely spaced compound eyes on head that is capable of twisting.
General Behavior and Ecology: Ambush predators of other insects.
Mating and Reproduction: Sexual. Eggs are produced as a mass within a protective case (ootheca). Males sometimes are consumed during mating.
Unique or Unusual Features: Adults often exhibit a startle response to avoid predation. Flying mantids can detect echolocating insectivorous bats with a specialized "ear."

Order Psocoptera (Barklice and Booklice)

Approximate Number of Species: 4,400.
Development: Hemimetabolous.
Mouthparts: Chewing, but mouthparts are used in a scraper-like fashion.
Wing Type: Winged and wingless adult forms may be produced. Winged individuals hold membranous wings rooflike over the body.
Leg Adaptations: Walking (cursorial).
Other Morphological Features: Very small size. Long filamentous antennae. Silk formed from labial glands.
General Behavior and Ecology: Feed on fungi, yeast, algae, and lichen. Inhabit dark and moist areas under rocks, in decaying vegetation, on bark of plants. Aggregations common.

Mating and Reproduction: Sexual. Female lays eggs in clusters often covered in silk or digested material.

Unique or Unusual Features: Able to extract atmospheric moisture through foregut.

Barklouse

Booklouse

Order Phthiraptera (Lice)

Approximate Number of Species: 4,900.
Development: Hemimetabolous.
Mouthparts: Two very different mouthparts are present in this order. Some have chewing mouthparts; others possess mouthparts that allow sucking blood.
Wing Type: Wingless.
Leg Adaptations: Uniquely modified for grasping hair or feathers of host.
Other Morphological Features: Body is flattened dorsoventrally and heavily sclerotized. In chewing lice, the head is wider than the prothorax; the head is narrower than the prothorax among sucking lice.

General Behavior and Ecology: All species are ectoparasites of vertebrates spending most, if not all, of their lives on the host. Depending on species, they will feed on blood, skin, hair, and feathers. Movement from among hosts requires close contact between hosts.

Mating and Reproduction: Sexual. Females of many species glue eggs to host hair or feathers.

Unique or Unusual Features: Several species are serious pests of domestic animals. Two species of sucking lice are of special concern to humans: the pubic louse and the head/body louse. The later transmits epidemic typhus.

Sucking louse

Chewing louse

Order Hemiptera (True Bugs, Aphids, Whiteflies, Scale Insects, Hoppers, Cicadas, and Others)

Approximate Number of Species: 82,000.
Development: Hemimetabolous. Some have nonmobile stages.
Mouthparts: Piercing–sucking.
Wing Types: Wing modifications vary between suborders. Heteroptera have front wings that are hemelytra and membranous hindwings. The wings are held flat over the body. Auchenorrhyncha have two pair of membranous wings that are held rooflike over body. Many Sternorrhyncha have forewings thickened (tegmina) and wings are held rooflike over the body.
Leg Adaptations: Typically walking (cursorial). Hindlegs of some species allow jumping. In many species, legs are lacking in juveniles and females.
Other Morphological Features: Digestive system modified to handle liquid diet and excreted waste is usually liquid. Some groups secrete wax to produce protective structures or excrete waxy waste.
General Behavior and Ecology: All are liquid feeders. Majority feed on plants, including many that are significant plant pests. Many are predators of other arthropods. A few are parasites of vertebrates.
Mating and Reproduction: Sexual for the most part. Some species are parthenogenetic.
Unique or Unusual Features: The largest order of hemimetabolous insects. Species are vectors for most viruses and all phytoplasmas and xylem-limited bacteria that produce plant disease. Wax produced by some species for protection.

Aphid (winged form)

Stink bug (feeding on sawfly)

Cicada

Oystershell scales

Order Thysanoptera (Thrips)

Approximate Number of Species: 5,500.
Development: Hemimetabolous. Last two immature stages (prepupa and pupa) are immobile and do not feed.
Mouthparts: Unique form of sucking mouthparts. A single mandible is present and is used to penetrate the surface of the food source. The mandibles are thin and elongate to puncture underlying cells. The labium forms a cone through which released fluids are drawn into the mouth.
Wing Type: Fringed.
Legs Adaptations: Walking, with tarsi having bladderlike adaptations to allow clinging.
Other Morphological Features: Very small insect (<3mm). Elongate head with distinct cone that contains mouthparts directed downward.
General Behavior and Ecology: Most are herbivorous, feeding on foliage and flowers. Pollen feeding is common and some feed on fungi. A few species are predaceous on other small insects.
Mating and Reproduction: Most exhibit haplodiploid reproduction (unfertilized eggs become males and fertilized eggs become females). Eggs are usually inserted into plants.
Unique or Unusual Features: Vectors of plant tospoviruses, including tomato spotted wilt. Some species are semisocial.

Thrips (adults and nymphs)

Order Coleoptera (Beetles)

Approximate Number of Species: 350,000+.
Development: Holometabolous. Hypermetamorphosis occurs within some families.
Mouthparts: Chewing.
Wing Type: Forewings elytra, hindwings membranous. Some species wingless.
Leg Adaptations: Adults of most species have walking legs (cursorial), some can run rapidly. Hindlegs of some adapted for jumping (saltatorial). Aquatic species have legs modified for swimming (natatorial). Species that dig have enlarged forelegs. Larvae usually have thoracic legs that allow crawling. Legs are absent or vestigial in groups that have larval forms that develop within plants.
Other Morphological Features: A very large, species-rich order that shows tremendous variety of size, shape, and habit.
General Behavior and Ecology: Generalizations are difficult. All terrestrial and freshwater environments support some species of beetles. A wide range of feeding behaviors is present within members of this enormous order.
Mating and Reproduction: Sexual. Maternal and paternal care present in a few families.
Unique or Unusual Features: Many species capable of spinning silk (for pupal cocoons) from modified Malpighian tubules. Some are capable of light production.

Thrips

Click beetle

Wireworm (larval click beetle)

White grub (larval scarab beetle)

Rhinoceros beetle (scarab beetle)

Weevil

Cottonwood leaf beetle (adult and larvae)

Order Neuroptera (Alderflies, Dobsonflies, Snakeflies, Lacewings, Antlions, and Owlflies)

Approximate Number of Species: 6,000.
Development: Holometabolous.
Hypermetamorphosis occurs with mantidflies.
Mouthparts: Adults have chewing mouthparts.
Larvae of most species have forward-projecting, sickle-shaped mouthparts with an internal groove allowing them to drain body fluids from their prey.
Wing Type: Membranous with extensive network of veins and wings held rooflike over the body.
Leg Adaptations: Adults' legs allow walking and crawling. Some predatory species have enlarged raptorial prothoracic legs. Larvae have legs for crawling or walking.
Other Morphological Features: Most are soft-bodied with delicate wings.
General Behavior and Ecology: Most are predatory. Hunting behaviors vary and many are ambush predators. Larvae of some groups develop as parasites of spider eggs or feed on sponges.
Mating and Reproduction: Sexual. Some species oviposit eggs on silken stalks.
Unique or Unusual Features: Silk produced from modified Malpighian tubules. Soil pits to assist prey capture used by some antlions. Digestive system in suborder Planipennia not fully connected until pupal stage.

Dobsonfly (male)

Antlion (larva)

Green lacewing (larva)

Antlion

Green lacewing

Bumble bee

Order Hymenoptera (Ants, Bees, Wasps, and Sawflies)

Approximate Number of Species: 125,000–150,000.

Development: Holometabolous.

Mouthparts: Chewing.

Wing Type: Membranous. Pro- and mesothoracic wings linked together with hamuli. Only reproductive form adults have wings; workers are wingless.

Leg Adaptations: Walking (cursorial) in adults. Larvae of sawflies have thoracic legs and abdominal prolegs. Other larvae within the order are legless.

Other Morphological Features: Females possess prominent ovipositor, which in some species has been modified to form a sting. Abdomen joined very narrowly to the thorax (sawflies are an exception).

General Behavior and Ecology: Feeding habits across the order are quite varied, including species that are herbivorous, predators of other arthropods, and scavengers. Many families of the order are truly social and may produce large colonies.

Mating and Reproduction: Sexual with haplodiploidy.

Unique or Unusual Features: Many species of bees, wasps, and *all* ants are eusocial, forming large colonies often with millions of individuals. Gall wasps produce unusual determinate galls on plants. All stinging insects are in the order Hymenoptera.

Carpenter ant (winged male and female)

Cicada killer

Conifer sawfly (male)

Armyworm

Order Lepidoptera (Butterflies and Moths)

Approximate Number of Species: 150,000.
Development: Holometabolous.
Mouthparts: Adults have mouthparts that allow siphoning of liquids or are vestigial. Larvae have chewing mouthparts.
Wing Type: Membranous, covered with scales.
Leg Adaptations: Adults have walking legs (some species with reduced prothoracic legs). Larvae have thoracic legs and 2–5 pairs of prolegs on abdomen.
Other Morphological Features: Adult body covered with scales or hair (setae). Wing scales often brightly colored and arranged in patterns on the wings.
General Behavior and Ecology: Adults feed on various fluids, with many feeding on nectar. Larvae of almost all species are herbivorous and many are plant pests. Some are scavengers and a few are predators.
Reproduction: Sexual, often with elaborate courting behaviors and extensive use of pheromones.
Unique or Unusual Features: Larval prolegs possess crochets. Silk produced through modified Malpighian tubules and, in many species, used to produce a cocoon within which the animal pupates. Some are commercially reared for silk production. Several species capable of long-distance migration as adults.

Sphinx moth (larva)

Twotailed swallowtail

Io moth (larva)

Silverspotted skipper

Order Trichoptera (Caddisflies)

Approximate Number of Species: 12,000.
Development: Holometabolous. Larvae are aquatic, adults terrestrial.
Mouthparts: Chewing.
Wing Type: Membranous and covered with short hairs. Wings held rooflike over the abdomen.
Leg Adaptations: Walking (cursorial) in adults. Larvae with thoracic legs and one pair of abdominal prolegs.
Other Morphological Features: Larvae of most species weave a protective silken "case" tying together pebbles or small sticks. Larval gills on the abdomen. Adults appear somewhat like moths with very long antennae.

General Behavior and Ecology: Larvae are aquatic. Larvae may feed on plant material and organic detritus or are predators. Larval gills found predominantly on the abdomen. Pupation occurs in water within silk cocoon and adult emergence is often synchronized. Adults are fluid feeders or do not feed.
Reproduction: Sexual. Oviposition occurs above water.
Unique or Unusual Features: Larvae produce silk from modified labial glands and can be used to form extremely complex protective cases. Some larvae use silk to form nets that capture food in moving water.

Caddisfly

Caddisfly

Caddisfly (larva in case)

Caddisfly (larva in case)

Order Mecoptera
(Scorpionflies)

Approximate Number of Species: 600.
Development: Holometabolous.
Mouthparts: Chewing.
Wing Type: Membranous of similar size.
Leg Adaptations: Walking (cursorial). Larval legs allow crawling.
Other Morphological Features: Adults possess elongated beak. Within the dominant family, male genitalia are greatly enlarged, appearing as a stinger, hence the name of the order.
General Behavior and Ecology: Mostly scavengers and predators.
Reproduction: Sexual. Mating behavior may involve nuptial gifts presented by the male to the female.
Unique or Unusual Features: Some species are active during winter. The order appears to be the ancestor to the true flies (Diptera).

Scorpionfly

Hangingfly

Order Siphonaptera
(Fleas)

Approximate Number of Species: 1,900.
Development: Holometabolous.
Mouthparts: Adults with uniquely modified piercing–sucking mouthpart used for blood feeding. Larvae have chewing mouthparts.
Wing Type: Wingless.
Leg Adaptations: Rear legs greatly enlarged to allow jumping (saltatorial). Legs designed to grasp and move through hair. Larvae with legs for crawling.
Other Morphological Features: Adults have a body laterally flattened and heavily sclerotized. Pronotal comb is present. Larvae with an elongate body and pale-colored.
General Behavior and Ecology: Ectoparasites of mammals and birds. Adults are blood feeders. Larvae feed on feces of adults or host debris. All stages closely associated with host animal. Pupation in silk cocoon within the host nest.
Reproduction: Sexual.
Unique or Unusual Features: Silk produced from modified labial glands. Able to jump relatively long distances. Responsible for transmitting plague.

Flea

Flea (larva)

Unique or Unusual Features: Responsible for transmission of numerous important diseases to mammals and birds.

Blow fly

Horse fly (larva)

Order Diptera (True Flies)

Approximate Number of Species: 154,000.
Development: Holometabolous.
Mouthparts: Sponging–lapping or piercing in adults. Larvae often have mouth hooks for chewing/slashing; some aquatic species have filter-feeding mouthparts.
Wing Type: Single pair of wings. Front wings are membranous and mesothorax is well developed. Rear wings modified into knob-like halteres.
Leg Adaptations: Walking in adults. Larvae legless.
Other Morphological Features: Two major suborders consisting of Nematocera with long antennae and Brachycera with short antennae. In Brachycera, the last larval skin covers the pupa (puparia).
General Behavior and Ecology: Wide variation in life histories and behaviors. Many species are scavengers. Many are plant feeders, including several agricultural pests. Some are predators of other invertebrates and some feed on vertebrate blood.
Reproduction: Sexual. Many engage in elaborate mating rituals.

Chironomid midge

Chironomid midge (larva)

Order Strepsiptera (Twisted-Wing Parasites)

Approximate Number of Species: 600.
Development: Holometabolous with hypermetamorphosis. First instars originally develop within the mother and then emerge to seek hosts.
Mouthparts: Mouthparts of adult males are usually nonfunctional; some have chewing mouthparts. Larvae feed within the host of body fluids. First instar larvae have modifications that allow them to soften and penetrate the host.
Wing type: Males winged with single pair of fanlike hindwings on metathorax. Front wings on mesothorax are in the form of antler-like clubs. Females are wingless.
Leg Adaptations: Walking type (cursorial) in adult males. Adult females legless. Larvae legless except in first instar (triungulin).
Other Morphological Features: Eyes of males are unique. Several large individual eyes are present rather than a single compound eye, giving the eye a raspberrylike appearance.
General Behavior and Ecology: All are internal parasites of other insects. When development is complete, females remain within the host insect; males emerge and fly to locate new parasitized hosts containing females.
Mating and Reproduction: Sexual. Mating occurs via the female's brood canal, located between the head and abdomen.
Unique or Unusual Features: Larvae develop as internal parasites of mother immediately following egg hatch. Eye structures and reduced metathoracic

wing unique to the order. Relationship of Strepsiptera with other orders unclear but they appear to be most closely related to some Diptera.

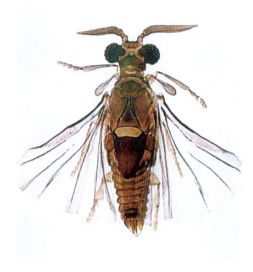

Strepsipteran (adult male)

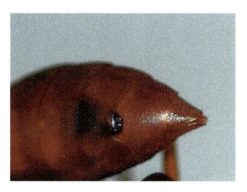

Strepsipteran (female in host abdomen)

Photo Credits for Appendix III:

Gary Alpert/Bugwood.org. Subterannean termites (soldier and workers)
Joseph Berger/Bugwood.org. Weevil, Antlion (larva), Common silverfish
Elke Buschman/University of Cincinnati. Strepsipteran (adult male)
David Cappaert/Michigan State University. Subterranean termite (winged reproductive)
Clemson University/Bugwood.org. Firebrat
Whitney Cranshaw/Colorado State University. European earwig, Blow fly, Cicada, Carpenter ants, Armyworm, Sphinx moth (larva), Twotailed

swallowtail, Io moth (larva), Silverspotted skipper, European mantid (nymph), Heelwalker, Jumping bristletail, Antlion, Grasshopper, Walkingstick (vertical), Thrips (mixed life stages)

Ken Gray/Oregon State University. Sucking louse, Chewing louse, Flea (larva)

Jim Kalisch/University of Nebraska. Death's head cockroach, American cockroach, Click beetle, White grub, Wireworm, Rhinoceros beetle, Cottonwood leaf beetle, Bumble bee, Chinese mantid

David Leatherman. Cicada killer

Sturgis McKeever/Georgia Southern University. Horse fly (larva)

Tom Murray. Entomobryid springtail, Globose springtail, Doru earwig, Dipluran, Chironomid midge, Chironomid midge (larva), Mayfly (female), Mayfly (subimago male), Mayfly (nymphs), Stink bug, Scorpionfly, Hangingfly, Damselfly (nymph), Dragonfly, Dragonfly (nymph), Jerusalem cricket, Field cricket, Walkingstick (horizontal), Stonefly (2 pictures), Stonefly (nymph), Flea, Barklouse, Booklouse, Caddisfly (2 adults), Caddisfly (2 larvae)

Jack Reed/Mississippi State University. Thrips

S. Dean Rider. Webspinner, Webspinner (winged male)

David Shetlar/The Ohio State University. Japygid dipluran, Oystershell scales, Conifer sawfly, Zorapteran

Brian Valentine. Aphid, Green lacewing (larva and adult), Damselflies

Alex Wild. Rock crawler

Glossary

Abdomen   The posterior of the three main body divisions in an insect; the posterior of the two main body regions in a mite, spider, or other arachnid.

Abscond   The behavior of honey bees when whole colonies abruptly abandon and disperse from a hive.

Active ventilation   Actively pumping air through the tracheal system.

Acari   The order of arachnids that contains the mites and ticks.

Acaricide   A pesticide used to control a member of the Arachnid family Acari (the mites and ticks). These are sometimes called **miticides**.

Acetylcholine   A chemical neurotransmitter. It is broken down by acetylcholinesterase enzymes.

Accessory gland   The reproductive structure found in both males and females responsible for producing various compounds and structures associated with reproduction. In males, the accessory gland produces much of the structure and material associated with the spermatophore. In females, it produces the structure and material associated with ootheca, venom, adhesive to attach the eggs, etc.

Aedeagus   The insect penis.

Air sac   The widening of the diameter of a trachea to form a saclike structure used to hold and store inhaled air and to occupy space that is latter used for growth.

Alate   With wings. Forms of aphids that possess wings in the adult stage known as **alatae**.

Alternate host   One of one or more hosts on which a parasite develops.

Altruistic behavior (altruism)   Nonselfish behavior. Helping others to the apparent (not necessarily actual) detriment of one's own fitness.

Amblypygi   The order of arachnids known as tailless whipscorpions, whipspiders, and amblypigids.

Ametabolous metamorphosis   A pattern of metamorphosis where there is little differentiation in external physical appearance between adult and immature stages, other than increased size. This metamorphosis occurs among the more primitive wingless insect orders (Microcoryphia and Thysanura) and Collembola.

Amphipoda   An order of crustaceans that includes scuds, beach fleas, and sand fleas.

Anamorphic development   A pattern of metamorphosis where developing stages of insects may increase the number of segments, legs, or other features following a molt. Crustaceans, millipedes, and some other noninsect arthropods show anamorphic development.

Annual   Having a life cycle that is completed in a single year.

Anostraca   An order of crustaceans that includes the fairy and brine shrimp.

Anus   The external opening of the hindgut through which waste passes to the outside.

Antennae   The sensory structure found on the head of an insect and certain other arthropods. It is often responsible for the collection of sensory information related to olfaction (smell), taste, sound, wind speed, etc.

Antenniform   In the form of antennae. For example, the forelegs of the tailless whipscorpions (Arachnida: Amblypygi) are long and thin and serve to sense by touch.

Apodeme   A platelike infolding of the exoskeleton to which muscles and other tissues are attached.

Aphophysis   A spinelike infolding of the exoskeleton to which muscles and other tissues are attached.

Apiculture   Beekeeping.

Apolysis   The stage in an insect molt when the older exoskeleton separates from the epidermis.

Aposematic coloration   The coloration of the exoskeleton such that it is extremely noticeable usually with bright and contrasting colors. Such coloration is used (ironically) as a predator avoidance mechanism as insects so colored are usually toxic when consumed.

Appendotomy   When grasped, the ability to lose legs to a predator.

Apterous   Without wings. Forms of aphids that are without wings known as **apterae**.

Apterygote   Insects that are wingless and never had ancestors with wings. The apterygote insects are those in the orders Microcoryphia and Thysanura.

Appendotomy   The ability to readily lose an appendage if it is pulled by some external force (e.g., a predator or rival male in a fight).

Aquatic   Associated with freshwater.

Arachnida   A class of arthropods that contain the spiders, scorpions, mites, and ticks, among others.

Arachnophobia   A disproportionate and unnatural fear (phobia) of spiders.

Araneae   The order of arachnids known as the spiders.

Arboviruses   Viruses of animals that are transmitted by an arthropod.

Archaeognatha   The order name often used for the bristletails. The alternate name Microcoryphia is used in this book.

Arthropoda   Phylum of invertebrate animals that includes the insects, arachnids, crustaceans, and other animals that possess an exoskeleton.

Autotomy   The ability to drop an appendage (also see **appendotomy**).

Asexual   Any type of reproduction not involving a union in which fertilization and meiosis occurs.

Asynchronous muscle   Muscles that can contract multiple times following a single nerve impulse.

Autocidal   Refers to either a genetically modified or sterile individual that when mated leads to the production of no viable offspring.

Axon   The long slender portion of a nerve cell.

Ballooning   A means of airborne dispersal used by spiders, spider mites, and very small Lepidoptera larvae where individuals release silk thread in order to catch and ride the wind as a mechanism for dispersal.

Base temperature   The lowest temperature at which an arthropod can develop.

Basement membrane   The cellular membrane underlying the epidermis of the insect's exoskeleton.

Batesian mimicry   A form of mimicry in which a harmless or palatable species mimics a colorful and usually toxic or venomous species.

Bilaterally symmetrical   Description of a body plan that can be divided by an imaginary plane such that the body is divided into two mirror images.

Bioluminescence   The ability of living organisms to produce light.

Binomial nomenclature   A system for describing a living organism that involves a unique two-part name including a genus name followed by a species name.

Biological control   The use of predators and parasitoids to manage pestiferous insect populations.

Bivoltine   Possessing two generations per year.

Bivouac   A temporary nest established by migrating army ants.

Black fly fever   Disease caused by numerous black fly bites and is characterized by fever, nausea, headache, and swollen lymph nodes.

Blattodea   The order of insects that include cockroaches.

Book lungs   The respiratory structures found in spiders and some other related arachnids. Book lungs are the site of oxygen inspiration.

Brachypterous   With wings that are short and usually nonfunctional for flight.

Brackish   Water having a low (relative to sea water) concentration of salt such as that occurs where rivers flow into oceans.

Brain   In insects, the anterior fusion of the first three ganglia of the central nervous system.
In insects, the brain is responsible for processing most (not all) of the sensory information regarding vision, taste, and odor.

Branchiopoda   Class of crustaceans that include the tadpole, fairy, and brine shrimp.

Broadleaf   A plant having broad leaves rather than needlelike or scalelike leaves. Most cultivated plants in gardens are **broadleaved**, except grasses and the conifers.

Bt   An abbreviation for *Bacillus thuringiensis*, a type of bacterium used for control of certain insects.

Bug   A common name most appropriately only applied to insects that occur within the suborder Heteroptera of the Hemiptera order, known as the "true bugs." They possess a piercing–sucking mouthpart that arises from the front of the head, and winged forms have a unique wing type that has the base of the forewing more thickened than the tip (hemelytra). In popular use, "bug" applies much more broadly to cover almost any "wee little creature."

Buzz pollination   Pollination carried out by bumble bees in which they vibrate the flower causing pollen to be dispersed onto the bee's body.

Caelifera   The shorthorned Orthoptera.

Calamistrum   Comblike projections on the hindleg of certain web-building spiders (e.g., combfooted spiders of the family Theridiidae).

Cambium   A layer of cells between the bark and wood that reproduce and divide into the phloem outward and xylem (wood) inward. If completely destroyed or exposed, the plant dies.

Cantharidin   Toxic compound capable of creating severe blistering. Produced by blister beetles.

Carapace   Typically, the large dorsal plate section of the exoskeleton that covers a portion of the cephalothorax in crustaceans.

Carrion   Dead animal or animals, may also refer to the flesh of dead animals.

Caste   In social insects, a group of related individuals performing a specific set of duties (e.g., the reproductive caste, the soldier caste, and the worker caste).

Catalepsy   Playing dead as a predator defense mechanism.

Caterpillar   The larva of a butterfly, moth, or sawfly.

Catfacing   A dimpling of fruit produce from injury during early development. Such injuries may result from feeding by various insects with sucking mouthparts, by egg laying wounds, or some that chew the surface of young fruit.

Cephalon   The area that includes the head of a crustacean.

Cephalothorax   One of two body sections found in arachnids and many crustaceans. It is formed by the fusion of the thorax and the head of the animal. The remaining tagma is the abdomen.

Cerci   Fingerlike projections extending from the rear of the abdomen of an insect. Are typically covered in sensory hairs.

Chelicerae   A pair of fang-like appendages used for penetrating tissues used by mites, spiders, and other arachnids (singular, **chelicera**).

Chagas disease   A trypanosome-caused disease of humans transmitted by species in the genus *Triatoma* (family Reduviidae, assassin bugs).

Chelae   The enlarged clawlike tips of the pedipalps of scorpions and some other arthropods.

Chemoreceptor   A sensory nerve evolved to detect a specific chemical or class of chemicals.

Chilopoda   A class of arthropods known as the centipedes.

Chitin   The tough, hard, polysaccharides that comprises much of an arthropod's exoskeleton.

Chorion   The eggshell of an arthropod.

Chrysalis   The pupal stage of a butterfly. It is not covered with silk and often contains spines and patterned colors. **Chrysalid** is an alternatively accepted term; **chrysalids** or **chrysalides** is plural.

Claspers   Modified cerci extending from the tip of the abdomen, usually in male insects. Claspers are used in mating to grasp and hold the opposite sex while in copula.

Class A taxonomic rank   A class is less inclusive than a phylum, and several classes may comprise a phylum. Insects are placed in the class Hexapoda and arachnids are in the class Arachnida.

Coarctate   Refers to certain dipteran pupae (known as puparia) that are smooth and rounded in appearance.

Cochineal   A type of red dye derived from cochineal scales. This is also known as **carmine**.

Cocoon   Silken case within which the pupal stage of many insects is formed.

Coevolution   Reciprocal evolutionary change in one species as a result of evolutionary change in another species.

Collophore   A straw-like structure located on the ventral side of the first abdominal segment of a collembolan. The collophore is used for imbibing water.

Coleoptera   The order of insects that contains the beetles.

Colon   In insects, part of the digestive system responsible for carrying waste products away from the midgut. The colon is considered part of the hindgut and is also partially responsible for water and salt regulation.

Comb   A row of stout hairs or spines located on the rear of the head of a flea. Prevents the flea from being scratched off or removed from the host.

Common name   Familiar, commonly used name for an insect species. Common names of many insects and arachnids are formally established by professional organizations.

Complete metamorphosis   A pattern of metamorphosis used by many insects (e.g., beetles, moths/butterflies, sawflies, and flies) that involves eggs, followed by immature larvae, a transition pupal stage, and finally adults. Adult and immature stages often have very different habits and appearance among insect groups with this metamorphosis. Also called holometabolous metamorphosis.

Compound eye   The many faceted eye of most adult insects. Depending on the species, the compound eye is comprised of hundreds to thousands of ommatidia.

Conifer   A general term for cone-bearing plants that include pines, cypresses, junipers, spruces, arborvitae, and yews.

Coprophage   An animal that feeds on animal dung.

Cornicle   A tubular structure (paired) on the posterior of aphids through which alarm pheromones are released.

Corpora allata   Neurosecretory cells responsible for releasing juvenile hormone.

Coxa   The first segment of the insect leg that is adjacent to the body of the thorax.

Crawler   The first stage after egg hatch of a scale or mealybug that is the most dispersive form of the insect.

Cremaster   Silken sticky structure by which a lepidopteran chrysalis is attached to a substrate.

Crepuscular   Active during dusk and early evening.

Cribellum   The silk-spinning organ found in spiders.

Crochets   Small hooks found on the tip of the prolegs of the larvae of moths and butterflies.

Crop   In insects, the portion of the digestive system responsible for partially storing consumed food prior to releasing it to the midgut for digestion. The crop is considered part of the foregut.

Cross-resistance   The ability to simultaneously develop resistance to several insecticides of similar modes of action.

Cryoprotectant   A substance used to protect tissues and cells from the effects of freezing.

Crypsis   The ability to blend in with the background; camouflaged.

Cryptic species   A species that is indistinguishable from another solely from examination of the use of external physical features.

Cursorial   Referring to legs modified for running or walking.

Cytotoxin   A toxin that damages the cells.

Decapoda   An order of crustaceans that includes the crayfish, lobster, shrimp, and crabs.

Deciduous   A plant that annually sheds its leaves.

DEET   *N,N*-diethyl-*m*-toluamide. A highly effective and widely used repellent for mosquitoes, ticks and some other biting arthropods.

Degree-day   A degree of temperature that is above the base temperature for development and persists for a 24-h period.

Defoliation   The loss of leaves as that occurring in natural shedding or from the feeding activities of insects and other plant feeders.

Dendrites   Portion of a neuron that receives the nervous impulse and sends it down to the neuron toward the soma and axon.

Dermaptera   The order of insects that includes the earwigs.

Detritivore   An animal that feeds on dead organic material.

Deutocerebrum   The second distinct lobe of the insect's brain responsible for processing sensory information coming in from the antennae.

Deutogyne   An overwintering form of an adult female eriophyid mite.

Deutonymph   The third and final immature stage of a mite or tick.

Diapause   A period of dormancy in which many insects undergo to avoid adverse conditions (e.g., winter cold). Diapause is more involved than simple dormancy or hibernation and can only be terminated by certain stimuli such as day length or a prescribed length of exposure to cold.

Dictyoptera   An order name used in some taxonomic arrangements that includes cockroaches, termites, and mantids.

Dieback   The decline and dying of the upper or terminal growth of a plant.

Diffraction   A phenomenon when light waves are scattered differentially in different directions making the object diffracting the light appear to change colors depending on the angle of view.

Diploid   Possessing two sets of chromosomes. May refer to a cell or an organism.

Diplopoda   Class of arthropods known as the millipedes.

Diplosegment   A pair of segments that are fused so they appear as a single segment. Such segments make up most of the trunk of millipedes.

Diptera   The order of insects that include the true flies.

Direct flight muscles   Muscles that produce wingbeat and are directly attached to the base of the wing.

Diurnal   Active during the day.

Distal   Refers to an object, structure, or appendage being away from the reference point (usually the organism's main body). Contrast to proximal.

Dorsal   The back or upper side.

Dorsal vessel   Large tubelike vessel located on the dorsal portion of the insect and is responsible for pumping hemolymph throughout the insect.

Dorsoventrally   Top to bottom.

Drone   Male honeybee.

Dulosis   Slave making in ants.

Ecdysis   During a molt, the actual splitting of the old exoskeleton resulting in the animal physically shedding the old covering.

Ecdysone   Molting hormone that is partially responsible for triggering a molt.

Eclosion   Escape of an immature insect from the eggshell (egg hatch); escape of an adult insect from the pupa.

Ectoparasite   A parasite that develops on the outside of its host.

Elytra   The hardened mesothoracic wings of beetles.

Embiidina   The order of insects that includes the webspinners.

Endotoxin   Toxin present within an intact bacterial cell. Such toxins are not secreted and therefore are not found in the fluids outside the cell.

Endoparasite   A parasite that develops on the inside of its host.

Ensifera   A suborder of Orthoptera generally marked with very long filament antennae.

Endemic   Refers to a species being native to a particular area in its distribution or can refer to the presence of a disease at a relatively low level of incidence (contrast with epidemic).

Endocuticle   The interior layer of the procuticle that is not sclerotized and relatively soft.

Entognathous   Having mouthparts sunken within the head.

Entomophagous   Feeding on insects.

Entomophagy   Eating insects by humans.

Ephemeroptera   The order of insects known as mayflies.

Epicuticle   The thin outermost layer of the insect's exoskeleton. The epicuticle is made of an external waterproof cement layer on top of a layer (cuticulin) made of a matrix of lipids and proteins.

Epidemic   Refers to a disease that occurs at an exceptionally high incidence.

Epigynum   The external genital opening of a female spider.

Epidermis   The outermost layer of cells of leaves, young stems, roots, flowers, fruits, and seeds. In insects, the epidermis is the single, outermost layer of cells that secretes the cuticle.

Epipharynx   A structure associated with the insect mouthparts that runs down the center of the interior surface of the labrum.

Epicuticle   The outermost layer of the exoskeleton.

Erineum   A change in plant growth, or gall, where plant hairs are produced in great abundance creating a felty patch on the leaf surface. Many eriophyid mites cause such changes in leaf growth.

Eriophyids   Mites in the superfamily Eriophyoidea that have two pairs of legs, generally elongate body form, are minute in size, and usually feed on plants. Many produce galls on plants.

Esophagus   A section of the insect foregut that runs between the pharynx and the crop.

Eusocial   Having the highest level of social organization. Most definitions involving arthropods would include reproductive division of labor, overlapping generations, and cooperative care of young as being attributes showing eusocial behavior.

Eversible vesicle   A small pouch-like structure found on the body of certain insects, often containing toxic or foul-smelling materials, capable of turning inside out, and dumping its contents outward.

Exocuticle   The exterior layer of the procuticle that is sclerotized and hardened.

Exodigestion   Digestion of a prey organism occurs external to the predator as a result of injection of digestive fluids into the prey.

Exoskeleton   A skeletal structure that is formed on the external surface of an animal, such as occurs with insects, mites, and other arthropods.

Extant   Refers to species, or traits, that are still in existence.

Extinct   Typically refers to a species that no longer exists anywhere on the planet.

Exuvium   The cast skin of an arthropod. The term does not exist in the singular and is also correct as **exuviae**.

Eyespots   Prominent markings resembling the eyes on the wings or body of certain insects.

Facultative   Optional or nonobligatory. The opposite of facultative is obligate in which there is no choice.

Family   A midlevel taxonomic classification or rank. Several genera may be grouped into a single family, and several families may be grouped into a single order.

Femur   Leading away from the body, this is the third segment of the insect leg. It is often large and contains significant musculature.

Filarial worm   A type of endoparasitic nematode.

File and scraper   Outgrowths or modifications of the exoskeleton found in some grasshoppers, crickets, and other orthopterans that when rubbed together are responsible for sound production.

Filiform   Threadlike.

Fingergalls   Abnormal growth forms in the shape of fingers, such as are produced by eriophyid mites on certain plants (e.g., wild plum and cherry).

Flagellum   The second antennal segment away from the body. Also the taillike structure responsible for motility found on sperm, and a variety of single-celled organisms.

Flagging   Yellowing, wilting of foliage, and/or twig breakage on a single branch.

Foregut   That portion of the insect's digestive track that is responsible for taking food into the body. The foregut consists of the mouth and associated glands, esophagus, crop, and proventriculus.

Forewings   The front pair of wings that are found on the mesothorax.

Fossorial   Referring to legs modified for digging through the soil, also referring to living within the soil.

Frass   Solid insect excrement typically consisting of a mixture of chewed plant fragments.

Fundatrix   The female aphid emerging from the overwintered egg that initiates colonies in spring.

Fungus gardens   Growths of specific types of fungus occurring in the colonies of leaf-cutter ants. The ants grow the fungus on the leaf material brought into the nest.

Furcula   A fork-shaped appendage found in Collembolans on the ventral side of the abdomen. Together with the retinaculum, it allows the animal to jump a considerable distance.

Gall   The abnormal growth of plant tissues, caused by the stimulus of an animal, microorganism, or wound.

Ganglia   A mass of nervous tissue that in insect is usually responsible for control of an individual segment of the body.

Gallery   A tunnel produced by insects during a life stage. For example, webspinners produce interconnected silklined tunnels in which the insects live. Bark beetles construct an egg gallery and larvae subsequently feed tunnel from this point.

Gastric caeca   Membranous outpocketing of the midgut where digestion and absorption of food occurs. Is one of the only places on an insect that is not lined by exoskeleton.

Genus   A relatively low-level taxonomic classification or rank. Several species are grouped into a single genus, and several genera (plural of genus) are grouped into a single family.

Gill   A respiratory structure responsible for extracting oxygen from water.

Gnathosome   In mites, the area of the head that surrounds the mouthparts.

Gonopod   Specially modified legs in millipedes that carry spermatophores.

Gossamer   The type of silk used by spiders to the balloon.

Gregarious   Living and feeding in groups.

Grub   The immature form of many beetles.

Grylloblattodea   The order of insects that include the rock or ice crawlers.

Haltere   The club-like metathoracic wing of a fly. Halteres (plural) are used to assist in flight stabilization.

Hamuli   A series of small hooks that connect together the front and rear wings of Hymenoptera.

Haploid   Possessing only a single set of chromosomes. May refer to a cell or organism.

Haplodiploid   Typically, a characteristic found in Hymenoptera in which females arise from diploid fertilized eggs and males arise from haploid unfertilized eggs.

Harborage   A group of cockroaches.

Hemimetabolous development   A pattern of metamorphosis used by many insects (e.g., Orthoptera, Isoptera, and Hemiptera) that involves eggs, followed by immature nymphs, and finally adults. Adult and immature stages often have similar different habits and appearance. Also called incomplete metamorphosis.

Hemelytron   Mesothoracic wing of the true bugs (suborder Heteroptera in the order Hemiptera) (plural, **hemelytra**).

Hemocyte   A blood cell.

Hemolymph   Arthropod blood.

Herbaceous   Plants that lack woody tissues and that annually die back.

Herbivore   An animal that feeds exclusively on plants.

Hemiptera   The order of insects that includes the true bugs, aphids, cicadas, leafhoppers, whiteflies, and scales.

Hexapods   Animals within the phylum Arthropoda that possess three pairs of legs. This includes the entognathous Hexapods (orders Collembola, Protura, and Diplura) and the insects.

Hibernaculum   A tiny silk cocoon spun by the first or second instar caterpillars within which they spend winter or some other period (plural, **hibernaculae**).

Hilltopping   Behavior in which individuals aggregate on hilltops.

Hindgut   The portion of the insect's digestive track that is responsible for transferring undigested material and waste products out of the body. The hindgut consists of the Malpighian tubules, the colon, rectum, and anus.

Hindwings   The back pair of wings that are associated with the metathorax.

Holometabolous development   A pattern of metamorphosis used by many insects (e.g., beetles, moths/butterflies, sawflies, and flies) that involves eggs, followed by immature larvae, a transition pupal stage, and finally adults. Adult and immature stages often have very different habits and appearance among insect groups with this metamorphosis. Also called complete metamorphosis.

Honeydew   The sugary, liquid excrement produced by certain aphids, scales, and other insects that feed in the phloem of the plant.

Hormones   Chemical messengers released within the body by one cell type and affect a different cell type. Hormones are typically transported by the bloodstream or hemolymph.

Host   The plant or animal on which an insect or pathogen feeds.

Hygroreceptor   A sensory receptor that is sensitive to humidity.

Hymenoptera   The order of insects that includes sawflies, bees, wasps, and ants.

Hypermetamorphosis   In development, the extreme change in body form as the development progresses from a highly mobile early instar to a grub-like immobile late-stage instar.

Hyperparasite   A parasite of a parasite.

Hypopharynx   A structure found in insects possessing chewing mouthparts. The hypopharynx functions as a tongue.

Hypostome   A structure typically associated with the mouthparts of mites and ticks. It is the hardened harpoon-like structure used to jab into their hosts or prey.

Imago   The adult.

Incomplete metamorphosis   A pattern of metamorphosis used by relatively primitive insects (e.g., roaches, termites, grasshoppers, and mantids) that involves eggs, followed by immature nymphs, and finally adults. Adult and immature stages often have very similar habits and appearance among insect groups with this metamorphosis. Also called hemimetabolous metamorphosis.

Indirect flight muscles   Muscles that produce wingbeats by alternate compression of the abdomen rather than being directly attached to the wing base.

Indirect mating   The passage of sperm to the female without direct coupling of the genitals.

Insecticide   A chemical produced and sold for the purpose of killing an insect.

Instar   The stage of an insect between the periods when the skin is shed.

Integrated pest management   A form of arthropod and weed pest control that attempts to minimize chemical controls through appropriate monitoring and alternative control practices.

Integument   The exoskeleton of an arthropod.

Interneuron   A nerve cell that typically connects a sensory nerve to a motor nerve or to a ganglia.

IPM   Acronym for integrated pest management.

Interference (light)   Refers to different wavelengths of light that interact to produce yet a different wavelength of light that is reflected from an object.

Isopoda   Order of crustaceans that includes the terrestrial sow bugs and pill bugs and a variety of marine species.

Isoptera   The order of insects that includes the termites.

Johnston's organ   A sensory structure found in the antennae of many insects that is exceptionally sensitive to vibrations in the air. As such, it is used to detect sound.

Juvenile hormone   A hormone used to control what type of molt occurs within an insect: larval–larval molt, larval–pupal molt, or pupal–adult molt.

Kin selection   An evolutionary strategy in which one's reproductive success is based on insuring the reproductive success of one's relatives (typically brothers or sisters).

Kingdom   The highest level of taxonomic classification or rank. Several phyla are grouped into a single kingdom.

Labium   A structure that is part of the insect's mouthparts. It functions as the lower lip to the mouth. The labium may often possess associated sensory palps.

Labrum   A structure that is part of the insect's mouthparts. It functions as the upper lip to the mouth.

Larva (plural, larvae)   The immature form, between egg and pupa, of an insect with complete metamorphosis, i.e., caterpillars, maggots, and grubs. The first immature stage of a mite or tick, which is six-legged.

Lateral   Directed to the side. Used to describe markings on the sides of an insect.

Laterally   Side to side.

Leafminer   An insect that has the habit of developing by feeding on internal leaf tissues, which it chews as it mines the leaf. Insects that tunnel needles in a similar manner are called **needleminers**.

Lepidoptera   The order of insects that includes the butterflies and moths.

Lerps   Wax-covered sugary excrement produced by some insects that suck sap, notably psyllids.

Lipids   A type of compound typically comprised of one or more long chains of carbon and includes waxes, fats, oils, certain vitamins, and sterols.

Luciferin   Compound, along with the enzyme luciferase, that is responsible for the light production in fireflies.

Macrodecomposer   An animal that feeds on dead animals, plants, or their by-products and, in the process, assists in the breakdown of biologic material into smaller and smaller fractions.

Maggot   The immature form of some flies (Diptera, suborder Brachycera) that do not have a distinct head capsule and feed using mouth hooks.

Maggot therapy   The use of larval blow flies to remove dead, necrotic, and gangrenous tissues from wounds.

Malacostraca   Class of crustaceans that includes the lobsters, crabs, shrimp, crayfish, pill bugs, and krill.

Malaria   Mosquito-transmitted diseases caused by species in the protozoan genus *Plasmodium*.

Melanistic   Dark colored.

Malpighian tubules   Excretory structures located at the junction of the midgut and foregut that float in the hemolymph of the body cavity. Malpighian tubules are filamentous blind sacs that filter the hemolymph of metabolic wastes and transfer the wastes to the hindgut.

Mandible   Paired appendages of the mouthparts located behind the labrum. In chewing insects the mandibles are hardened and used to cut and gnaw. In some insects that suck fluids the mandibles are highly elongated and may be used to penetrate.

Mantophasmatodea   The most recently discovered order of extant insects known variously as heelwalkers, gladiators and African rock crawlers.

Mantodea   The order of insects known as the mantids.

Marine   Associated with ocean waters.

Marsupium   Egg pouch in Isopoda.

Mating disruption   An insect management technique to disrupt insect mating by releasing high levels of sex pheromones.

Maxillae   A paired set of appendages of the insect mouthparts located immediately behind the mandibles. In chewing insects the maxillae are usually used to help manipulate food and often have sensory structures. In some insects that suck fluids the maxillae are highly elongated and variously modified.

Maxilliped   In crustaceans, this is an appendage that has evolved to function as a mouthpart.

Meconium   The waste material excreted by a moth shortly after it has emerged from the pupa. Also the excretory waste sac in some Neuroptera.

Mechanoreceptor   A sensory structure sensitive to touch, pressure, stretching, or contraction.

Mecoptera   The order of insects that includes the scorpion flies.

Merozoite   The stage of the disease organism-causing malaria that infects human blood cells.

Mesophyll   The cell tissues within a leaf or other part of the plant where photosynthesis occurs.

Mesothorax   The middle segment of the thorax of an insect. The forewings of the insects and the middle pair of legs are attached to the mesothorax.

Metamorphosis   The change in body form that occurs in insects as they grow and develop.

Metathorax   The last of the three segments of the thorax of an insect. A pair of legs and the hindwings are attached to this segment.

Micropyle   Opening of the eggshell (chorion) that allows sperm to enter for fertilization.

Midgut   The portion of the insect's digestive track that is responsible for digesting food. The midgut includes the gastric caeca and the peritrophic membrane.

"Miller moth"   Any species of moth that is locally very abundant. The term is derived from the scales of moths and their resemblance to the flour on the clothing of one who mills grain; the scales of a locally abundant moth may also similarly accumulate.

Mimicry   Appearing as something else. Usually, one species will mimic another species or unrelated object (e.g., walking sticks mimic sticks).

Mine   To form a burrow or excavate a tunnel. Used to describe the activities of insects that live or feed within a leaf or needle.

Mode of action   Refers to the functioning of an insecticide. Different compounds interfere in different ways (modes of action) with different aspects of an insect's physiology.

Molt   The shedding of the exoskeleton by an insect in the process of development.

Molting fluid   A fluid secreted by the cells of the epidermis into the space following separation of the old procuticle during the apolysis stage of molting. The molting fluid contains enzymes that are inactive until a new exoskeleton is formed. The enzymes are then activated and digest parts of the old exoskeleton so that it can be reabsorbed.

Motor neurons   Nerve cells that stimulate muscles or other organs to function.

Mottephobia   A phobia involving excessive fear or dread of moths or butterflies.

Müllerian mimicry   A type of mimicry in which several species share a common model.

Multivoltine   Many generations per year.

Mushroom bodies   Portions of the insect brain responsible for learning and processing sensory signals related to smell.

Mutualistic Providing benefit to both species that are in some association.

mya Million years ago.

Mycetocytes A specialized cell that harbors symbiotic bacteria.

Mycetome A organelle or group of cells consisting of mycetocytes.

Myiasis Infestation with larvae of flies (Diptera).

Myriapoda A term describing the classes of arthropods that have the body generally divided into two body regions, a head and elongated trunk. On the trunk are legs associated with most every segment. These many-legged animals include centipedes, millipedes, pauropods, and symphylans.

Nasus Nozzle-like structure on the head of some species of soldier termites. From the nasus, the soldier can squirt noxious chemicals as a means of defending the colony.

Natatorial Referring to legs modified for swimming.

Nauplii The first instar fairy shrimp.

Nectar guides The lines of pigments in the petals of certain flowers that are seen in ultraviolet light.

Necrotic arachnidism A type of tissue damage caused by a venomous spider bite.

Nematode Generally microscopic, wormlike animals that live saprophytically in water or soil or as parasites of plants and animals. They are classified in the phylum Nematoda.

Neuroendocrine system Secretory cells of the nervous system that release hormones when stimulated.

Neuroptera The order of insects that contain lacewings, dobsonflies, ant lions, owlflies, and snakeflies.

Neurotoxin A toxin that affects nerve function.

Nit Egg of a louse.

Nociceptor A pain receptor of the nervous system (not found in insects).

Nocturnal Active at night.

Nomadic phase A colony-based behavior of legionary ants in which the colony moves through the habitat a considerable distance each day.

Nontarget organism An organism incidentally affected by a control practice applied against another organism. The term is usually applied to mammals, birds, beneficial arthropods, and other animals that are affected by insecticide applications directed against a certain pest insect.

Notoptera An order of insects used in some taxomonic arrangements that includes both the rock crawlers (Grylloblattodea) and heelwalkers (Mantophasmatodea).

Notostraca An order of crustaceans that includes the tadpole shrimp.

Notum A dorsal sclerite of a thoracic segment.

Nuptial gift Some food or material (dead insect, glandular secretion) provided by a male insect to entice a female to mate.

Nymph An immature stage of an insect with simple metamorphosis, e.g., aphids, bugs, and grasshoppers.

Obligate Opposite of facultative. A behavior or developmental stage that must occur.

Ocelli Simple, single-lensed eyes of adult insects.

Odonata The order that contains damselflies and dragonflies.

Oligophagous A type of feeding behavior in which the types of food utilized are partially restricted (as opposed to polyphagous). A limited diet in terms of variety.

Ommatidium One of the hexagonally faced individual units that comprise the adult insect compound eye. Ommatidia (plural). Many thousands of ommatidia may comprise a compound eye.

Omnivore An animal that feeds on a wide variety of materials include animal products and plants.

Onchocerciasis A blackfly-transmitted nematode disease of humans. Symptoms include depigmentation of the skin, loss of skin elasticity, and blindness. Also known as river blindness.

Ootheca Egg pods of roaches, mantids, primitive termites, and a few other groups of insects. Eggs are held together and encased in a hardened protein capsule that is formed by the female accessory gland during oviposition. Plural is **oothecae.**

Open tracheal system The common respiratory system in insects. It consists of spiracles that open to the atmosphere allowing oxygen to diffuse into a series of tubes (trachea and tracheoles) that ultimately come in direct contact with the tissues.

Opiliones Order of arachnids known as the daddy longlegs, harvestmen, and opilionids.

Opisthosoma The abdomen of some arachnids.

Order A midlevel taxonomic classification or rank. Several families can be grouped into a single order and several orders can be grouped into a single class. The class Insecta consists of many orders (e.g., Ephemeroptera, Odonata, Isoptera, Phasmida, Thysanoptera, and Coleoptera).

Orthoptera   The order of insects that includes grasshoppers, katydids, and crickets.

Osmeterium   A fleshy Y-shaped gland that is extended by swallowtail larvae in response to disturbance (plural, **osmeteria**).

Ovaries   Reproductive organs in females that produce the eggs.

Ovarioles   Organelles within the ovary that are ultimately responsible for egg production.

Overwinter   To spend the winter.

Oviparous   Term used to describe females that lay eggs. While developing, offspring are nourished entirely by the yolk of the egg.

Oviposition   The process of laying an egg by an insect. The verb is **oviposit**.

Ovipositor   The egg-laying apparatus of a female insect.

Ovoviviparous   Term used to describe females that give birth to live offspring (as opposed to eggs). Eggs are retained within the female and hatched within her body. While developing, offspring are nourished by the yolk of the egg.

Ovisac   The large waxy sack into which eggs are laid by mealybugs, some soft scales, and related insects.

Palpigradi   The order of arachnids commonly known as the microwhip scorpions.

Pandemic   A disease epidemic that has spread, or is spreading, across a large portion of the earth.

Parasite   An organism that lives at the expense of another. (The term **parasitoid** is often used to describe parasitic insects that kill the host in which they develop.)

Parthenogenesis   Reproduction in which an unfertilized egg develops into an individual.

Passive ventilation   Allowing air (oxygen) to passively enter the tracheal system by simple diffusion.

Pathogen   An entity that is responsible for causing a disease.

Pauropoda   Class of arthropods known as pauropods, may be confused with millipedes.

Pedicel   (1) The basal segment of the insect antennae; (2) a body constriction between the abdomen and the thorax or cephalothorax.

Pediculosis   Infestation with lice.

Pedipalps   In spiders and related arthropods, these are the fingerlike appendages associated with the mouthparts.

Perennial   Long-lived, a plant that grows every year (e.g., a tree).

Periodical cicada   Either the 13- or 17-year periodical cicada. Both of which are characterized by predictable and repeatable life cycles resulting in mass emergence of adults at specified intervals (13 or 17 years).

Periphyton   The mixture of algae, cyanobacteria, other microbes, and organic debris that typically covers the surfaces of submerged objects.

Peritrophic membrane   Noncellular membrane produced by the gastric caeca. The peritrophic membrane envelops swallowed food and functions to protect the insect from pathogens that may have been consumed.

Pesticide   A chemical produced and sold for the purpose of controlling some sort of pest. Examples of pesticides include herbicides (used to control plants), fungicides (used to control fungi), and insecticides (used to control insects).

Petioles   The stem or stalk of a leaf stem that attaches to a twig.

Phase polymorphism   In Orthoptera, the ability of the individuals within a population to cycle in synchrony over multiple generations between two distinct morphs, each associated with distinct behaviors.

Phasmatodea   The order of insects that includes walking sticks and leaf insects.

Pheromone   A chemical used to communicate between members of the same species. For example, many female moths produce sex pheromones to attract mates.

Phloem   Food-conducting tissue located in the bark of woody plants that consist of sieve tubes, companion cells, phloem parenchyma, and fibers. The phloem is largely responsible for conducting carbohydrates from the leaves to the rest of the plant.

Phthiraptera   The order of insects that includes the lice.

Phyllopods   Leaflike appendages on the underside of some aquatic crustaceans that help to sweep food particles toward the mouth.

Phytophagous   Feeding on plants.

Phytoplasma   A group of bacteria that lack a cell wall and develop within the phloem tissues of plants.

Phylum   A high-level taxonomic classification or rank. Several classes of organisms are grouped into a single phylum, and several phyla are grouped into a kingdom. The class Insecta, along with other classes, belongs to the phylum Arthropoda.

Pitch   A resinous material exuded by conifers either naturally or in response to a wound.

Pharynx   An area of the foregut immediately behind the oral opening (mouth) of an insect.

Phoresy   One animal using another for dispersal.

Photoperiod   The relative amount of time during the day that is light (or dark).

Plague   Usually refers to the flea-transmitted disease caused by the bacteria *Yersinia pestis*. Three forms of plague are recognized: bubonic plague, septicemic plague, and pneumonic plague; the latter may be transmitted without the aid of a flea.

Plasma   The liquid portion of the hemolymph.

Plastron   A bubble attached to or surrounding the abdomen of an insect. The plastron is held in place by a combination of hydrophobic forces and small hairs located on the animal's body. The plastron functions as a gill.

Plecoptera   The order of insects that includes the stone flies.

Pleuron   A lateral sclerite of a thoracic segment.

Poikilothermic   Refers to "cold-blooded" animals that cannot regulate their body temperature.

Polarized light   Light with light waves all oriented in the same direction.

Polyphagous   Feeding on a very wide variety of food items.

Predator   An animal that moves and hunts other animals (**prey**). Predators require multiple prey to complete their life cycle.

Prepupae   The period during the last larval stage when insects undergo changes in behavior and, sometimes, form prior to pupation.

Pretarsus   The very tip of the tarsus, usually modified to be a small claw or pad.

Prey   An animal that is hunted and killed for food.

Primitive   Early or poorly evolved. Opposite of advanced.

Procuticle   The dominant layer of the multilayered exoskeleton, located underneath the thin epicuticle. The procuticle often is subdivided into the endocuticle and the exocuticle, the latter of which undergoes chemical reactions that produce strengthening, hardening, and water resistance.

Proleg   Fleshy leglike extensions of the abdomen found in the larval stages of larvae of sawflies and all moths and butterflies.

Pronotum   The dorsal plate on the prothorax.

Propolis   Plant resins collected by honeybees and used to help create the structure of the hive.

Prosoma   The front region of the arachnid body that contains the head and thorax (cephalothorax).

Prothorax   The first segment of the thorax, just behind the head.

Prothoracic gland   A gland composed of neurosecretory cells that are a major source of ecdysone.

Protocerebrum   The large anterior portion or lobe of the insect brain responsible for the processing of visual sensory input from the organs of vision (compound eyes, ocelli).

Protogyne   The form of an eriophyid mite female that occurs during the growing season (vs. overwintering form known as a deutogyne).

Protonymph   The second immature stage of a mite or tick.

Protozoan   Often one- or few-celled animallike creatures often equipped with flagella or cilia that provide movement.

Proventriculus   Part of the foregut of an insect and is responsible for further grinding and processing food before it enters the midgut.

Proximal   Adjacent or near the point of reference (usually the main body). Opposite of distal.

Pseudoscorpiones   The order of arachnids that contain the pseudoscorpions.

Psocoptera   The order of insects that includes the booklice or psocids.

Pterygote   Insects that are winged or have had ancestors that are winged.

Pupa   The transitional stage, between larva and adult, of insects with complete metamorphosis.

Puparia   A type of pupa characterized by a subgroup of flies in which the larva pupate within the last cast larval exoskeleton. The pupa is smooth and rounded.

Questing   Dispersal behavior exhibited by ticks in which the animals climb up on top of vegetation and rhythmically wave their legs in order to cling to a passing host.

Rad   A unit of the absorbed dose of ionizing radiation.

Raptorial   Referring to legs modified for grasping prey.

Rectum   In insects, a portion of the hindgut partially responsible for maintaining water balance and transporting wastes out of the body.

Reflexive bleeding   Oozing, using noxious or toxic, hemolymph through special pores or joints on the exoskeleton as a mechanism to avoid predation.

Refugia   Areas typically left untreated in a pest management program. Also, those areas in which populations or individuals may find safety from predation or harsh environmental conditions.

Repagula   Structure produced by adult female owlflies to protect the recently deposited egg mass.

Repugnatorial gland   A type of gland found in many arthropods (especially daddy longlegs and hemipterans) that produces toxic or repellent chemicals when the animal is threatened or disturbed.

Resilin   An elastic protein found in the exoskeleton of fleas and some other species of insects that allows them to store and release mechanical energy.

Resistance   The ability of an organism to exclude or overcome, completely or in some degree, the effect of a pathogen, insecticide, insect, or other damaging factor. For example, some plants have resistance to insect attack, and some insects can have resistance to the effects of an insecticide.

Retinaculum   A hooklike appendage found in Collembolans on the ventral side of the abdomen. Together with the retinaculum, it allows the animal to jump a considerable distance.

Ricinulei   The order of arachnids known as the hooded tick spiders.

River blindness   See onchocerciasis.

Rostrum   Pointed portion of decapods (lobsters, shrimp, etc.) head.

Royal jelly   Highly nutritious food produced in special glands located in the heads of certain worker honey bees. It is fed to developing bee larvae to produce queens.

Russeting   A bronzy and somewhat thickened surface of a fruit or leaf that may be induced by feeding of certain mites and from other causes of injury.

Saline   Salt. Saline waters have high concentrations of salt.

Saltatorial   Referring to legs modified for jumping.

Sapwood   Young, physiologically active zone of wood; outermost growth layers of xylem in woody plants that conduct water.

Scale   A flattened seta (hair) of an insect.

Scale insect   Insect members of the order Hemiptera that occur within the superfamily Coccoidea. These are generally marked by production of a waxy covering over the body.

Schizomida   The order of arachnids know as short-tailed whip scorpions.

Scientific name   A formal name of a species recognized under the international codes of zoological nomenclature. The scientific name is unique and made of a genus name and a species name. For example, the scientific name of the house fly is *Musca domestica*.

Sclerite   A visible plate of the exoskeleton.

Sclerotin   A structural protein found within the exoskeleton.

Sclerotization   The chemical bonding of chitin and sclerotin resulting in the hardening of the exoskeleton after a molt.

Scorpiones   The order of arachnids that is known commonly as the scorpions.

Scutellum   Large triangular plate of exoskeleton located between the wing bases just behind the thorax in hemipterans.

Secondary pest   A mite or insect that only becomes a pest due to improper insecticide use.

Sedentary   Moving little, if at all, as it develops.

Sensory neurons   Nerve cells that sense light, chemicals, vibration, humidity or other signals in the environment surrounding an organism.

Sericulture   The culture of silkworms to produce silk.

Serpentine   Winding, twisting pattern typically used to describe the shape of certain leaf mines.

Seta (plural, setae)   A bristle or hair.

Sexual   Produced as a result of a union in which fertilization and meiosis occurs.

Sexual dimorphism   Substantial difference in size and/or appearance between adult males and females of the same species.

Shellac   Resin produced by certain species of lac scales.

Shelter tubes   Tunnels of soil and debris built by subterranean termites to access food distant from the colony.

Shothole   Small holes in leaves. These are most characteristics of flea beetles and some other insects that chew pits in the interior of leaves.

Silk   Filamentous protein fibers extruded from specialized glands.

Simple metamorphosis   A pattern of metamorphosis used by many insects (e.g., true bugs, aphids, grasshoppers, and earwigs) that involves eggs, followed by immature nymphs, and finally adults. Adult and immature stages usually feed in the same manner and are primarily differentiated by the adult features of sexual maturity and (usually) wings. Also called gradual metamorphosis or hemimetabolous.

Simuliotoxicosis   Death or illness as a result of multiple black fly bites and the injected of saliva.

Siphonaptera   Order of insects that includes the fleas.

Skeletonize   The feeding pattern of certain leaf-feeding insects that avoid feeding on larger veins of the leaf producing a lacy "skeleton" of the leaf.

Skep   Primitive artificial housing sites provided to honeybees.

Solitary   Living along, not in groups.

Solifugae   Order of arachnids known as windscorpions, solpugids, sunspiders, and camel spiders.

Soma   The main body of a nerve.

Sooty mold   A dark, typically black, fungus growing on the insect honeydew.

Species   The lowest level of taxonomic classification or rank. One or more species are grouped into a single genus. A species is also a group of interbreeding individuals that do not breed with other groups or individuals.

sp.   A contraction for species in the singular; plural is **spp.** It is typically used in making a scientific name where the genus is known but the species name(s) is (are) generalized. For example, the bedbug and its close relatives are in the genus *Cimex* and together they might be referred to as *Cimex* spp.

Sperm   The reproductive cells produced by the male of a species. They are haploid.

Spermatheca   A reproductive structure found in female insects in which sperm is stored after mating.

Spermatophore   A reproductive structure constructed by the male in which a packet of sperm is surrounded by a protective protein matrix.

Spermaflege   A weakened area on the female abdomen of some true bugs (e.g., bed bugs) that the male pierces with his aedeagus during the type of mating known as traumatic insemination.

Spiderling   A newly hatched spider.

Spinnerets   External structures that are used to spin silk. These occur at the tip of the abdomen in spiders and near the mouth in caterpillars.

Spiracle   The opening in the body through which arthropods breathe.

Spirochete   Spiral-shaped gram negative bacteria.

Spore   The reproductive unit of fungi consisting of one or more cells; it is analogous to the seed of green plants.

Sporozoite   The haploid stage of the disease organism-causing malaria. The sporozoite is the stage transferred from the insect to a human.

Stabilimentum   A thick band of silk, often laid in zigzag pattern, that is produced in the center of webs by some orb-weaving spiders.

Stadium   (Plural, Stadia), the duration of an instar.

Statary phase   A colony-based behavior of legionary ants in which the colony remains in place while the queen produces thousands of eggs.

Stemmata   A type of single-lensed simple eye often found in immature holometabolous insects.

Sternum   A ventral sclerite of a thoracic segment.

Stinger   Most often used to describe a modification of the ovipositor in certain Hymenoptera that has associated venom sacs and is used for defense. The term is also used to describe the ovipositor of parasitic Hymenoptera that oviposit in/on other insects. The tip of the abdomen of scorpions is also associated with venom and is described as a stinger.

Stippling   Small, white flecking injuries on foliage produced by certain insects (some leafhoppers, lace bugs, thrips) and spider mites resulting from removal of cell contents.

Strepsiptera   The order of insect that includes the twisted-wing parasites.

Stridulatory organs   Areas of the exoskeleton modified to produce sound by rubbing two parts together. For example, the on the leading edge of the forewings of a cricket are structures known as a file and scraper that are used to produce songs.

Stylets   When applied to insects, a reference to mouthparts designed to penetrate and to suck fluids. In Hemiptera, the stylet bundle consists of paired mandibles and maxillae that are extremely elongated and very fine.

Stylostome   The feeding tube produced within the host skin as a chigger feeds.

Subesophageal ganglion   The ganglion that connects the insect brain (specifically the tritocerebrum) to the rest of the central nervous system.

Subimago   The first of two-winged stages of insects in the order Ephemeroptera (Mayflies). They are often somewhat dark-colored and are called the "dun" stage by fisherman.

Supercooling   Ability to avoid the formation of ice crystallization at temperatures below freezing.

Supersedure   The replacement of the queen in a honey bee colony with a new queen.

Sylvatic plague   Plague occurring among wild animals.

Symbionts   From symbiotic relationships in which two species may be obligately or facultatively dependent on each other for survival.

Symphyla   Class of arthropods related to the centipedes and millipedes. They are commonly known as symphylans.

Synanthropic   Describes a species associated with humans.

Synapse   The small gap that defines the area where two nerve cells come together.

Systematics   The area of biology that determines the evolutionary relationships among individual species.

Systemic insecticide   An insecticide that is capable of moving internally through a plant. Systemic insecticides may be sprayed on leaves, applied to soil, or injected and moved from the point of application. Acephate and imidacloprid are examples of systemic insecticides.

Taenidia   Reinforcing rings of exoskeleton found in the trachea.

Tagma   A main body segment. In insects, there are three tagma: head, thorax, and abdomen. The plural is **tagmata**.

Tail-wag dance   The complex dance of the honeybee that provides colony nest mates with both direction and distance to resources located by foragers.

Tapetum   A layer of tissue in the eyes of some spiders, moths and other animals that is highly reflective and assists with vision in dimly lit conditions.

Tarsus   The most distal segment of an insect leg. Commonly referred to as the insect's foot.

Taxonomy   The science of classifying organisms using a standard nomenclature that accounts for systematic relationships.

Tegmen   The thickened front (mesothoracic) wing of grasshoppers, crickets, mantids, and earwigs (plural, **tegmina**).

Tents   Protective shelter constructed of silk and spun by certain caterpillars.

Tergite heteronomy   In centipedes, alternate body segments are shortened so that body sway is dampened when running.

Terminal growth   Typically, new growth or buds at the end of a branch or twig.

Terrestrial   Living on land.

Test   The waxy covering produced by a scale insect.

Thelyphonida   The order of arachnids commonly referred to as whipscorpions or vinegarroons.

Thigmotactic   Movement toward (positive thigmotaxis) or away from (negative thigmotaxis) a mechanical stimulus such as touch or vibration.

Thorax   The middle section of an insect body where the legs and wings are attached.

Thysanoptera   The order of insects that includes the thrips.

Tibia   Leading outward from the body, the tibia is the fourth segment of the insect leg.

Tick paralysis   A disease caused by the injection of neurotoxin during a tick bite.

Toxin   Poisonous secretion produced by a living organism.

Trachea   A hollow tube that is a basic structure of the respiratory system that delivers oxygen to the tissues of insects. Plural is tracheae.

Transgenic   An organism that has its genetic material deliberately altered with genes derived from another type of organism.

Transovarial transmission   Passage of a microorganism from one generation to the other via the egg.

Traumatic insemination   Reproductive behavior exhibited by bed bugs by which the male pierces the female's abdomen with his aedeagus and deposits sperm within her hemolymph.

Trichogen cell   Specialized cells within the cellular epidermis of the exoskeleton that are responsible for producing exoskeletal setae.

Trichoptera   The order of insects that includes the caddisflies.

Tritocerebrum   The third and most anterior portion of the insect's brain, responsible for receiving and processing sensory information from the mouthparts.

Triungulins   First instar larvae of blister beetles and some species of Neuroptera that are highly active and seek host insects on which they later develop.

Trochanter   Leading outward from the body, the trochanter is the second segment of the insect leg.

Trophallaxis   The mouth-to-mouth or anus-to-mouth transfer of chemicals (often contained in liquids or food) among individuals typically in a social colony. It is a form of chemical communication.

True bugs   Insects in the hemipteran suborder Heteroptera. A hemelytra forewing is characteristic of these insects.

Trunk   The elongated region and leg-bearing region of the body behind the head in myriapods (millipedes, centipedes, pauropods, and symphylans).

Tubercles   A rounded protuberance found on many insects, particularly caterpillars.

Tumbler   A mosquito pupa.

Tymbal   Sound-producing structure within the abdomen of cicadas.

Tympanum   Sound-detecting organs in insects (plural, tympana).

Typhus   A bacterial disease transmitted to humans by lice. Symptoms include high fever, headache, disorientation, red spots, and death.

Undescribed species   A species that exists but has not yet been formally described so that it has a recognized scientific name.

Uropods   Taillike processes on terrestrial arthropods (e.g., sowbugs and pill bugs) that are used in absorbing water and maintaining water balance.

Urticating   Stinging or irritating.

Vector   A living organism (i.e., insect, bird, and higher animal) able to carry and transmit a pathogen.

Venomous   The ability to deliver a sting or bite that contains a poisonous venom of some type.

Ventral   The underside of the body.

Vestigial   Remnant features (e.g., legs) that are greatly reduced in size from that of ancestral forms.

Virus   A submicroscopic obligate parasite consisting of nucleic acid and protein.

Viviparous   Term used to describe females that give birth to live offspring (as opposed to eggs). While developing, offspring are nourished by the mother.

Wheel position   Describes behavior and positioning of male and female Odonata while mating. Males and females fly in copula forming a heart-shaped mating wheel.

Wing pads   Incompletely developed, nonfunctional wings typically seen on immature insects with simple metamorphosis.

Wingspan   The measurement between tips of the extended forewings of an insect.

Wireworm   Larva of a click beetle (Elateridae).

Witches'-broom   Broom-like growth or massed proliferation caused by the dense clustering of branches of woody plants.

Wriggler   A mosquito larva.

Xylem   The complex supporting, water- and mineral-conducting tissue of vascular plants that makes up sapwood and heartwood.

Xylem-limited bacteria   Bacteria that live within the xylem vessels of a plant and are spread by xylem-feeding insects such as sharpshooter leafhoppers.

Yellow fever   Mosquito transmitted viral disease of humans. Symptoms include pain, internal bleeding, vomiting of blood, and jaundice.

Zoraptera   The order of insects that includes the zorapterans and angel insects.

Zygentoma   A name sometimes given to the insect order that contains the silverfish. The alternate order name Thysanura is used in this book.

Index